MATHEMATICS RESEARCH DEVELOPMENTS

BINARY PERIODIC SIGNALS AND FLOWS

Mathematics Research Developments

Additional books in this series can be found on Nova's website under the Series tab.

Additional E-books in this series can be found on Nova's website under the E-book tab.

MATHEMATICS RESEARCH DEVELOPMENTS

BINARY PERIODIC SIGNALS AND FLOWS

SERBAN E. VLAD

nova publishers
New York

Library of Congress Cataloging-in-Publication Data

Names: Vlad, Serban E., 1959- editor.
Title: Binary periodic signals and flows / Serban E. Vlad (The Romanian Society of Applied and Industrial Mathematics, Oradea, Romania), editor.
Description: Hauppauge, New York : Nova Science Publishers, Inc., [2016] | Series: Mathematics research developments | Includes bibliographical references and index.
Identifiers: LCCN 2016008513 (print) | LCCN 2016016866 (ebook) | ISBN 9781634844529 (hardcover) | ISBN 9781634849586 ()
Subjects: LCSH: Signal processing. | Signal detection. | Binary system (Mathematics) | Modulation (Electronics) | Frequencies of oscillating systems.
Classification: LCC TK5102.9 .B5354 2016 (print) | LCC TK5102.9 (ebook) | DDC 621.382/23--dc23
LC record available at https://lccn.loc.gov/2016008513

Published by Nova Science Publishers, Inc. † New York

Contents

Preface

The Boolean autonomous deterministic regular asynchronous systems have been defined for the first time in our work [16] and a deeper study of such systems can be found in [18]. The concept has its origin in switching theory, the theory of modeling the switching circuits from the digital electrical engineering. The attribute Boolean vaguely refers to the Boole algebra with two elements; autonomous means that there is no input; determinism means the existence of a unique state function; and regular indicates the existence of a function $\Phi : \{0,1\}^n \to \{0,1\}^n, \Phi = (\Phi_1, ..., \Phi_n)$ that 'generates' the system. The time set is discrete: $\{-1,0,1,...\}$ or continuous: $\mathbf{R}$. The system, which is analogue to the (real, usual) dynamical systems, iterates (asynchronously) on each coordinate $i \in \{1,...,n\}$, one of

- Φ_i : we say that Φ is computed, at that time instant, on that coordinate;

- $\{0,1\}^n \ni (\mu_1, ..., \mu_i, ..., \mu_n) \longmapsto \mu_i \in \{0,1\}$: we use to say that Φ is not computed, at that time instant, on that coordinate.

The flows are these that result by analogy with the dynamical systems.

The 'nice' discrete time and real time functions that the (Boolean) asynchronous systems work with are called signals and periodicity is a very important feature in Nature.

In the first two chapters we give the most important concepts concerning the signals and periodicity. The periodicity properties are used to characterize the eventually constant signals in Chapter 3 and the constant signals in Chapter 4. Chapters 5,...,8 are dedicated to the eventually periodic points, eventually periodic signals, periodic points and periodic signals.

Chapter 9 shows constructions that, given an (eventually) periodic point, by changing some values of the signal, change the periodicity properties of the point.

The monograph continues with flows. Chapter 10 is dedicated to the computation functions, i.e. to the functions that show when and how the function Φ is iterated (asynchronously). Chapter 11 introduces the flows and Chapter 12 gives a wider point of view on the flows, which are interpreted as deterministic asynchronous systems. Chapters 13,...,15 restate the topics from Chapters 3,...,8 in the special case when the signals are flows and the main interest is periodicity.

The bibliography consists in general in works of (real, usual) dynamical systems and we use analogies.

The book ends with a list of notations, an index of notions and an appendix with lemmas. These lemmas are frequently used in the exposure and some of them are interesting by themselves.

The book is structured in chapters, each chapter consists in several sections and each section is structured in paragraphs. The chapters begin with an abstract. The paragraphs

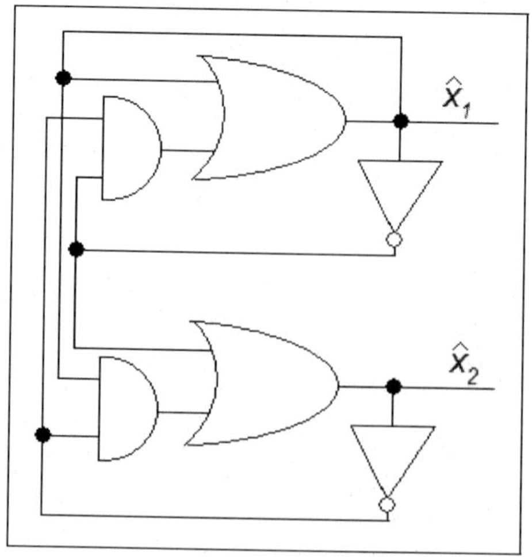

Figure 1. Asynchronous circuit.

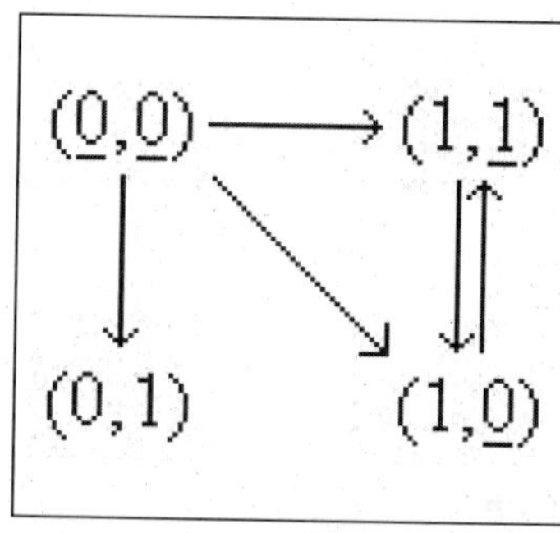

Figure 2. The state diagram of the circuit from Figure 1.

are of the following kinds: definitions, notations, remarks, theorems, corollaries, lemmas, examples and propositions. Each kind of paragraph is numbered separately on the others. Inside the paragraphs, the equations and, more generally, the most important statements are numbered also. When we refer to the statement (x,y) this means the $y-th$ statement of the $x-th$ section of the current chapter. Sometimes we write $(x,y)_{page\ z}$ in order to indicate the page where the statement occurs.

We refer to a definition, theorem, example,... by indicating its number and, when necessary, its page.

In order to point out our source of inspiration, we give the example of the circuit from Figure 1, where $\widehat{x}:\{-1,0,1,...\}\longrightarrow\{0,1\}^2$ is the signal representing the state of the system, and the initial state is $\widehat{x}(-1)=(0,0)$. The function that generates the system is $\Phi:\{0,1\}^2\longrightarrow\{0,1\}^2$, $\forall\mu\in\{0,1\}^2$,

$$\Phi(\mu)=(\mu_1\cup\overline{\mu_1}\cdot\overline{\mu_2},\overline{\mu_1}\cup\mu_1\cdot\overline{\mu_2}).$$

The evolution of the system is given by its state portrait from Figure 2, where the arrows indicate the increase of time and we have underlined these coordinates $\mu_i, i=\overline{1,2}$ that, by the computation of Φ, change their value: $\Phi_i(\mu) = \overline{\mu_i}$. Let $\alpha : \{0,1,2,...\} \longrightarrow \{0,1\}^2$ be the computation function whose values α_i^k show that Φ_i is computed at the time instant k if $\alpha_i^k = 1$, respectively that it is not computed at the time instant k if $\alpha_i^k = 0$, where $i = \overline{1,2}$ and $k \in \{0,1,2,...\}$. The uncertainty related with the circuit, depending in general on the technology, the temperature, etc. manifests in the fact that the order and the time of computation of each coordinate function Φ_i are not known. If the second coordinate is computed at the time instant 0, then $\alpha^0 = (0,1)$ indicates the transfer from $(0,0)$ to $(0,1)$, where the system remains indefinitely long for any values of $\alpha^1, \alpha^2, \alpha^3, ...,$ since $\Phi(0,1) = (0,1)$. Such a signal $\widehat{x}$ is called eventually constant and it corresponds to a stable system. The eventually constant discrete time signals are eventually periodic with an arbitrary period $p \geq 1$.

Another possibility is that the first coordinate of Φ is computed at the time instant 0, thus $\alpha^0 = (1,0)$. Figure 2 indicates the transfer from $(0,0)$ to $(1,0)$, while $\alpha^0 = (1,1)$ indicates the transfer from $(0,0)$ to $(1,1)$, as resulted by the simultaneous computation of $\Phi_1(0,0)$ and $\Phi_2(0,0)$. And if $\alpha^k = (1,1), k \in \{1,2,...\}$, then $\widehat{x}$ is eventually periodic with the period $p \in \{2,4,6,...\}$, as it switches from $(1,1)$ to $(1,0)$ and from $(1,0)$ to $(1,1)$. This last possibility represents an unstable system.

An interesting open problem that we have reached during these studies (mentioned in Chapter 6) is the following one[1]. We suppose that time is real $t \in \mathbf{R}$. If a signal x is periodic, then its points[2] $x(t)$ are periodic

$$\begin{cases} \exists T > 0, \forall t \in \mathbf{R}, \forall z \in \mathbf{Z}, x(t) = x(t+zT) \\ \Longrightarrow \forall t \in \mathbf{R}, \exists T > 0, \forall z \in \mathbf{Z}, x(t) = x(t+zT), \end{cases}$$

and the set of periods of the signal is the intersection of the sets of periods of its points. Is the inverse statement true?

$$\begin{cases} \forall t \in \mathbf{R}, \exists T > 0, \forall z \in \mathbf{Z}, x(t) = x(t+zT) \\ \overset{?}{\Longrightarrow} \exists T > 0, \forall t \in \mathbf{R}, \forall z \in \mathbf{Z}, x(t) = x(t+zT) \end{cases}$$

In other words, if all the points $x(t)$ are periodic, the possibility exists that the intersection of the sets of their periods is empty? If so, the signal is not periodic.

The book addresses to researchers in systems theory and computer science, but it is also interesting to those that study periodicity itself. From this last perspective, the binary signals may be thought of as functions with finitely many values.

The author is aware of the fact that the exposure could have been occasionally shorter, or simpler, or perhaps more correct. For this reason, he thanks in advance to those that would accept to make suggestions of improvements for the next editions of the book.

Serban E. Vlad
E-mail address: serban_e_vlad@yahoo.com
Oradea, December 2015

[1]Because we want to make ourselves understood, we state this open problem under a more general form than the needs of the exposure of the monograph.

[2]point of x = value $x(t)$ of x in some t.

Chapter 1

Preliminaries

The signals from digital electrical engineering are modeled by 'nice' discrete time and real time functions, which are also called signals and their introduction is the purpose of this chapter. We define the left and the right limits of the real time signals, the initial and the final values of the signals, the initial and the final time of the signals, the forgetful function, the orbits, the omega limit sets and the support sets. The last section refers to the images of the signals via a function.

1. The Definition of the Signals

Notation 1. *We denote by* $\mathbf{B} = \{0,1\}$ *the binary Boole algebra. Its laws are the usual ones:*

$$
\begin{array}{cc} \overline{} & \\ 0 & 1 \\ 1 & 0 \end{array} ; \quad
\begin{array}{c|cc} \cdot & 0 & 1 \\ \hline 0 & 0 & 0 \\ 1 & 0 & 1 \end{array} ; \quad
\begin{array}{c|cc} \cup & 0 & 1 \\ \hline 0 & 0 & 1 \\ 1 & 1 & 1 \end{array} ; \quad
\begin{array}{c|cc} \oplus & 0 & 1 \\ \hline 0 & 0 & 1 \\ 1 & 1 & 0 \end{array}
$$

Table 1

called negation, not, or (logical) complement; product or intersection; sum or union; and modulo 2 sum or disjoint union. These laws induce laws that are denoted with the same symbols on $\mathbf{B}^n, n \geq 1$.

Definition 1. *Both sets* $\mathbf{B}$ *and* $\mathbf{B}^n$ *are organized as topological spaces by the discrete topology.*

Notation 2. $\mathbf{N}$, $\mathbf{Z}$, $\mathbf{R}$ *denote the sets of the non negative integers, of the integers and of the real numbers.* $\mathbf{N}_{\_} = \mathbf{N} \cup \{-1\}$ *is the notation of the discrete time set.*

Notation 3. *We denote*

$$\widehat{Seq} = \{(k_j) | k_j \in \mathbf{N}_{\_}, j \in \mathbf{N}_{\_} \text{ and } k_{-1} < k_0 < k_1 < ...\},$$

$$Seq = \{(t_k) | t_k \in \mathbf{R}, k \in \mathbf{N} \text{ and } t_0 < t_1 < t_2 < ... \text{ superiorly unbounded}\}.$$

Example 1. *A typical example of element of* $\widehat{Seq}$ *is the sequence* $k_j = j, j \in \mathbf{N}_{\_}$ *and typical examples of elements of* Seq *are given by the sequences* $z, z+1, z+2, ..., z \in \mathbf{Z}$.

Proposition 1. *Let* $(t_k) \in Seq$ *and* $t \in \mathbf{R}$ *be arbitrary. Then*

$$\exists \varepsilon > 0, \{k | k \in \mathbf{N}, t_k \in (t-\varepsilon, t+\varepsilon)\} = \begin{cases} \{k'\}, if\ t = t_{k'}, \\ \varnothing, if\ \forall k \in \mathbf{N}, t \neq t_k. \end{cases}$$

Proof. We have the following possibilities.

Case $t < t_0$; we take $\varepsilon \in (0, t_0 - t)$, for which $\{k | k \in \mathbf{N}, t_k \in (t-\varepsilon, t+\varepsilon)\} = \varnothing$.

Case $t = t_0$; for $\varepsilon \in (0, t_1 - t)$ we have $\{k | k \in \mathbf{N}, t_k \in (t-\varepsilon, t+\varepsilon)\} = \{0\}$.

Case $t \in (t_{k'-1}, t_{k'}), k' \geq 1$; $\varepsilon \in (0, \min\{t - t_{k'-1}, t_{k'} - t\})$ gives $\{k | k \in \mathbf{N}, t_k \in (t-\varepsilon, t+\varepsilon)\} = \varnothing$.

Case $t = t_{k'}, k' \geq 1$; in this situation any $\varepsilon \in (0, \min\{t - t_{k'-1}, t_{k'+1} - t\})$ gives $\{k | k \in \mathbf{N}, t_k \in (t-\varepsilon, t+\varepsilon)\} = \{k'\}$. □

Remark 1. *The previous* ε *obviously depends on* t. *On the other hand, let us consider the sequence* $t_k = \frac{1}{0+1} + \frac{1}{1+1} + ... + \frac{1}{k+1}, k \in \mathbf{N}$. *We notice that* $(t_k) \in Seq$ *and*

$$\forall \varepsilon > 0, \exists t \in \mathbf{R}, card(\{k | k \in \mathbf{N}, t_k \in (t-\varepsilon, t+\varepsilon)\}) > 1$$

holds.

Notation 4. $\chi_A : \mathbf{R} \to \mathbf{B}$ *is the notation of the characteristic function of the set* $A \subset \mathbf{R} : \forall t \in \mathbf{R}$,

$$\chi_A(t) = \begin{cases} 1, if\ t \in A, \\ 0, otherwise. \end{cases}$$

Definition 2. *The* ***discrete time signals*** *are by definition the functions* $\widehat{x} : \mathbf{N}_{\_} \to \mathbf{B}^n$. *Their set is denoted with* $\widehat{S}^{(n)}$.

The ***continuous time signals*** *are the functions* $x : \mathbf{R} \to \mathbf{B}^n$ *of the form* $\forall t \in \mathbf{R}$,

$$x(t) = \mu \cdot \chi_{(-\infty, t_0)}(t) \oplus x(t_0) \cdot \chi_{[t_0, t_1)}(t) \oplus ... \oplus x(t_k) \cdot \chi_{[t_k, t_{k+1})}(t) \oplus ... \tag{1.1}$$

where $\mu \in \mathbf{B}^n$ *and* $(t_k) \in Seq$. *Their set is denoted by* $S^{(n)}$.

Example 2. *The constant functions* $\widehat{x} \in \widehat{S}^{(1)}, x \in S^{(1)}$ *equal with* $\mu \in \mathbf{B}$:

$$\forall k \in \mathbf{N}_{\_}, \widehat{x}(k) = \mu, \tag{1.2}$$

$$\forall t \in \mathbf{R}, x(t) = \mu \tag{1.3}$$

are typical examples of signals. Here are some other examples:

$$\forall k \in \mathbf{N}_{\_}, \widehat{x}(k) = \begin{cases} 1, if\ k\ is\ odd, \\ 0,\ if\ k\ is\ even, \end{cases} \tag{1.4}$$

$$\forall t \in \mathbf{R}, x(t) = \chi_{[0,\infty)}(t), \tag{1.5}$$

$$\forall t \in \mathbf{R}, x(t) = \chi_{[0,1)}(t) \oplus \chi_{[2,3)}(t) \oplus ... \oplus \chi_{[2k,2k+1)}(t) \oplus ... \tag{1.6}$$

The signal from (1.5) is called the (unitary) step function (of Heaviside).

Remark 2. *At Definition 2 a convention of notation has occurred for the first time, namely a hat '$\widehat{\ }$' is used to show that we have discrete time. The hat will make the difference between, for example, the notation of the discrete time signals $\widehat{x}, \widehat{y}, ..$ and the notation of the real time signals $x, y, ...$*

Remark 3. *The discrete time signals are sequences. The real time signals are piecewise constant functions.*

Remark 4. *As we shall see in the rest of the book, the study of the periodicity of the signals does not use essentially the fact that they take values in $\mathbf{B}^n$, but the fact that they take finitely many values. For example, instead of using '$\cdot$' and '$\oplus$' in (1.1), we can write equivalently*

$$x(t) = \begin{cases} \mu, t < t_0, \\ x(t_0), t \in [t_0, t_1), \\ \dots \\ x(t_k), t \in [t_k, t_{k+1}), \\ \dots \end{cases}$$

Remark 5. *The signals model the electrical signals of the circuits from the digital electrical engineering.*

2. Left and Right Limits

Theorem 1. *For any $x \in S^{(n)}$ and any $t \in \mathbf{R}$, there exist $x(t-0), x(t+0) \in \mathbf{B}^n$ with the property*

$$\exists \varepsilon > 0, \forall \xi \in (t-\varepsilon, t), x(\xi) = x(t-0), \tag{2.1}$$

$$\exists \varepsilon > 0, \forall \xi \in (t, t+\varepsilon), x(\xi) = x(t+0). \tag{2.2}$$

Proof. We presume that x, t are arbitrary and fixed and that x is of the form

$$x(t) = \mu \cdot \chi_{(-\infty, t_0)}(t) \oplus x(t_0) \cdot \chi_{[t_0, t_1)}(t) \oplus \dots \oplus x(t_k) \cdot \chi_{[t_k, t_{k+1})}(t) \oplus \dots \tag{2.3}$$

with $\mu \in \mathbf{B}^n$ and $(t_k) \in Seq$. We take $\varepsilon > 0$ small enough, see Proposition 1, page 2 such that

$$\{k | k \in \mathbf{N}, t_k \in (t-\varepsilon, t+\varepsilon)\} = \begin{cases} \{k'\}, if\ t = t_{k'}, \\ \varnothing, if\ \forall k \in \mathbf{N}, t \neq t_k. \end{cases}$$

We have the following possibilities:

Case $t < t_0$;

$$\forall \xi \in (t-\varepsilon, t), x(\xi) = \mu,$$
$$\forall \xi \in (t, t+\varepsilon), x(\xi) = \mu.$$

Case $t = t_0$;

$$\forall \xi \in (t-\varepsilon, t), x(\xi) = \mu,$$
$$\forall \xi \in (t, t+\varepsilon), x(\xi) = x(t_0).$$

Case $t \in (t_{k'-1}, t_{k'}), k' \geq 1$;

$$\forall \xi \in (t-\varepsilon, t), x(\xi) = x(t_{k'-1}),$$

$$\forall \xi \in (t, t+\varepsilon), x(\xi) = x(t_{k'-1}).$$

Case $t = t_{k'}, k' \geq 1$;

$$\forall \xi \in (t-\varepsilon, t), x(\xi) = x(t_{k'-1}),$$

$$\forall \xi \in (t, t+\varepsilon), x(\xi) = x(t_{k'}).$$

□

Definition 3. *The functions* $\mathbf{R} \ni t \to x(t-0) \in \mathbf{B}^n, \mathbf{R} \ni t \to x(t+0) \in \mathbf{B}^n$ *are called the* ***left limit*** *function of x and the* ***right limit*** *function of x.*

Remark 6. *Theorem 1 states that the signals* $x \in S^{(n)}$ *have a left limit function* $x(t-0)$ *and a right limit function* $x(t+0)$*. Moreover, if (2.3) is true, then*

$$x(t-0) = \mu \cdot \chi_{(-\infty,t_0]}(t) \oplus x(t_0) \cdot \chi_{(t_0,t_1]}(t) \oplus \ldots \oplus x(t_k) \cdot \chi_{(t_k,t_{k+1}]}(t) \oplus \ldots, \tag{2.4}$$

$$x(t+0) = x(t) \tag{2.5}$$

hold, meaning in particular that $x(t-0)$ *is not a signal and that* $x(t+0)$ *coincides with* $x(t)$.

Remark 7. *The property (2.5) stating in fact that the real time signals x are right continuous will be used later under the form*

$$\forall t \in \mathbf{R}, \exists \varepsilon > 0, \forall \xi \in [t, t+\varepsilon), x(\xi) = x(t). \tag{2.6}$$

3. Initial and Final Values, Initial and Final Time

Definition 4. *The* ***initial value*** *of* $\widehat{x} \in \widehat{S}^{(n)}$ *is* $\widehat{x}(-1) \in \mathbf{B}^n$.

For $x \in S^{(n)}$,

$$x(t) = \mu \cdot \chi_{(-\infty,t_0)}(t) \oplus x(t_0) \cdot \chi_{[t_0,t_1)}(t) \oplus \ldots \oplus x(t_k) \cdot \chi_{[t_k,t_{k+1})}(t) \oplus \ldots, \tag{3.1}$$

where $\mu \in \mathbf{B}^n$ *and* $(t_k) \in Seq$*, the* ***initial value*** *is* μ.

Notation 5. *There is no special notation for the initial value of* $\widehat{x}$.

The initial value of x has two usual notations, $x(-\infty+0)$ *and* $\lim\limits_{t \to -\infty} x(t)$.

Definition 5. *By definition, the* ***initial time (instant)*** *of* $\widehat{x}$ *is* $k = -1$.

The ***initial time (instant)*** *of x is any number* $t_0 \in \mathbf{R}$ *that fulfills*

$$\forall t \leq t_0, x(t) = x(-\infty+0). \tag{3.2}$$

Notation 6. *The set of the initial time instants of x is denoted by* I^x.

Definition 6. *The **final value** $\mu \in \mathbf{B}^n$ of $\widehat{x} \in \widehat{S}^{(n)}$ is defined by $\exists k' \in \mathbf{N}_{\_}$,*

$$\forall k \geq k', \widehat{x}(k) = \mu \tag{3.3}$$

*and the **final value** $\mu \in \mathbf{B}^n$ of $x \in S^{(n)}$ is defined by $\exists t' \in \mathbf{R}$,*

$$\forall t \geq t', x(t) = \mu. \tag{3.4}$$

Notation 7. *The usual notations for μ in (3.3) are $\widehat{x}(\infty - 0)$ and $\lim_{k \to \infty} \widehat{x}(k)$.*

The final value μ from (3.4) is denoted with either of $x(\infty - 0)$ and $\lim_{t \to \infty} x(t)$.

Definition 7. *If the final value μ of $\widehat{x}$ exists, then any $k' \in \mathbf{N}_{\_}$ like in (3.3) is called **final time (instant)** of $\widehat{x}$.*

*Similarly, if the final value μ of x exists and (3.4) holds, then any such $t' \in \mathbf{R}$ is called **final time (instant)** of x.*

Notation 8. *The set of the final time instants of $\widehat{x}$ is denoted by $\widehat{F^{x}}$.*

The set of the final time instants of x has the notation F^x.

Example 3. *The signals from (1.2), (1.3) fulfill $\lim_{k \to \infty} \widehat{x}(k) = \lim_{t \to -\infty} x(t) = \lim_{t \to \infty} x(t) = \mu$, $\widehat{F^{x}} = \mathbf{N}_{\_}, I^x = F^x = \mathbf{R}$ and the signal from (1.5) fulfills $\lim_{t \to -\infty} x(t) = 0$, $\lim_{t \to \infty} x(t) = 1, I^x = (-\infty, 0), F^x = [0, \infty)$. The signals from (1.4), (1.6) have no final value: $\widehat{F^{x}} = F^x = \emptyset$.*

Remark 8. *For arbitrary $\widehat{x}, x$ the initial value exists and it is unique; the initial time of $\widehat{x}$ is unique and the initial time of x is not unique.*

The final value of $\widehat{x}, x$ might not exist, but if it exists, it is unique. The final time of $\widehat{x}, x$ might not exist, but if it exists, it is not unique.

Theorem 2. *a) Let $\widehat{x} \in \widehat{S}^{(n)}$ and $k_0 \in \mathbf{N}_{\_}$. The following equivalencies hold:*

$$\begin{cases} \forall k \geq k_0, \widehat{x}(k) = \widehat{x}(\infty - 0), \\ k_0 \geq 0 \Longrightarrow \widehat{x}(k_0 - 1) \neq \widehat{x}(\infty - 0) \end{cases} \Longleftrightarrow \widehat{F^{x}} = \{k_0, k_0 + 1, k_0 + 2, ...\}, \tag{3.5}$$

$$\forall k \in \mathbf{N}_{\_}, \widehat{x}(k) = \widehat{x}(\infty - 0) \Longleftrightarrow \widehat{F^{x}} = \mathbf{N}_{\_}. \tag{3.6}$$

b) Let $x \in S^{(n)}, t_0 \in \mathbf{R}$. The following equivalencies take place:

$$\begin{cases} \forall t < t_0, x(t) = x(-\infty + 0), \\ x(t_0) \neq x(-\infty + 0) \end{cases} \Longleftrightarrow I^x = (-\infty, t_0), \tag{3.7}$$

$$\begin{cases} \forall t \geq t_0, x(t) = x(\infty - 0), \\ x(t_0 - 0) \neq x(t_0) \end{cases} \Longleftrightarrow F^x = [t_0, \infty), \tag{3.8}$$

$$\forall t \in \mathbf{R}, x(t) = x(-\infty + 0) \Longleftrightarrow I^x = \mathbf{R}, \tag{3.9}$$

$$\forall t \in \mathbf{R}, x(t) = x(\infty - 0) \Longleftrightarrow F^x = \mathbf{R}. \tag{3.10}$$

Proof. a) Two possibilities exist.

Case $k_0 = -1$

The statements $\forall k \in \mathbf{N}_{\text{-}}, \widehat{x}(k) = \widehat{x}(\infty - 0)$ and $\{k'|\forall k \geq k', \widehat{x}(k) = \widehat{x}(\infty - 0)\} = \mathbf{N}_{\text{-}}$ are equivalent indeed and this special case of (3.5) coincides with (3.6).

Case $k_0 \geq 0$

We have that $(\forall k \geq k_0, \widehat{x}(k) = \widehat{x}(\infty - 0)$ and $\widehat{x}(k_0 - 1) \neq \widehat{x}(\infty - 0))$ is equivalent with $\{k'|\forall k \geq k', \widehat{x}(k) = \widehat{x}(\infty - 0)\} = \{k_0, k_0 + 1, k_0 + 2, ...\}$.

b) The statement $(\forall t < t_0, x(t) = x(-\infty + 0)$ and $x(t_0) \neq x(-\infty + 0))$ is equivalent with $\{t'|\forall t \leq t', x(t) = x(-\infty + 0)\} = (-\infty, t_0)$. This coincides with (3.7). The statement $\forall t \in \mathbf{R}, x(t) = x(-\infty + 0)$ is equivalent with $\{t'|\forall t \leq t', x(t) = x(-\infty + 0)\} = \mathbf{R}$, giving the truth of (3.9). Statements (3.8) and (3.10) are obvious now. □

Remark 9. *Versions of Theorem 2 exist, showing that $\widehat{x}$ is constant if and only if $\widehat{F^{x}} = \mathbf{N}_{\text{-}}$ and non constant otherwise, respectively the statements:*

i) x is not constant;

ii) $t_0 \in \mathbf{R}$ exists with

$$\forall t < t_0, x(t) = x(-\infty + 0),$$

$$x(t_0) \neq x(-\infty + 0);$$

iii) $t_0 \in \mathbf{R}$ exists with $I^x = (-\infty, t_0)$

are equivalent etc.

Theorem 3. *Let the signal $x \in S^{(n)}$ from (3.1). We define $\widehat{x} \in \widehat{S}^{(n)}$ by*

$$\widehat{x}(-1) = \mu,$$

$$\forall k \in \mathbf{N}, \widehat{x}(k) = x(t_k).$$

The following statements hold.

a) $\lim\limits_{k\to\infty} \widehat{x}(k)$ exists if and only if $\lim\limits_{t\to\infty} x(t)$ exists and in case that the previous limits exist we have $\lim\limits_{k\to\infty} \widehat{x}(k) = \lim\limits_{t\to\infty} x(t)$.

b) We suppose that $\lim\limits_{k\to\infty} \widehat{x}(k), \lim\limits_{t\to\infty} x(t)$ exist. Then -1 is final time of $\widehat{x}$ if and only if any $t' < t_0$ is final time of x and $\forall k' \geq 0$, k' is final time of $\widehat{x}$ if and only if $t_{k'}$ is final time of x.

Proof. a) From the hypothesis we infer that for any $k' \in \mathbf{N}$ we can write

$$\{\widehat{x}(k)|k \geq k'\} = \{x(t)|t \geq t_{k'}\}. \tag{3.11}$$

Then

$$\lim_{k\to\infty} \widehat{x}(k) \text{ exists} \Longleftrightarrow \exists k' \in \mathbf{N}, card(\{\widehat{x}(k)|k \geq k'\}) = 1 \Longleftrightarrow$$

$$\Longleftrightarrow \exists k' \in \mathbf{N}, card(\{x(t)|t \geq t_{k'}\}) = 1 \Longleftrightarrow \lim_{t\to\infty} x(t) \text{ exists}$$

and if one of the previous equivalent statements is true, we obtain the existence of $\mu' \in \mathbf{B}^n, k' \in \mathbf{N}$ such that

$$\{\widehat{x}(k)|k \geq k'\} = \{\mu'\} = \{x(t)|t \geq t_{k'}\},$$

i.e.

$$\lim_{k\to\infty}\widehat{x}(k)=\mu'=\lim_{t\to\infty}x(t). \tag{3.12}$$

b) Let us presume that $\lim_{k\to\infty}\widehat{x}(k),\lim_{t\to\infty}x(t)$ exist. We have

$$-1\in\widehat{F^{\widehat{x}}}\Longleftrightarrow\forall k\in\mathbf{N}_{\_},\widehat{x}(k)=\mu\Longleftrightarrow\forall t\in\mathbf{R},x(t)=\mu\Longleftrightarrow$$

$$\Longleftrightarrow\forall t'<t_0,\forall t\geq t',x(t)=\mu\Longleftrightarrow\forall t'<t_0,t'\in F^x$$

and similarly for any $k'\geq 0$, we infer from (3.12) that

$$k'\in\widehat{F^{\widehat{x}}}\Longleftrightarrow\forall k\geq k',\widehat{x}(k)=\mu'\Longleftrightarrow\forall t\geq t_{k'},x(t)=\mu'\Longleftrightarrow t_{k'}\in F^x.$$

□

4. The Forgetful Function

Definition 8. *The* ***discrete time forgetful function*** $\widehat{\sigma}^{k'}:\widehat{S}^{(n)}\to\widehat{S}^{(n)}$ *is defined for* $k'\in\mathbf{N}$ *by*

$$\forall\widehat{x}\in\widehat{S}^{(n)},\forall k\in\mathbf{N}_{\_},\widehat{\sigma}^{k'}(\widehat{x})(k)=\widehat{x}(k+k')$$

and the ***real time forgetful function*** $\sigma^{t'}:S^{(n)}\to S^{(n)}$ *is defined for* $t'\in\mathbf{R}$ *in the following manner*

$$\forall x\in S^{(n)},\forall t\in\mathbf{R},\sigma^{t'}(x)(t)=\begin{cases}x(t),t\geq t',\\x(t'-0),t<t'.\end{cases}$$

Theorem 4. *The signals* $\widehat{x}\in\widehat{S}^{(n)},x\in S^{(n)}$ *are given. The following statements hold:*

a) $\widehat{\sigma}^0(\widehat{x})=\widehat{x}$*; if* $I^x=\mathbf{R}$*, then* $\forall t'\in\mathbf{R},\sigma^{t'}(x)=x$ *and if* $\exists t_0\in\mathbf{R}$, $I^x=(-\infty,t_0)$*, then* $\forall t'\leq t_0,\sigma^{t'}(x)=x$*;*

b) for $k',k''\in\mathbf{N}$ *we have* $(\widehat{\sigma}^{k'}\circ\widehat{\sigma}^{k''})(\widehat{x})=\widehat{\sigma}^{k'+k''}(\widehat{x})$*; for any* $t',t''\in\mathbf{R}$ *we have* $(\sigma^{t'}\circ\sigma^{t''})(x)=\sigma^{\max\{t',t''\}}(x)$.

Proof. a) The discrete time statement is obvious. In order to prove the real time statement, we notice that $I^x=\mathbf{R}$ is true only if x is constant, see Theorem 2, page 5, so that we can suppose now that x is not constant and some t_0 exists such that $I^x=(-\infty,t_0)$:

$$\forall t<t_0,x(t)=x(-\infty+0),$$

$$x(t_0)\neq x(-\infty+0).$$

Let $t'\leq t_0$ arbitrary. We have $\forall t\in\mathbf{R}$,

$$\sigma^{t'}(x)(t)=\begin{cases}x(t),t\geq t'\\x(t'-0),t<t'\end{cases}=\begin{cases}x(t),t\geq t'\\x(-\infty+0),t<t'\end{cases}=x(t).$$

b) We fix arbitrarily $k',k''\in\mathbf{N}$. We can write for any $k\in\mathbf{N}$ that

$$(\widehat{\sigma}^{k'}\circ\widehat{\sigma}^{k''})(\widehat{x})(k)=\widehat{x}(k+k'+k'')=\widehat{\sigma}^{k'+k''}(\widehat{x})(k).$$

Let us take now $t',t''\in\mathbf{R}$ arbitrarily. We get the existence of the next possibilities.

Case $t'' \leq t'$
For any $t \in \mathbf{R}$ we infer

$$(\sigma^{t'} \circ \sigma^{t''})(x)(t) = \sigma^{t'}(\sigma^{t''}(x))(t) = \begin{cases} \sigma^{t''}(x)(t), t \geq t' \\ \sigma^{t''}(x)(t'-0), t < t' \end{cases}$$

$$= \begin{cases} x(t), t \geq t' \\ x(t'-0), t < t' \end{cases} = \sigma^{t'}(x)(t).$$

Case $t'' > t'$
We get for arbitrary $t \in \mathbf{R}$ that

$$(\sigma^{t'} \circ \sigma^{t''})(x)(t) = \sigma^{t'}(\sigma^{t''}(x))(t) = \begin{cases} \sigma^{t''}(x)(t), t \geq t' \\ \sigma^{t''}(x)(t'-0), t < t' \end{cases}$$

$$= \begin{cases} \sigma^{t''}(x)(t), t \geq t' \\ x(t''-0), t < t' \end{cases} = \sigma^{t''}(x)(t).$$

□

Remark 10. *Let us give $\widehat{x}$ by its values $\widehat{x} = x^{-1}, x^0, x^1, \ldots$ where $x^k \in \mathbf{B}^n, k \in \mathbf{N}_{\_}$. Then $\widehat{\sigma}^1(\widehat{x}) = x^0, x^1, \ldots$ i.e. $\widehat{x}$ has forgotten its first value. Furthermore, $\widehat{\sigma}^0(\widehat{x})$ makes $\widehat{x}$ forget nothing and $\widehat{\sigma}^{k'}(\widehat{x})$ makes $\widehat{x}$ forget its first $k' \geq 1$ values.*

Remark 11. *$\sigma^{t'}(x)$ makes x forget its values prior to t' : no value if $\forall t < t', x(t) = x(-\infty + 0)$ and some values otherwise.*

5. Orbits, Omega Limit Sets and Support Sets

Definition 9. *The **orbits** of $\widehat{x} \in \widehat{S}^{(n)}, x \in S^{(n)}$ are the sets of the values of these functions:*

$$\widehat{Or}(\widehat{x}) = \{\widehat{x}(k) | k \in \mathbf{N}_{\_}\},$$

$$Or(x) = \{x(t) | t \in \mathbf{R}\}.$$

Definition 10. *The **omega limit set** $\widehat{\omega}(\widehat{x})$ of $\widehat{x}$ is defined as*

$$\widehat{\omega}(\widehat{x}) = \{\mu | \mu \in \mathbf{B}^n, \exists (k_j) \in \widehat{Seq}, \forall j \in \mathbf{N}_{\_}, \widehat{x}(k_j) = \mu\}$$

*and the **omega limit set** $\omega(x)$ of x is defined by*

$$\omega(x) = \{\mu | \mu \in \mathbf{B}^n, \exists (t_k) \in Seq, \forall k \in \mathbf{N}, x(t_k) = \mu\}.$$

*The points of $\widehat{\omega}(\widehat{x}), \omega(x)$ are called **omega limit points**.*[1]

[1] In a real time construction, in [18], when x represents the state of a (control, nondeterministic, asynchronous) system, the value μ of x is called (accessible) recurrent if $\forall t_0 \in \mathbf{R}, \exists t > t_0, x(t) = \mu$, i.e. if $\mu \in \omega(x)$.

Example 4. *We define* $\widehat{x} \in \widehat{S}^{(2)}$ *by*

$$\widehat{x}(k) = \begin{cases} (0,0), k = -1, \\ (1,1), k = 3k', k' \geq 0, \\ (0,1), k = 3k'+1, k' \geq 0, \\ (1,0), k = 3k'+2, k' \geq 0 \end{cases}$$

and $x \in S^{(2)}$ *by*

$$x(t) = \widehat{x}(-1) \cdot \chi_{(-\infty,0)}(t) \oplus \widehat{x}(0) \cdot \chi_{[0,1)}(t) \oplus \ldots \oplus \widehat{x}(k) \cdot \chi_{[k,k+1)}(t) \oplus \ldots$$

We see that $\widehat{Or}(\widehat{x}) = Or(x) = \mathbf{B}^2$ *and* $\widehat{\omega}(\widehat{x}) = \omega(x) = \{(1,1),(0,1),(1,0)\}$.

Definition 11. *For* $\widehat{x} \in \widehat{S}^{(n)}, x \in S^{(n)}$ *and* $\mu \in \mathbf{B}^n$, *we define the* ***support sets*** *of* μ *by*

$$\widehat{\mathbf{T}}_{\mu}^{\widehat{x}} = \{k | k \in \mathbf{N}_{\_}, \widehat{x}(k) = \mu\},$$

$$\mathbf{T}_{\mu}^{x} = \{t | t \in \mathbf{R}, x(t) = \mu\}.$$

Remark 12. *The previous definition allows us to express the fact that t is an initial time instant of x,* $t \in I^x$ *under the equivalent form* $(-\infty, t] \subset \mathbf{T}_{x(-\infty+0)}^{x}$. *We shall use sometimes this possibility in the rest of the exposure.*

Theorem 5. *Let* $\widehat{x} \in \widehat{S}^{(n)}, x \in S^{(n)}$. *We have that*
a) $\widehat{\omega}(\widehat{x}) = \{\mu | \mu \in \mathbf{B}^n, \widehat{\mathbf{T}}_{\mu}^{\widehat{x}}$ *is infinite*$\}$, $\omega(x) = \{\mu | \mu \in \mathbf{B}^n, \mathbf{T}_{\mu}^{x}$ *is unbounded from above*$\}$;
b) $\widehat{\omega}(\widehat{x}) \neq \varnothing, \omega(x) \neq \varnothing$;
c) for any $\widetilde{k} \in \mathbf{N}, \widetilde{t} \in \mathbf{R}$ *the following diagrams commute*

$$\begin{array}{ccc} \widehat{Or}(\widehat{\sigma}^{\widetilde{k}}(\widehat{x})) & \subset & \widehat{Or}(\widehat{x}) \\ \cup & & \cup \\ \widehat{\omega}(\widehat{\sigma}^{\widetilde{k}}(\widehat{x})) & = & \widehat{\omega}(\widehat{x}) \end{array}, \quad \begin{array}{ccc} Or(\sigma^{\widetilde{t}}(x)) & \subset & Or(x) \\ \cup & & \cup \\ \omega(\sigma^{\widetilde{t}}(x)) & = & \omega(x) \end{array}$$

Proof. a) Indeed, for any $\mu \in \mathbf{B}^n$, the fact that $\mu \in \widehat{\omega}(\widehat{x})$ is equivalent with any of:

a sequence $k_{-1} < k_0 < k_1 < \ldots$ exists such that $\forall j \in \mathbf{N}_{\_}, \widehat{x}(k_j) = \mu$,

the set $\{k | k \in \mathbf{N}_{\_}, \widehat{x}(k) = \mu\}$ is infinite

and the fact that $\mu \in \omega(x)$ is equivalent with any of

an unbounded from above sequence $t_0 < t_1 < t_2 < \ldots$ exists with $\forall j \in \mathbf{N}, x(t_j) = \mu$,

the set $\{t | t \in \mathbf{R}, x(t) = \mu\}$ is unbounded from above.

b) The sets $\widehat{\mathbf{T}}_{\mu}^{\widehat{x}}$, $\mu \in \mathbf{B}^n$ are either empty, or finite non-empty, or infinite. We put $\mathbf{B}^n$ under the form $\mathbf{B}^n = \{\mu^1, \mu^2, \ldots, \mu^{2^n}\}$. Because in the equation

$$\widehat{\mathbf{T}}_{\mu^1}^{\widehat{x}} \cup \ldots \cup \widehat{\mathbf{T}}_{\mu^{2^n}}^{\widehat{x}} = \mathbf{N}_{\_}$$

the right hand set is infinite, we infer that infinite sets $\widehat{\mathbf{T}}^{\widehat{x}}_{\mu^i}$ always exist, let them be, without loosing the generality, $\widehat{\mathbf{T}}^{\widehat{x}}_{\mu^1},...,\widehat{\mathbf{T}}^{\widehat{x}}_{\mu^p}$. We have from a)

$$\widehat{\omega}(\widehat{x}) = \{\mu^1,...,\mu^p\}.$$

Similarly, we consider the equation

$$\mathbf{T}^x_{\mu^1} \cup ... \cup \mathbf{T}^x_{\mu^{2^n}} = \mathbf{R}$$

where the right hand set is unbounded from above. We infer that the left hand term contains sets $\mathbf{T}^x_{\mu^i}$ which are unbounded from above and let them be, without loosing the generality, $\mathbf{T}^x_{\mu^1},...,\mathbf{T}^x_{\mu^p}$. We infer from a) that

$$\omega(x) = \{\mu^1,...,\mu^p\}.$$

c) We prove that $\widehat{\omega}(\widehat{x}) \subset \widehat{Or}(x)$. Some sets $\widehat{\mathbf{T}}^{\widehat{x}}_{\mu^i}$ may exist which are finite non-empty, let them be without loosing the generality $\widehat{\mathbf{T}}^{\widehat{x}}_{\mu^{p+1}},...,\widehat{\mathbf{T}}^{\widehat{x}}_{\mu^s}$, where $p \leq s \leq 2^n$. Then

$$\widehat{\omega}(\widehat{x}) = \{\mu^1,...,\mu^p\} \subset \{\mu^1,...,\mu^s\} = \widehat{Or}(\widehat{x}).$$

The previous inclusion is true as equality if finite non-empty sets $\widehat{\mathbf{T}}^{\widehat{x}}_{\mu^i}$ do not exist and $p = s$.

The proof of $\omega(x) \subset Or(x)$ is similar, we presume that the non-empty, bounded sets $\mathbf{T}^x_{\mu^i}$ are $\mathbf{T}^x_{\mu^{p+1}},...,\mathbf{T}^x_{\mu^s}$, with $p \leq s \leq 2^n$. Then

$$\omega(x) = \{\mu^1,...,\mu^p\} \subset \{\mu^1,...,\mu^s\} = Or(x)$$

and the previous inclusion holds as equality in the situation when all the non-empty sets $\mathbf{T}^x_{\mu^i}$ are unbounded from above, i.e. when $p = s$.

Equation $\widehat{\omega}(\widehat{\sigma}^{\widetilde{k}}(\widehat{x})) = \widehat{\omega}(\widehat{x})$ is a consequence of the fact that for any $\mu \in \mathbf{B}^n$, the sets $\widehat{\mathbf{T}}^{\widehat{x}}_{\mu} = \{k | k \geq -1, \widehat{x}(k) = \mu\}$ and $\widehat{\mathbf{T}}^{\widehat{\sigma}^{\widetilde{k}}(\widehat{x})}_{\mu} = \{k | k \geq \widetilde{k} - 1, \widehat{x}(k) = \mu\}$ are both finite (the empty sets are in this situation) or infinite.

Equation $\omega(\sigma^{\widetilde{t}}(x)) = \omega(x)$ results from the fact that for any $\mu \in \mathbf{B}^n$, the sets $\mathbf{T}^x_{\mu}$ and $\mathbf{T}^{\sigma^{\widetilde{t}}(x)}_{\mu}$ fulfilling $\mathbf{T}^x_{\mu} \cap [\widetilde{t}, \infty) = \mathbf{T}^{\sigma^{\widetilde{t}}(x)}_{\mu} \cap [\widetilde{t}, \infty)$ are both superiorly bounded (including the empty sets, that are considered to have this property) or superiorly unbounded.

We prove $\widehat{Or}(\widehat{\sigma}^{\widetilde{k}}(\widehat{x})) \subset \widehat{Or}(x), Or(\sigma^{\widetilde{t}}(x)) \subset Or(x)$ in the following way:

$$\widehat{Or}(\widehat{\sigma}^{\widetilde{k}}(\widehat{x})) = \{\widehat{\sigma}^{\widetilde{k}}(\widehat{x})(k) | k \in \mathbf{N}_{\_}\} = \{\widehat{x}(k + \widetilde{k}) | k \in \mathbf{N}_{\_}\} =$$

$$= \{\widehat{x}(k) | k \geq \widetilde{k} - 1\} \subset \{\widehat{x}(k) | k \in \mathbf{N}_{\_}\} = \widehat{Or}(x)$$

and on the other hand let $\varepsilon > 0$ with $\forall \xi \in (\widetilde{t} - \varepsilon, \widetilde{t}), x(\xi) = x(\widetilde{t} - 0)$; then

$$Or(\sigma^{\widetilde{t}}(x)) = \{\sigma^{\widetilde{t}}(x)(t) | t \in \mathbf{R}\} = \{x(t) | t > \widetilde{t} - \varepsilon\} \subset \{x(t) | t \in \mathbf{R}\} = Or(x).$$

□

Theorem 6. *The signals $\widehat{x}, x$ are given and we suppose that the sequence $(t_k) \in Seq$ exists such that*

$$x(t) = \widehat{x}(-1) \cdot \chi_{(-\infty,t_0)}(t) \oplus \widehat{x}(0) \cdot \chi_{[t_0,t_1)}(t) \oplus \ldots \oplus \widehat{x}(k) \cdot \chi_{[t_k,t_{k+1})}(t) \oplus \ldots \quad (5.13)$$

a) We have $\widehat{Or}(\widehat{x}) = Or(x)$ and $\widehat{\omega}(\widehat{x}) = \omega(x)$.

b) For any $\widetilde{k} \in \mathbf{N}, \widetilde{t} \in \mathbf{R}$ we infer $\widehat{\omega}(\widehat{\sigma}^{\widetilde{k}}(\widehat{x})) = \omega(\sigma^{\widetilde{t}}(x))$; if either $\widetilde{k} = 0, \widetilde{t} \leq t_0$, or $\widetilde{k} \geq 1$, $\widetilde{t} \in (t_{\widetilde{k}-1}, t_{\widetilde{k}}]$, then $\widehat{Or}(\widehat{\sigma}^{\widetilde{k}}(\widehat{x})) = Or(\sigma^{\widetilde{t}}(x))$.

Proof. a) We have

$$\widehat{Or}(\widehat{x}) = \{\widehat{x}(k) | k \in \mathbf{N}_{\_}\} \overset{(5.13)}{=} \{x(t) | t \in \mathbf{R}\} = Or(x).$$

In order to prove the second equality, let some arbitrary $\mu \in \widehat{\omega}(\widehat{x})$, thus the sequence $(k_j) \in \widehat{Seq}$ exists with the property that

$$\forall j \in \mathbf{N}_{\_}, \widehat{x}(k_j) = \mu.$$

For x given by (5.13), we can define the unbounded from above sequence

$$\forall j \in \mathbf{N}_{\_}, t'_{j+1} \overset{def}{=} t_{k_j},$$

for which we get

$$\forall j \in \mathbf{N}_{\_}, x(t'_{j+1}) = x(t_{k_j}) = \widehat{x}(k_j) = \mu,$$

thus $\mu \in \omega(x)$ and $\widehat{\omega}(\widehat{x}) \subset \omega(x)$. The inverse inclusion is proved similarly.

b) We fix $\widetilde{k} \in \mathbf{N}, \widetilde{t} \in \mathbf{R}$ arbitrarily. The first statement results from

$$\widehat{\omega}(\widehat{\sigma}^{\widetilde{k}}(\widehat{x})) \overset{Theorem\,5}{=} \widehat{\omega}(\widehat{x}) \overset{a)}{=} \omega(x) \overset{Theorem\,5}{=} \omega(\sigma^{\widetilde{t}}(x)).$$

We prove the second statement. If $\widetilde{k} = 0, \widetilde{t} \leq t_0$, then $\widehat{\sigma}^{\widetilde{k}}(\widehat{x}) = \widehat{x}$ and $\sigma^{\widetilde{t}}(x) = x$, thus

$$\widehat{Or}(\widehat{\sigma}^{\widetilde{k}}(\widehat{x})) = \widehat{Or}(\widehat{x}) \overset{a)}{=} Or(x) = Or(\sigma^{\widetilde{t}}(x)).$$

We suppose from this moment that $\widetilde{k} \geq 1, \widetilde{t} \in (t_{\widetilde{k}-1}, t_{\widetilde{k}}]$ hold. We conclude that

$$\widehat{Or}(\widehat{\sigma}^{\widetilde{k}}(\widehat{x})) = \{\widehat{x}(k) | k \geq \widetilde{k} - 1\} \overset{(5.13)}{=} \{x(t) | t \geq t_{\widetilde{k}-1}\} = \{\sigma^{\widetilde{t}}(x)(t) | t \in \mathbf{R}\} = Or(\sigma^{\widetilde{t}}(x)).$$

□

Theorem 7. *For any $\widehat{x} \in \widehat{S}^{(n)}, x \in S^{(n)}$ we have*

$$\exists k' \in \mathbf{N}_{\_}, \forall k'' \geq k', \widehat{\omega}(\widehat{x}) = \{\widehat{x}(k) | k \geq k''\},$$

$$\exists t' \in \mathbf{R}, \forall t'' \geq t', \omega(x) = \{x(t) | t \geq t''\}.$$

Proof. We denote once again the elements of $\mathbf{B}^n$ with $\mu^1,...,\mu^{2^n}$. From the proof of Theorem 5, if $\widehat{\mathbf{T}}^{\widehat{x}}_{\mu^1},...,\widehat{\mathbf{T}}^{\widehat{x}}_{\mu^p}$ are the infinite sets $\widehat{\mathbf{T}}^{\widehat{x}}_{\mu^i}, i=\overline{1,2^n}$ then $\widehat{\omega}(\widehat{x})=\{\mu^1,...,\mu^p\}$. The number

$$k'=\begin{cases} 1+\max\{k|k\in\mathbf{N}_{\_},\widehat{x}(k)\in\widehat{Or}(\widehat{x})\setminus\widehat{\omega}(\widehat{x})\},\ if\ \widehat{Or}(\widehat{x})\setminus\widehat{\omega}(\widehat{x})\neq\varnothing \\ -1,\ otherwise\end{cases}$$

satisfies the property that $\forall i\in\{1,...,p\},\forall k''\geq k',\{k'',k''+1,k''+2,...\}\cap\widehat{\mathbf{T}}^{\widehat{x}}_{\mu^i}\neq\varnothing$ and $\forall i\in\{p+1,...,n\},\forall k''\geq k',\{k'',k''+1,k''+2,...\}\cap\widehat{\mathbf{T}}^{\widehat{x}}_{\mu^i}=\varnothing$, thus $\forall k''\geq k'$,

$$\{\mu^1,...,\mu^p\}=\{\widehat{x}(k)|k\in\widehat{\mathbf{T}}^{\widehat{x}}_{\mu^1}\cup...\cup\widehat{\mathbf{T}}^{\widehat{x}}_{\mu^p}\}$$

$$=\{\widehat{x}(k)|k\in(\widehat{\mathbf{T}}^{\widehat{x}}_{\mu^1}\cup...\cup\widehat{\mathbf{T}}^{\widehat{x}}_{\mu^p})\cap\{k'',k''+1,k''+2,...\}\}=\{\widehat{x}(k)|k\geq k''\}.$$

Similarly, if $\mathbf{T}^x_{\mu^1},...,\mathbf{T}^x_{\mu^p}$ are the unbounded from above sets $\mathbf{T}^x_{\mu^i}, i=\overline{1,2^n}$ then $\omega(x)=\{\mu^1,...,\mu^p\}$ and, with the notation

$$t'=\begin{cases} \sup\{t|t\in\mathbf{R},x(t)\in Or(x)\setminus\omega(x)\},\ if\ Or(x)\setminus\omega(x)\neq\varnothing \\ 0,\ otherwise\end{cases}$$

we have that $\forall i\in\{1,...,p\},\forall t''\geq t',[t'',\infty)\cap\mathbf{T}^x_{\mu^i}\neq\varnothing$ and $\forall i\in\{p+1,...,n\},\forall t''\geq t',[t'',\infty)\cap\mathbf{T}^x_{\mu^i}=\varnothing$, thus $\forall t''\geq t'$,

$$\{\mu^1,...,\mu^p\}=\{x(t)|t\in\mathbf{T}^x_{\mu^1}\cup...\cup\mathbf{T}^x_{\mu^p}\}$$

$$=\{x(t)|t\in(\mathbf{T}^x_{\mu^1}\cup...\cup\mathbf{T}^x_{\mu^p})\cap[t'',\infty)\}=\{x(t)|t\geq t''\}.$$

□

Remark 13. *Let the signals $\widehat{x}$ and x. If $\widehat{Or}(\widehat{x})\neq\widehat{\omega}(\widehat{x})$, the time instant $k'\in\mathbf{N}$ exists that determines two time intervals for $\widehat{x}$: $\{-1,0,...,k'-1\}$ when $\widehat{x}$ can take values in any of $\widehat{Or}(\widehat{x})\setminus\widehat{\omega}(\widehat{x}),\widehat{\omega}(\widehat{x})$ and $\{k',k'+1,...\}$ when $\widehat{x}$ takes values in $\widehat{\omega}(\widehat{x})$ only. Similarly for x, if $Or(x)\neq\omega(x)$, the time instant $t'\in\mathbf{R}$ exists that determines two time intervals for x: $(-\infty,t')$ when x can take values in both sets $Or(x)\setminus\omega(x),\omega(x)$ and $[t',\infty)$ when x takes values in $\omega(x)$ only.*

6. The Image of a Signal via a Function

Notation 9. *Let $f:\mathbf{B}^n\to\mathbf{B}^m$ and the signals $\widehat{x}\in\widehat{S}^{(n)},x\in S^{(n)}$. We denote with $\widehat{f}(\widehat{x}):\mathbf{N}_{\_}\to\mathbf{B}^m, f(x):\mathbf{R}\to\mathbf{B}^m$ the functions*

$$\forall k\in\mathbf{N}_{\_},\widehat{f}(\widehat{x})(k)=f(\widehat{x}(k)),\tag{6.1}$$

$$\forall t\in\mathbf{R},f(x)(t)=f(x(t)).\tag{6.2}$$

Theorem 8. *The function f defines by (6.1), (6.2) two functions $\widehat{f}:\widehat{S}^{(n)}\to\widehat{S}^{(m)},f:S^{(n)}\to S^{(m)}$.*

Proof. Indeed, we suppose in the real time case that x is given by

$$x(t) = \mu \cdot \chi_{(-\infty,t_0)}(t) \oplus x(t_0) \cdot \chi_{[t_0,t_1)}(t) \oplus \dots \oplus x(t_k) \cdot \chi_{[t_k,t_{k+1})}(t) \oplus \dots, \tag{6.3}$$

where $\mu \in \mathbf{B}^n$ and $(t_k) \in Seq$. Then we have

$$\begin{aligned} f(x)(t) = f(\mu) \cdot \chi_{(-\infty,t_0)}(t) \oplus f(x(t_0)) \cdot \chi_{[t_0,t_1)}(t) \oplus \\ \dots \oplus f(x(t_k)) \cdot \chi_{[t_k,t_{k+1})}(t) \oplus \dots \end{aligned} \tag{6.4}$$

We infer that $f(x) \in S^{(m)}$. □

Definition 12. *The signals $\widehat{f}(\widehat{x}), f(x)$ are called the* ***images of*** *$\widehat{x}, x$* ***via*** *f.*

Example 5. *We take $i \in \{1, ..., n\}$ and $f : \mathbf{B}^n \to \mathbf{B}$ defined by*

$$\forall \mu \in \mathbf{B}^n, f(\mu) = \mu_i.$$

The images of $\widehat{x} \in \widehat{S}^{(n)}, x \in S^{(n)}$ via f are the coordinates $\widehat{x}_i \in \widehat{S}^{(1)}, x_i \in S^{(1)}$.

Theorem 9. *For $f : \mathbf{B}^n \to \mathbf{B}^m$ and $x \in S^{(n)}$, we have $\forall t \in \mathbf{R}$,*

$$f(x)(t-0) = f(x(t-0)), \tag{6.5}$$

$$f(x)(t+0) = f(x(t+0)). \tag{6.6}$$

Proof. We suppose that x is of the form (6.3), with $\mu \in \mathbf{B}^n$ and $(t_k) \in Seq$, thus $x(t-0)$ and $x(t+0)$ are given by (2.4), (2.5). We have

$$\begin{aligned} f(x)(t-0) &\overset{(6.4)}{=} f(\mu) \cdot \chi_{(-\infty,t_0]}(t) \oplus f(x(t_0)) \cdot \chi_{(t_0,t_1]}(t) \oplus \\ &\quad \dots \oplus f(x(t_k)) \cdot \chi_{(t_k,t_{k+1}]}(t) \oplus \dots \\ &\overset{(2.4)}{=} f(x(t-0)), \end{aligned}$$

$$f(x)(t+0) = f(x)(t) = f(x(t)) \overset{(2.5)}{=} f(x(t+0)).$$

□

Theorem 10. *The initial and the final time sets of $\widehat{f}(\widehat{x}), f(x)$ fulfill $\widehat{F}^{\widehat{x}} \subset \widehat{F}^{\widehat{f}(\widehat{x})}$, $I^x \subset I^{f(x)}$, $F^x \subset F^{f(x)}$.*

Proof. In order to prove the first inclusion, we take an arbitrary $k' \in \widehat{F}^{\widehat{x}}$, thus $\mu \in \mathbf{B}^n$ exists with $\forall k \geq k', \widehat{x}(k) = \mu$. We infer the existence of $\mu'(= f(\mu))$ such that $\forall k \geq k', \widehat{f}(\widehat{x})(k) = f(\widehat{x}(k)) = f(\mu)$, thus $k' \in \widehat{F}^{\widehat{f}(\widehat{x})}$.

We prove the second inclusion. If x is constant, then $f(x)$ is constant and the inclusion holds with $I^x = I^{f(x)} = \mathbf{R}$. We can suppose now without loosing the generality that x is given by (6.3) and that $\mu \neq x(t_0)$. Let us take an arbitrary $t' \in (-\infty, t_0) = I^x$. As $f(x)(t)$ is given by (6.4), we infer that $\forall t \leq t', f(x)(t) = f(\mu)$, in other words $t' \in I^{f(x)}$. □

Theorem 11. *We have*

$$f(\lim_{t\to-\infty} x(t)) = \lim_{t\to-\infty} f(x)(t)$$

and if $\lim\limits_{k\to\infty}\widehat{x}(k), \lim\limits_{t\to\infty}x(t)$ *exist, then*

$$f(\lim_{k\to\infty}\widehat{x}(k)) = \lim_{k\to\infty}\widehat{f}(\widehat{x})(k),$$

$$f(\lim_{t\to\infty}x(t)) = \lim_{t\to\infty}f(x)(t).$$

Proof. We prove the last equality in the following manner. If $\lim\limits_{t\to\infty}x(t)$ exists, then $\mu\in\mathbf{B}^n$ and $t'\in\mathbf{R}$ exist such that $\forall t\geq t', x(t)=\mu$, thus $\forall t\geq t', f(x)(t)=f(x(t))=f(\mu)$. We have shown that $\lim\limits_{t\to\infty}f(x)(t)$ exists and it is equal with $f(\mu)$. We infer:

$$f(\lim_{t\to\infty}x(t)) = f(\mu) = \lim_{t\to\infty}f(x)(t).$$

□

Theorem 12. *The forgetful function acts on the images of* $\widehat{x}, x$ *via* f *like this:*

$$\forall k'\in\mathbf{N}, \widehat{\sigma}^{k'}(\widehat{f}(\widehat{x})) = \widehat{f}(\widehat{\sigma}^{k'}(\widehat{x})), \tag{6.7}$$

$$\forall t'\in\mathbf{R}, \sigma^{t'}(f(x)) = f(\sigma^{t'}(x)). \tag{6.8}$$

Proof. We prove the first equality and let $k'\in\mathbf{N}, k\in\mathbf{N}\_$ arbitrary. We get:

$$\widehat{\sigma}^{k'}(\widehat{f}(\widehat{x}))(k) = \widehat{f}(\widehat{x})(k+k') = \widehat{f}(\widehat{x}(k+k')) = \widehat{f}(\widehat{\sigma}^{k'}(\widehat{x})(k)) = \widehat{f}(\widehat{\sigma}^{k'}(\widehat{x}))(k).$$

We prove the second equality now and we take for this $t'\in\mathbf{R}, t\in\mathbf{R}$ arbitrary. We obtain:

$$\sigma^{t'}(f(x))(t) = \begin{cases} f(x)(t), t\geq t', \\ f(x)(t'-0), t<t' \end{cases} \overset{Theorem\,9}{=} \begin{cases} f(x(t)), t\geq t', \\ f(x(t'-0)), t<t' \end{cases}$$

$$= f(\begin{cases} x(t), t\geq t', \\ x(t'-0), t<t' \end{cases}) = f(\sigma^{t'}(x)(t)) = f(\sigma^{t'}(x))(t).$$

□

Theorem 13. *If* $f, \widehat{x}, x$ *are like previously, then for any* $\mu\in\mathbf{B}^n$ *we have* $\widehat{\mathbf{T}}_{\mu}^{\widehat{x}} \subset \widehat{\mathbf{T}}_{f(\mu)}^{\widehat{f}(\widehat{x})}, \mathbf{T}_{\mu}^{x} \subset \mathbf{T}_{f(\mu)}^{f(x)}$.

Proof. Let $k\in\widehat{\mathbf{T}}_{\mu}^{\widehat{x}}$ arbitrary, thus $\widehat{x}(k)=\mu$. This means that $f(\widehat{x}(k))=f(\mu)$, i.e. $\widehat{f}(\widehat{x})(k)=f(\mu)$, in other words $k\in\widehat{\mathbf{T}}_{f(\mu)}^{\widehat{f}(\widehat{x})}$. □

Theorem 14. *We have*

$$\widehat{Or}(\widehat{f}(\widehat{x})) = f(\widehat{Or}(\widehat{x})), \tag{6.9}$$

$$Or(f(x)) = f(Or(x)), \tag{6.10}$$

$$\widehat{\omega}(\widehat{f}(\widehat{x})) = f(\widehat{\omega}(\widehat{x})), \tag{6.11}$$

$$\omega(f(x)) = f(\omega(x)). \tag{6.12}$$

Proof. We prove the first equation:

$$\mu \in \widehat{Or}(\widehat{f}(\widehat{x})) \Longleftrightarrow \exists k \in \mathbf{N}_{\_}, \widehat{f}(\widehat{x})(k) = \mu$$

$$\Longleftrightarrow \exists k \in \mathbf{N}_{\_}, f(\widehat{x}(k)) = \mu \Longleftrightarrow \mu \in f(\widehat{Or}(\widehat{x})).$$

We prove the last equation now: $\omega(f(x)) \subset f(\omega(x))$. Let $\mu' \in \omega(f(x))$ arbitrary, meaning that $\mathbf{T}_{\mu'}^{f(x)}$ is superiorly unbounded. A sequence $(t_k) \in Seq$ exists then such that $\forall k \in \mathbf{N}, f(x(t_k)) = \mu'$. In this situation, a subsequence $(t_{i_k}) \subset (t_k)$ from Seq exists, together with $\mu \in \mathbf{B}^n$, having the property that $\forall k \in \mathbf{N}, x(t_{i_k}) = \mu$. We have $\mu \in \omega(x)$ and $f(\mu) = \mu'$, thus $\mu' \in f(\omega(x))$.

$f(\omega(x)) \subset \omega(f(x))$. We take $\mu' \in f(\omega(x))$ arbitrary. This fact means that $\mu \in \omega(x)$ exists with $\mathbf{T}_{\mu}^{x}$ superiorly unbounded and $\mu' = f(\mu)$. But the inclusion $\mathbf{T}_{\mu}^{x} \subset \mathbf{T}_{f(\mu)}^{f(x)}$ shows that $\mathbf{T}_{f(\mu)}^{f(x)}$ is superiorly unbounded, in other words $\mu' = f(\mu) \in \omega(f(x))$. □

Chapter 2

The Main Definitions on Periodicity

In this chapter we list the main definitions on periodicity, that are necessary in order to understand the rest of the exposure: the eventually periodic points and the eventually periodic signals, the periodic points and the periodic signals.

1. Eventually Periodic Points

Definition 13. *In case that, for $\mu \in \widehat{Or}(\widehat{x})$, $p \geq 1$, some $k' \in \mathbf{N}_{\_}$ exists such that we have*

$$\begin{cases} \widehat{\mathbf{T}}_{\mu}^{\widehat{x}} \cap \{k', k'+1, k'+2, ...\} \neq \varnothing \text{ and } \forall k \in \widehat{\mathbf{T}}_{\mu}^{\widehat{x}} \cap \{k', k'+1, k'+2, ...\}, \\ \{k + zp | z \in \mathbf{Z}\} \cap \{k', k'+1, k'+2, ...\} \subset \widehat{\mathbf{T}}_{\mu}^{\widehat{x}}, \end{cases} \tag{1.1}$$

*then μ is said to be **eventually periodic** (an **eventually periodic point of** $\widehat{x}$, or **of** $\widehat{Or}(\widehat{x})$) with the **period** p and with the **limit of periodicity** k'.*

Let $\mu \in Or(x)$ and $T > 0$ such that $t' \in \mathbf{R}$ exists with

$$\mathbf{T}_{\mu}^{x} \cap [t', \infty) \neq \varnothing \text{ and } \forall t \in \mathbf{T}_{\mu}^{x} \cap [t', \infty), \{t + zT | z \in \mathbf{Z}\} \cap [t', \infty) \subset \mathbf{T}_{\mu}^{x}. \tag{1.2}$$

*Then μ is said to be **eventually periodic** (an **eventually periodic point of** x, or **of** $Or(x)$) with the **period** T and with the **limit of periodicity** t'.*

Notation 10. *We use the notation $\widehat{P}_{\mu}^{\widehat{x}}$ for the set of the periods of $\mu \in \widehat{Or}(\widehat{x})$:*

$$\widehat{P}_{\mu}^{\widehat{x}} = \{p | p \geq 1, \exists k' \in \mathbf{N}_{\_}, (1.1) \text{ holds}\}.$$

The notation P_{μ}^{x} is used for the analogue set of the periods of $\mu \in Or(x)$:

$$P_{\mu}^{x} = \{T | T > 0, \exists t' \in \mathbf{R}, (1.2) \text{ holds}\}.$$

Notation 11. *We denote with $\widehat{L}_{\mu}^{\widehat{x}}$ the set of the limits of periodicity of $\mu \in \widehat{Or}(\widehat{x})$:*

$$\widehat{L}_{\mu}^{\widehat{x}} = \{k' | k' \in \mathbf{N}_{\_}, \exists p \geq 1, (1.1) \text{ holds}\}$$

and L_{μ}^{x} denotes the set of the limits of periodicity of $\mu \in Or(x)$:

$$L_{\mu}^{x} = \{t' | t' \in \mathbf{R}, \exists T > 0, (1.2) \text{ is true}\}.$$

Definition 14. *The numbers*

$$\widetilde{p} = \min \widehat{P}^{\widehat{x}}_{\mu},$$

$$\widetilde{T} = \min P^{x}_{\mu}$$

are called ***prime periods*** *(of μ) and the numbers*

$$\widetilde{k'} = \min \widehat{L}^{\widehat{x}}_{\mu},$$

$$\widetilde{t'} = \min L^{x}_{\mu}$$

are called ***prime limits of periodicity*** *(of μ).*

Remark 14. *The periodic and the eventually periodic points of a vector field $f: M \to M$ occur in the real numbers dynamical systems theory and, by translation to binary concepts, they should refer to a function $\Phi: \mathbf{B}^n \to \mathbf{B}^n$. This is the framework of the last chapters of this monograph. Here, periodicity is adapted to signals and, instead of the eventual periodicity of a* ***point*** *of $\widehat{x}, x$, we can speak sometimes about the eventual periodicity of a* ***value*** *of $\widehat{x}, x$.*

Remark 15. *The eventual periodicity of $\mu \in \widehat{Or}(\widehat{x})$ with the period p and the limit of periodicity k' means a periodic behavior that starts from k': for any $k \in \widehat{\mathbf{T}}^{\widehat{x}}_{\mu} \cap \{k', k'+1, k'+2, ...\}$, we can go upwards and downwards with multiples of p to $k + zp, z \in \mathbf{Z}$ without getting out of the 'final' time set $\{k', k'+1, k'+2, ...\}$ and we still remain in $\widehat{\mathbf{T}}^{\widehat{x}}_{\mu}$. In other words*

$$\mu = \widehat{x}(k) = \widehat{x}(k-p) = \widehat{x}(k-2p) = ... = \widehat{x}(k-k_1 p),$$

where $k_1 \in \mathbf{N}$, fulfills $k - k_1 p \geq k'$, $k - (k_1+1)p < k'$ and also

$$\mu = \widehat{x}(k) = \widehat{x}(k+p) = \widehat{x}(k+2p) = ...$$

Remark 16. *The requirement $\widehat{\mathbf{T}}^{\widehat{x}}_{\mu} \cap \{k', k'+1, k'+2, ...\} \neq \varnothing$ is one of non-triviality. It is necessary, because for any point $\mu \in \widehat{Or}(\widehat{x}) \smallsetminus \widehat{\omega}(\widehat{x})$, the set $\widehat{\mathbf{T}}^{\widehat{x}}_{\mu}$ is finite, some $k' \in \mathbf{N}\_$ exists such that $\widehat{\mathbf{T}}^{\widehat{x}}_{\mu} \cap \{k', k'+1, k'+2, ...\} = \varnothing$ and*

$$\begin{cases} \forall k \in \widehat{\mathbf{T}}^{\widehat{x}}_{\mu} \cap \{k', k'+1, k'+2, ...\}, \\ \{k + zp | z \in \mathbf{Z}\} \cap \{k', k'+1, k'+2, ...\} \subset \widehat{\mathbf{T}}^{\widehat{x}}_{\mu}, \end{cases}$$

equivalent with

$$\forall k, k \in \varnothing \Longrightarrow \{k + zp | z \in \mathbf{Z}\} \cap \{k', k'+1, k'+2, ...\} \subset \widehat{\mathbf{T}}^{\widehat{x}}_{\mu},$$

is true, $\forall p \geq 1$ (the falsity implies anything).

Remark 17. *The eventually periodic points $\mu \in \widehat{Or}(\widehat{x})$ are omega limit points $\mu \in \widehat{\omega}(\widehat{x})$ because the set $\widehat{\mathbf{T}}^{\widehat{x}}_{\mu}$ is necessarily infinite.*

Remark 18. *Definition 13 avoids the triviality expressed by the possibility* $\widehat{\mathbf{T}}_{\mu}^{\widehat{x}} \cap \{k', k'+1, k'+2, ...\} = \varnothing$*, but a way of obtaining the same result is to ask* $\mu \in \widehat{\omega}(\widehat{x})$ *instead of* $\mu \in \widehat{Or}(\widehat{x})$*, see Lemma 1, page 223, since in that case we have that* $\widehat{\mathbf{T}}_{\mu}^{\widehat{x}}$ *is infinite and* $\forall k' \in \mathbf{N}_{\_}, \widehat{\mathbf{T}}_{\mu}^{\widehat{x}} \cap \{k', k'+1, k'+2, ...\} \neq \varnothing$*. With this note, the discrete time part of Definition 13 becomes, equivalently:* $\mu \in \widehat{\omega}(\widehat{x})$ *is eventually periodic with the period p and the limit of periodicity* k' *if*

$$\forall k \in \widehat{\mathbf{T}}_{\mu}^{\widehat{x}} \cap \{k', k'+1, k'+2, ...\}, \{k+zp | z \in \mathbf{Z}\} \cap \{k', k'+1, k'+2, ...\} \subset \widehat{\mathbf{T}}_{\mu}^{\widehat{x}}.$$

Remark 19. *The eventual periodicity of* $\mu \in Or(x)$ *with the period T and the limit of periodicity* t' *means periodicity that starts from* $t' \in \mathbf{R}$*: for any* $t \in \mathbf{T}_{\mu}^{x} \cap [t', \infty)$ *we can go arbitrarily upwards and downwards with multiples of* T*, to* $t + zT, z \in \mathbf{Z}$ *without leaving the 'final' time set* $[t', \infty)$ *and we still remain in* $\mathbf{T}_{\mu}^{x}$*.*

Remark 20. *The eventual periodicity of* $\mu \in Or(x)$ *obviously implies that* $\mu \in \omega(x)$*, because the set* $\mathbf{T}_{\mu}^{x}$ *is superiorly unbounded.*

Remark 21. *The requirement* $\mathbf{T}_{\mu}^{x} \cap [t', \infty) \neq \varnothing$ *in (1.2) is one of non-triviality. An equivalent way of obtaining non-triviality is to ask* $\mu \in \omega(x)$ *and to replace (1.2) with*

$$\forall t \in \mathbf{T}_{\mu}^{x} \cap [t', \infty), \{t + zT | z \in \mathbf{Z}\} \cap [t', \infty) \subset \mathbf{T}_{\mu}^{x}.$$

Remark 22. *We have* $\widehat{P}_{\mu}^{\widehat{x}} \neq \varnothing \Longleftrightarrow \widehat{L}_{\mu}^{\widehat{x}} \neq \varnothing$ *and* $P_{\mu}^{x} \neq \varnothing \Longleftrightarrow L_{\mu}^{x} \neq \varnothing$*.*

Example 6. *The signal* $\widehat{x} \in \widehat{S}^{(2)}$ *with* $\widehat{\mathbf{T}}_{(1,1)}^{\widehat{x}} = \{1, 3, 5, ...\}$ *fulfills the property that* $(1, 1)$ *is eventually periodic with the period* 2 *and the limit of periodicity* $k' = 0$*.*

Example 7. *Let* $\widehat{x} \in \widehat{S}^{(2)}$ *arbitrary with* $(1, 1) \notin \widehat{Or}(\widehat{x})$ *and* $\widehat{x}(-1) \neq \widehat{x}(0)$*.* $y \in S^{(2)}$ *is defined like this:*

$$y(t) = \widehat{x}(-1) \cdot \chi_{(-\infty,0)}(t) \oplus \widehat{x}(0) \cdot \chi_{[0,1)}(t) \oplus (1,1) \cdot \chi_{[1,2)}(t) \oplus \widehat{x}(2) \cdot \chi_{[2,3)}(t) \oplus$$

$$\oplus (1,1) \cdot \chi_{[3,4)}(t) \oplus \widehat{x}(4) \cdot \chi_{[4,5)}(t) \oplus (1,1) \cdot \chi_{[5,6)}(t) \oplus ...$$

The point $(1, 1)$ *is an eventually periodic point of* y *with the period* $T = 2$ *and any* $t' \in [0, \infty)$ *is a limit of periodicity. The situation* $\widehat{x}(-1) = \widehat{x}(0)$ *generates a special case of eventual periodicity called periodicity, that will be analyzed later and the situation* $(1, 1) \in \widehat{Or}(\widehat{x})$ *might generate several possibilities, for example* y *has the arbitrary period* $T > 0$ *or* y *changes its limit of periodicity.*

Remark 23. *Let the function* $f : \mathbf{B}^n \to \mathbf{B}^m$ *and we suppose that* $\mu \in \mathbf{B}^n, p \geq 1, k' \in \mathbf{N}_{\_}, T > 0, t' \in \mathbf{R}$ *are such that (1.1), (1.2) hold. We analyze the truth of the statements*

$$\widehat{\mathbf{T}}_{f(\mu)}^{\widehat{f}(\widehat{x})} \cap \{k', k'+1, k'+2, ...\} \neq \varnothing, \tag{1.3}$$

$$\begin{cases} \forall k \in \widehat{\mathbf{T}}_{f(\mu)}^{\widehat{f}(\widehat{x})} \cap \{k', k'+1, k'+2, ...\}, \\ \{k + zp | z \in \mathbf{Z}\} \cap \{k', k'+1, k'+2, ...\} \subset \widehat{\mathbf{T}}_{f(\mu)}^{\widehat{f}(\widehat{x})}. \end{cases} \tag{1.4}$$

From $\widehat{\mathbf{T}}_{\mu}^{\widehat{x}} \subset \widehat{\mathbf{T}}_{f(\mu)}^{\widehat{f}(\widehat{x})}$ *and from (1.1) we infer that*

$$\varnothing \neq \widehat{\mathbf{T}}_{\mu}^{\widehat{x}} \cap \{k', k'+1, k'+2, ...\} \subset \widehat{\mathbf{T}}_{f(\mu)}^{\widehat{f}(\widehat{x})} \cap \{k', k'+1, k'+2, ...\},$$

thus (1.3) is true. We cannot say anything about the truth of (1.4) however, because $k \geq k'$ *with* $\widehat{f}(\widehat{x})(k) = f(\mu)$ *and* $\widehat{x}(k) \neq \mu$ *might exist and then (1.1) cannot be used. The discussion is similar for the real time case, when the eventual periodicity of the point* $f(\mu)$ *of* $f(x)$ *is uncertain.*

Example 8. *We continue Example 7 in the following manner. Let* $f : \mathbf{B}^2 \to \mathbf{B}^2$ *with* $f(1,1) = (1,1)$ *and we consider the signal* $z \in S^{(2)}, z = f(y)$ *fulfilling the property:*

$$\begin{aligned} & f(\widehat{x}(2)), f(\widehat{x}(4)), f(\widehat{x}(6)), ... = \\ & = \underbrace{(0,0)}_{1}, (1,1), \underbrace{(0,0), (0,0)}_{2}, (1,1), \underbrace{(0,0), (0,0), (0,0)}_{3}, (1,1), ... \end{aligned}$$

Then, even if $(1,1)$ *is an eventually periodic point of* y*,* $f(1,1) = (1,1)$ *is not an eventually periodic point of* $f(y) = z$.

2. Eventually Periodic Signals

Definition 15. *For* $p \geq 1$ *and* $k' \in \mathbf{N}_{\_}$, *if*

$$\forall k \geq k', \widehat{x}(k) = \widehat{x}(k+p), \tag{2.1}$$

we say that $\widehat{x}$ *is* **eventually periodic** *with the* **period** p *and the* **limit of periodicity** k'.

Let $T > 0$. *If* $t' \in \mathbf{R}$ *exists such that*

$$\forall t \geq t', x(t) = x(t+T) \tag{2.2}$$

is true, we say that x *is* **eventually periodic** *with the* **period** T *and the* **limit of periodicity** t'.

Notation 12. *We use the notation* $\widehat{P^x}$ *for the set of the periods of* $\widehat{x}$:

$$\widehat{P^x} = \{p | p \geq 1, \exists k' \in \mathbf{N}_{\_}, (2.1) \text{ holds}\}$$

and also the notation P^x *for the set of the periods of* x:

$$P^x = \{T | T > 0, \exists t' \in \mathbf{R}, (2.2) \text{ holds}\}.$$

Notation 13. *We use the notations*

$$\widehat{L^x} = \{k' | k' \in \mathbf{N}_{\_}, \exists p \geq 1, (2.1) \text{ holds}\},$$

$$L^x = \{t' | t' \in \mathbf{R}, \exists T > 0, (2.2) \text{ holds}\}.$$

Definition 16. *The numbers*

$$\widetilde{p} = \min \widehat{P^{\widehat{x}}},$$

$$\widetilde{T} = \min P^x$$

are called ***prime periods*** *(of $\widehat{x}, x$) and the numbers*

$$\widetilde{k'} = \min \widehat{L^{\widehat{x}}},$$

$$\widetilde{t'} = \min L^x$$

are called ***prime limits of periodicity*** *(of $\widehat{x}, x$).*

Remark 24. *The eventual periodicity of $\widehat{x}$ with the period p and the limit of periodicity k' means that all the values $\mu \in \widehat{\omega}(\widehat{x})$ are eventually periodic with the same period p and with the same limit of periodicity k'.*

Remark 25. *The signal x is eventually periodic with the period T and the limit of periodicity t' if all the values $\mu \in \omega(x)$ are eventually periodic with the same period T and with the same limit of periodicity t'.*

Remark 26. *We see that $\widehat{P^{\widehat{x}}} \neq \varnothing \Longleftrightarrow \widehat{L^{\widehat{x}}} \neq \varnothing$ and $P^x \neq \varnothing \Longleftrightarrow L^x \neq \varnothing$.*

Example 9. *The signal $\widehat{x} \in \widehat{S^{(1)}}$ defined by $\widehat{x} = 0, 1, 1, 1, \ldots$ is eventually constant with $\widehat{F^{\widehat{x}}} =$ $\mathbf{N}$. It is eventually periodic also with the period $p = 1$ and the limit of periodicity $k' = 0$.*

Example 10. *The real time analogue of the previous example is given by $x \in S^{(1)}, x(t) = \chi_{[0,\infty)}(t)$. The signal x is eventually constant and eventually periodic, with the arbitrary period $T > 0$. We have $I^x = (-\infty, 0)$ and $F^x = L^x = [0, \infty)$.*

Remark 27. *Let $f : \mathbf{B}^n \to \mathbf{B}^m$. If $p \geq 1, k' \in \mathbf{N}\_$ and $T > 0, t' \in \mathbf{R}$ are such that (2.1), (2.2) hold, then the fact that*

$$\forall k \geq k', \widehat{f}(\widehat{x})(k) = f(\widehat{x}(k)) = f(\widehat{x}(k+p)) = \widehat{f}(\widehat{x})(k+p),$$

$$\forall t \geq t', f(x)(t) = f(x(t)) = f(x(t+T)) = f(x)(t+T)$$

shows the validity of the inclusions $\widehat{P^{\widehat{x}}} \subset \widehat{P^{\widehat{f}(\widehat{x})}}, \widehat{L^{\widehat{x}}} \subset \widehat{L^{\widehat{f}(\widehat{x})}}$ and $P^x \subset P^{f(x)}, L^x \subset L^{f(x)}$.

3. Periodic Points

Definition 17. *We consider the signals $\widehat{x} \in \widehat{S^{(n)}}, x \in S^{(n)}$.*

Let $\mu \in \widehat{Or}(\widehat{x})$ and $p \geq 1$. If

$$\forall k \in \widehat{\mathbf{T}}_{\mu}^{\widehat{x}}, \{k + zp | z \in \mathbf{Z}\} \cap \mathbf{N}\_ \subset \widehat{\mathbf{T}}_{\mu}^{\widehat{x}}, \tag{3.1}$$

we say that μ is ***periodic*** *(a* ***periodic point of*** *$\widehat{x}$, or* ***of*** *$\widehat{Or}(\widehat{x})$) with the* ***period*** *p.*

Let $\mu \in Or(x)$ and $T > 0$ such that $t' \in I^x$ exists with

$$\forall t \in \mathbf{T}_{\mu}^{x} \cap [t', \infty), \{t + zT | z \in \mathbf{Z}\} \cap [t', \infty) \subset \mathbf{T}_{\mu}^{x}. \tag{3.2}$$

Then μ is called ***periodic*** *(a* ***periodic point of*** *x, or* ***of*** *$Or(x)$) with the* ***period*** *T.*

Remark 28. *The periodicity of the point $\mu \in \widehat{Or}(\widehat{x})$ (of the value μ of $\widehat{x}$) with the period $p \geq 1$ means eventual periodicity that starts at the limit of periodicity $k' = -1$. The property is non-trivial (in the sense that was mentioned at Remark 18, page 19), since $\mu \in \widehat{Or}(\widehat{x})$ implies $\emptyset \neq \widehat{\mathbf{T}}^{\widehat{x}}_{\mu} = \widehat{\mathbf{T}}^{\widehat{x}}_{\mu} \cap \{-1,0,1,...\}$.*

Remark 29. *The periodicity of $\mu \in Or(x)$ with the period $T > 0$ means eventual periodicity with the property that the limit of periodicity t' is an initial time instant of x also. Even if the initial time t' is not unique in the real time case, periodicity is an eventual periodicity that starts 'from the very beginning'. The property is non-trivial as far as $\mathbf{T}^{x}_{\mu} \cap [t',\infty) \neq \emptyset$ results from Lemma 2, page 223.*

Remark 30. *Because the periodicity of μ is a special case of eventual periodicity, the notations $\widehat{P}^{\widehat{x}}_{\mu}, P^{x}_{\mu}, \widehat{L}^{\widehat{x}}_{\mu}, L^{x}_{\mu}$ and the concepts of prime period, prime limit of periodicity are used for the periodic points also, with the remark that $\widehat{L}^{\widehat{x}}_{\mu} = \mathbf{N}_{\_}$, $L^{x}_{\mu} \cap I^{x} \neq \emptyset$.*

Remark 31. *The periodic points are omega limit points. On one hand even if there is a periodic point, omega limit points which are distinct from it might exist that are not periodic and on the other hand when stating periodicity we must not ask $\mu \in \widehat{\omega}(\widehat{x}), \mu \in \omega(x)$ because triviality is impossible.*

Remark 32. *Mentioning the limit of periodicity in case of periodicity is not necessary: in the discrete time case because $k' = -1$ is always clear and in the real time case because the property of periodicity does not depend on the choice of t', as we shall see later.*

Example 11. *Let $\widehat{x} \in \widehat{S}^{(n)}$, $\mu \in \widehat{Or}(\widehat{x})$ with $\widehat{\mathbf{T}}^{\widehat{x}}_{\mu} = \{-1,1,3,5,...\}$, thus the point μ is periodic with the period $p = 2$.*

We define $x \in S^{(n)}$ by

$$x(t) = \widehat{x}(-1) \cdot \chi_{(-\infty,0)}(t) \oplus \widehat{x}(0) \cdot \chi_{[0,1)}(t) \oplus ... \oplus \widehat{x}(k) \cdot \chi_{[k,k+1)}(t) \oplus ...$$

We have $x(-\infty+0) = \mu$ and $\mathbf{T}^{x}_{x(-\infty+0)} = \mathbf{T}^{x}_{\mu} = (-\infty,0) \cup [1,2) \cup [3,4) \cup ...$ For any $t' \in [-1,0)$, we infer the truth of $(-\infty,t'] \subset \mathbf{T}^{x}_{x(-\infty+0)}$, $\mathbf{T}^{x}_{\mu} \cap [t',\infty) = [t',0) \cup [1,2) \cup [3,4) \cup ...$ and

$$\forall t \in [t',0) \cup [1,2) \cup [3,4) \cup ...,$$

$$\{t + z2 | z \in \mathbf{Z}\} \cap [t',\infty) \subset (-\infty,0) \cup [1,2) \cup [3,4) \cup ...$$

μ has the period $T = 2$.

Remark 33. *When $f : \mathbf{B}^n \to \mathbf{B}^m$ is a function and $\mu \in \mathbf{B}^n$ is a periodic point of $\widehat{x}, x$, we do not infer that $f(\mu)$ is a periodic point of $\widehat{f}(\widehat{x}), f(x)$, see Remark 23 page 19 referring to the eventually periodic points.*

4. Periodic Signals

Definition 18. *Let $\widehat{x} \in \widehat{S}^{(n)}, x \in S^{(n)}$ and $p \geq 1, T > 0$.*

If

$$\forall k \in \mathbf{N}_{\_}, \widehat{x}(k) = \widehat{x}(k+p), \tag{4.1}$$

*we say that $\widehat{x}$ is **periodic** with the **period** p.*

In case that $\exists t' \in I^x$,

$$\forall t \geq t', x(t) = x(t+T) \tag{4.2}$$

*holds, we say that x is **periodic** with the **period** T.*

Remark 34. *If $\widehat{x}$ is periodic with the period p then all its values $\mu \in \widehat{Or}(\widehat{x})$ are periodic with the period p. Since the periodic points are omega limit points, this means in particular that the periodicity of $\widehat{x}$ implies $\widehat{Or}(\widehat{x}) = \widehat{\omega}(\widehat{x})$.*

Remark 35. *If the signal x is periodic with the period T then all the values $\mu \in Or(x)$ are periodic with the same period T. Note that $Or(x) = \omega(x)$.*

Remark 36. *The periodic signals are special cases of eventually periodic signals when $k' = -1$ instead of $k' \in \widehat{L^{\widehat{x}}}$, respectively when $t' \in I^x \cap L^x$, instead of $t' \in L^x$. In particular the notations $\widehat{P^{\widehat{x}}}, P^x, \widehat{L^{\widehat{x}}}, L^x$ and the concepts of prime period, prime limit of periodicity are used for the periodic signals too. We have $\widehat{L^{\widehat{x}}} = \mathbf{N}_{\_}, L^x \cap I^x \neq \varnothing$.*

Remark 37. *Mentioning the limit of periodicity k', t' in Definition 18 is not necessary, since the property itself does not depend on the choice of k', t'.*

Example 12. *The signal $\widehat{x} \in \widehat{S}^{(1)}$ given by $\widehat{x} = 1, 0, 1, 0, 1, ...$ is periodic with the period 2, $\widehat{Or}(\widehat{x}) = \{0, 1\}$ and both points 0, 1 are periodic with the period 2.*

Example 13. *The signal $x \in S^{(1)}$ that is defined in the following way:*

$$x(t) = \chi_{(-\infty,0)}(t) \oplus \chi_{[1,2)}(t) \oplus \chi_{[3,4)}(t) \oplus ...$$

has the period 2 if we take arbitrarily the initial time=limit of periodicity $t' \in [-1, 0)$. If we take $t' < -1$ then (4.2) does not hold, i.e. t' is not limit of periodicity; if we take $t' \geq 0$, then t' is not initial time.

Remark 38. *Let $f : \mathbf{B}^n \to \mathbf{B}^m$ some function. The existence of $p \geq 1, T > 0, t' \in I^x$ such that (4.1), (4.2) hold shows that $t' \in I^{f(x)}$ see Theorem 10, page 13 and*

$$\forall k \in \mathbf{N}_{\_}, \widehat{f}(\widehat{x})(k) = f(\widehat{x}(k)) = f(\widehat{x}(k+p)) = \widehat{f}(\widehat{x})(k+p),$$

$$\forall t \geq t', f(x)(t) = f(x(t)) = f(x(t+T)) = f(x)(t+T)$$

are true, thus the situation is similar with the one from Remark 27, page 21 with $\widehat{L^{\widehat{x}}} = \widehat{L^{\widehat{f}(\widehat{x})}} = \mathbf{N}_{\_}$. The inclusion $L^x \subset L^{f(x)}$ from Remark 27 does not necessarily become equality here, for example the situation when x is not constant and $f(x)$ is constant is possible, and in this case $L^x \subsetneqq L^{f(x)} = \mathbf{R}$.

Chapter 3

Eventually Constant Signals

The purpose of the chapter is that of giving properties that are equivalent with the eventual constancy of the signals, a concept that is anticipated in Chapter 1, Definition 6, page 5 and the following paragraphs and in Chapter 2, Example 9 and Example 10, page 21. The point is that eventual periodicity, which occurs in many of these properties, must hold with any period in order to be equivalent to eventual constancy. The importance of eventual constancy is that in the asynchronous systems theory it coincides with stability.

The first group of eventual constancy properties of Section 1 does not involve periodicity. Groups 2 and 3 are related with the eventual periodicity of the points and they are introduced in Sections 3, 4 and 5. Group 4 of eventual constancy properties is related with the eventual periodicity of the signals and it is introduced in Section 6. Section 7 shows the connection between discrete time and continuous time as far as eventual constancy is concerned and Section 8 contains a discussion.

1. The First Group of Eventual Constancy Properties

Remark 39. *The first group of eventual constancy properties of the signals contains these properties that are not related with periodicity.*

Theorem 15. *Let the signals $\widehat{x} \in \widehat{S}^{(n)}, x \in S^{(n)}$.*

a) The statements

$$\exists \mu \in \mathbf{B}^n, \exists k' \in \mathbf{N}_{\_}, \forall k \geq k', \widehat{x}(k) = \mu, \tag{1.1}$$

$$\exists \mu \in \mathbf{B}^n, \exists k' \in \mathbf{N}_{\_}, \{k', k'+1, k'+2, ...\} \subset \widehat{\mathbf{T}}_{\mu}^{\widehat{x}}, \tag{1.2}$$

$$\exists \mu \in \mathbf{B}^n, \widehat{\omega}(\widehat{x}) = \{\mu\} \tag{1.3}$$

are equivalent.

b) The statements

$$\exists \mu \in \mathbf{B}^n, \exists t' \in \mathbf{R}, \forall t \geq t', x(t) = \mu, \tag{1.4}$$

$$\exists \mu \in \mathbf{B}^n, \exists t' \in \mathbf{R}, [t', \infty) \subset \mathbf{T}_{\mu}^{x}, \tag{1.5}$$

$$\exists \mu \in \mathbf{B}^n, \omega(x) = \{\mu\} \tag{1.6}$$

are also equivalent.

Proof. a) (1.1)$\Longrightarrow$(1.2) $\mu \in \mathbf{B}^n$ and $k' \in \mathbf{N}_-$ exist with the property

$$\forall k \geq k', \widehat{x}(k) = \mu.$$

Then

$$\{k', k'+1, k'+2, ...\} \subset \{k | k \in \mathbf{N}_-, \widehat{x}(k) = \mu\} \tag{1.7}$$

holds.

(1.2)$\Longrightarrow$(1.3) $\mu \in \mathbf{B}^n$ and $k' \in \mathbf{N}_-$ exist such that (1.7) holds. We suppose, see Theorem 7, page 11, that $k'' \in \mathbf{N}_-$ fulfills

$$\{\widehat{x}(k) | k \geq k''\} = \widehat{\omega}(\widehat{x}). \tag{1.8}$$

For $k_1 = \max\{k', k''\}$ we can write that

$$\{\mu\} \overset{(1.7)}{=} \{\widehat{x}(k) | k \geq k_1\} \overset{(1.8)}{=} \widehat{\omega}(\widehat{x}).$$

(1.3)$\Longrightarrow$(1.1) From (1.3) and Theorem 7, page 11 we have the existence of $k' \in \mathbf{N}_-$ such that

$$\{\mu\} = \widehat{\omega}(\widehat{x}) = \{\widehat{x}(k) | k \geq k'\},$$

wherefrom the truth of (1.1).

b) (1.4)$\Longrightarrow$(1.5) We suppose that $\mu \in \mathbf{B}^n$ and $t' \in \mathbf{R}$ exist such that $\forall t \geq t', x(t) = \mu$. Then $[t', \infty) \subset \{t | t \in \mathbf{R}, x(t) = \mu\}$.

(1.5)$\Longrightarrow$(1.6) Some $t_1 \in \mathbf{R}$ exists satisfying $\{x(t) | t \geq t_1\} = \omega(x)$ and, from the hypothesis, $\mu \in \mathbf{B}^n$ and $t' \in \mathbf{R}$ exist such that $[t', \infty) \subset \{t | t \in \mathbf{R}, x(t) = \mu\}$. We use the notation $t'' = \max\{t_1, t'\}$ and we have

$$\{\mu\} = \{x(t) | t \geq t''\} = \omega(x).$$

(1.6)$\Longrightarrow$(1.4) The hypothesis (1.6) and Theorem 7 show the existence of $\mu \in \mathbf{B}^n, t' \in \mathbf{R}$ with

$$\{\mu\} = \omega(x) = \{x(t) | t \geq t'\},$$

wherefrom the truth of (1.4). □

2. Eventual Constancy

Definition 19. *If $\widehat{x} \in \widehat{S}^{(n)}$ fulfills one of (1.1),...,(1.3), it is called* ***eventually constant*** *and if $x \in S^{(n)}$ fulfills one of (1.4),...,(1.6), it is called* ***eventually constant.*** *In (1.1), (1.2), $k' \in \mathbf{N}_-$ is called the* ***limit of constancy****, or* ***limit of equilibrium****, or* ***final time*** *of $\widehat{x}$. Similarly in (1.4), (1.5), $t' \in \mathbf{R}$ is called the* ***limit of constancy****, or* ***limit of equilibrium****, or* ***final time*** *of x.*

Theorem 16. *a) If $\widehat{x}$ is constant, it is eventually constant.*

b) If x is constant, it is eventually constant.

Proof. a) The constancy of $\widehat{x}$ means the eventual constancy of $\widehat{x}$ with the limit of constancy $k' = -1$.

b) The constancy of x is its eventual constancy with the limit of constancy $t' \in I^x$. □

Remark 40. *The eventual constancy of a signal coincides with the existence of the final value, Definition 6, page 5. This is the reason why in Definition 19, k' and t' are also called final time.*

Remark 41. *If $f : \mathbf{B}^n \to \mathbf{B}^m$ is some function and $\widehat{x}, x$ are eventually constant, then $\widehat{f}(\widehat{x}), f(x)$ are eventually constant and we have*

$$\widehat{\omega}(\widehat{f}(\widehat{x})) \overset{Theorem\,14,\,page\,14}{=} f(\widehat{\omega}(\widehat{x})) \overset{(1.3)}{=} \{f(\mu)\},$$

$$\omega(f(x)) \overset{Theorem\,14,\,page\,14}{=} f(\omega(x)) \overset{(1.6)}{=} \{f(\mu)\}.$$

Remark 42. *Eventual constancy is important in systems theory since it is associated with stability: if modeling is deterministic and the signal is an asynchronous flow, then stability means exactly the eventual constancy of that flow; and if modeling is non-deterministic and we have a set of deterministic flows, then stability means the eventual constancy of all these flows.*

3. The Second Group of Eventual Constancy Properties

Remark 43. *This group of eventual constancy properties of the signals involves the eventual periodicity of all the points μ of the orbit, i.e. in (3.1),...,(3.4), (3.5),...,(3.12) to follow we ask $\forall \mu \in \widehat{Or}(\widehat{x}), \forall \mu \in Or(x)$.*

Remark 44. *In order to understand better the way that these properties were written, to be noticed the existence of the following symmetries:*

- *(3.1)-{(3.5),(3.6)}; (3.2)-{(3.7),(3.8)}; (3.3)-{(3.9),(3.10)}; (3.4)-{(3.11), (3.12)};*
- *(3.1)-(3.2), (3.3)-(3.4) and (3.5)-(3.6)-(3.7)-(3.8), (3.9)-(3.10)-(3.11)-(3.12);*
- *(3.1)-(3.3), (3.2)-(3.4) and (3.5)-(3.6)-(3.9)-(3.10), (3.7)-(3.8)-(3.11)-(3.12).*

Theorem 17. *Let the signals $\widehat{x} \in \widehat{S}^{(n)}, x \in S^{(n)}$.*

a) The following statements are equivalent with the eventual constancy of $\widehat{x}$:

$$\begin{cases} \forall p \geq 1, \forall \mu \in \widehat{Or}(\widehat{x}), \exists k' \in \mathbf{N}_{\_}, \forall k \in \widehat{\mathbf{T}}_{\mu}^{\widehat{x}} \cap \{k', k'+1, k'+2, ...\}, \\ \{k + zp | z \in \mathbf{Z}\} \cap \{k', k'+1, k'+2, ...\} \subset \widehat{\mathbf{T}}_{\mu}^{\widehat{x}}, \end{cases} \tag{3.1}$$

$$\begin{cases} \forall p \geq 1, \forall \mu \in \widehat{Or}(\widehat{x}), \exists k'' \in \mathbf{N}, \forall k \in \widehat{\mathbf{T}}_{\mu}^{\widehat{\sigma}^{k''}(\widehat{x})}, \\ \{k + zp | z \in \mathbf{Z}\} \cap \mathbf{N}_{\_} \subset \widehat{\mathbf{T}}_{\mu}^{\widehat{\sigma}^{k''}(\widehat{x})}, \end{cases} \tag{3.2}$$

$$\begin{cases} \forall p \geq 1, \forall \mu \in \widehat{Or}(\widehat{x}), \exists k' \in \mathbf{N}_{\_}, \forall k \geq k', \widehat{x}(k) = \mu \Longrightarrow \\ \Longrightarrow (\widehat{x}(k) = \widehat{x}(k+p) \text{ and } k - p \geq k' \Longrightarrow \widehat{x}(k) = \widehat{x}(k-p)), \end{cases} \tag{3.3}$$

$$\begin{cases} \forall p \geq 1, \forall \mu \in \widehat{Or}(\widehat{x}), \exists k'' \in \mathbf{N}, \forall k \in \mathbf{N}_{\_}, \widehat{\sigma}^{k''}(\widehat{x})(k) = \mu \Longrightarrow \\ \Longrightarrow (\widehat{\sigma}^{k''}(\widehat{x})(k) = \widehat{\sigma}^{k''}(\widehat{x})(k+p) \text{ and} \\ \text{and } k - p \geq -1 \Longrightarrow \widehat{\sigma}^{k''}(\widehat{x})(k) = \widehat{\sigma}^{k''}(\widehat{x})(k-p)). \end{cases} \tag{3.4}$$

b) The following statements are equivalent with the eventual constancy of x:

$$\begin{cases} \forall T>0, \forall \mu \in Or(x), \exists t' \in I^x, \\ \exists t'_1 \geq t', \forall t \in \mathbf{T}^x_\mu \cap [t'_1,\infty), \{t+zT|z\in\mathbf{Z}\}\cap[t'_1,\infty)\subset \mathbf{T}^x_\mu, \end{cases} \tag{3.5}$$

$$\begin{cases} \forall T>0, \forall \mu \in Or(x), \exists t'_1 \in \mathbf{R}, \\ \forall t \in \mathbf{T}^x_\mu \cap [t'_1,\infty), \{t+zT|z\in\mathbf{Z}\}\cap[t'_1,\infty)\subset \mathbf{T}^x_\mu, \end{cases} \tag{3.6}$$

$$\begin{cases} \forall T>0, \forall \mu \in Or(x), \exists t'' \in \mathbf{R}, \exists t' \in I^{\sigma^{t''}(x)}, \\ \forall t \in \mathbf{T}^{\sigma^{t''}(x)}_\mu \cap [t',\infty), \{t+zT|z\in\mathbf{Z}\}\cap[t',\infty)\subset \mathbf{T}^{\sigma^{t''}(x)}_\mu, \end{cases} \tag{3.7}$$

$$\begin{cases} \forall T>0, \forall \mu \in Or(x), \exists t'' \in \mathbf{R}, \exists t' \in \mathbf{R}, \\ \forall t \in \mathbf{T}^{\sigma^{t''}(x)}_\mu \cap [t',\infty), \{t+zT|z\in\mathbf{Z}\}\cap[t',\infty)\subset \mathbf{T}^{\sigma^{t''}(x)}_\mu, \end{cases} \tag{3.8}$$

$$\begin{cases} \forall T>0, \forall \mu \in Or(x), \exists t' \in I^x, \exists t'_1 \geq t', \forall t \geq t'_1, x(t)=\mu \Longrightarrow \\ \Longrightarrow (x(t)=x(t+T)\ and\ t-T\geq t'_1 \Longrightarrow x(t)=x(t-T)), \end{cases} \tag{3.9}$$

$$\begin{cases} \forall T>0, \forall \mu \in Or(x), \exists t'_1 \in \mathbf{R}, \forall t \geq t'_1, x(t)=\mu \Longrightarrow \\ \Longrightarrow (x(t)=x(t+T)\ and\ t-T\geq t'_1 \Longrightarrow x(t)=x(t-T)), \end{cases} \tag{3.10}$$

$$\begin{cases} \forall T>0, \forall \mu \in Or(x), \exists t'' \in \mathbf{R}, \exists t' \in I^{\sigma^{t''}(x)}, \\ \forall t \geq t', \sigma^{t''}(x)(t)=\mu \Longrightarrow (\sigma^{t''}(x)(t)=\sigma^{t''}(x)(t+T)\ and \\ and\ t-T\geq t' \Longrightarrow \sigma^{t''}(x)(t)=\sigma^{t''}(x)(t-T)), \end{cases} \tag{3.11}$$

$$\begin{cases} \forall T>0, \forall \mu \in Or(x), \exists t'' \in \mathbf{R}, \exists t' \in \mathbf{R}, \forall t \geq t', \sigma^{t''}(x)(t)=\mu \Longrightarrow \\ \Longrightarrow (\sigma^{t''}(x)(t)=\sigma^{t''}(x)(t+T)\ and \\ and\ t-T\geq t' \Longrightarrow \sigma^{t''}(x)(t)=\sigma^{t''}(x)(t-T)). \end{cases} \tag{3.12}$$

Proof. a) (1.3)$\Longrightarrow$(3.1) From Theorem 7, page 11, $k' \in \mathbf{N}\_$ exists such that $\{\widehat{x}(k)|k\geq k'\} = \widehat{\omega}(\widehat{x})$ and, if we take into account (1.3) also, $\mu \in \mathbf{B}^n$ exists with

$$\{\widehat{x}(k)|k\geq k'\} = \{\mu\}(=\widehat{\omega}(\widehat{x})). \tag{3.13}$$

Let $p\geq 1$ and $\mu' \in \widehat{Or}(\widehat{x})$ arbitrary. We have two possibilities.

Case $\mu' \neq \mu$

This corresponds from (3.13) to the situation when $\mu' \in \widehat{Or}(\widehat{x}) \setminus \widehat{\omega}(\widehat{x})$ and $\widehat{\mathbf{T}}^{\widehat{x}}_{\mu'} \cap \{k',k'+1,k'+2,...\} = \emptyset$. The statement

$$\begin{cases} \forall k \in \widehat{\mathbf{T}}^{\widehat{x}}_{\mu'} \cap \{k',k'+1,k'+2,...\}, \\ \{k+zp|z\in\mathbf{Z}\}\cap\{k',k'+1,k'+2,...\} \subset \widehat{\mathbf{T}}^{\widehat{x}}_{\mu'} \end{cases} \tag{3.14}$$

takes place trivially.

Case $\mu' = \mu$

In this case $\widehat{\mathbf{T}}^{\widehat{x}}_{\mu} \cap \{k',k'+1,k'+2,...\} \neq \emptyset$ and let $k \in \widehat{\mathbf{T}}^{\widehat{x}}_{\mu} \cap \{k',k'+1,k'+2,...\}$, $z\in\mathbf{Z}$ arbitrary such that $k+zp\geq k'$. Then from (3.13) we get $\widehat{x}(k+zp)=\mu$, thus $k+zp \in \widehat{\mathbf{T}}^{\widehat{x}}_{\mu}$.

(3.1)$\Longrightarrow$(3.2) Let $p\geq 1, \mu \in \widehat{Or}(\widehat{x})$ be arbitrary. From (3.1) we have the existence of $k' \in \mathbf{N}\_$ with

$$\begin{cases} \forall k \in \widehat{\mathbf{T}}^{\widehat{x}}_{\mu} \cap \{k',k'+1,k'+2,...\}, \\ \{k+zp|z\in\mathbf{Z}\}\cap\{k',k'+1,k'+2,...\} \subset \widehat{\mathbf{T}}^{\widehat{x}}_{\mu}. \end{cases} \tag{3.15}$$

We define $k'' = k' + 1$ and there are two possibilities.

Case $\widehat{\mathbf{T}}_{\mu}^{\widehat{\sigma}^{k''}(\widehat{x})} = \varnothing$

This situation occurs because $\widehat{Or}(\widehat{\sigma}^{k''}(\widehat{x})) \subset \widehat{Or}(\widehat{x})$, when $\mu \in \widehat{Or}(\widehat{x}) \setminus \widehat{Or}(\widehat{\sigma}^{k''}(\widehat{x}))$. The statement

$$\forall k \in \widehat{\mathbf{T}}_{\mu}^{\widehat{\sigma}^{k''}(\widehat{x})}, \{k + zp | z \in \mathbf{Z}\} \cap \mathbf{N}_{\_} \subset \widehat{\mathbf{T}}_{\mu}^{\widehat{\sigma}^{k''}(\widehat{x})} \tag{3.16}$$

takes place trivially.

Case $\widehat{\mathbf{T}}_{\mu}^{\widehat{\sigma}^{k''}(\widehat{x})} \neq \varnothing$

In this case $\mu \in \widehat{Or}(\widehat{\sigma}^{k''}(\widehat{x}))$. We take $k \in \widehat{\mathbf{T}}_{\mu}^{\widehat{\sigma}^{k''}(\widehat{x})}, z \in \mathbf{Z}$ arbitrary such that $k + zp \geq -1$. We conclude

$$\widehat{\sigma}^{k''}(\widehat{x})(k) = \mu = \widehat{x}(k + k'') = \widehat{x}(k + k' + 1),$$

in other words $k + k' + 1 \in \widehat{\mathbf{T}}_{\mu}^{\widehat{x}}, k + k' + 1 \geq k'$. Furthermore, $k + zp + k' + 1 \geq -1 + k' + 1 = k'$, thus we can apply (3.15), wherefrom $k + zp + k' + 1 \in \widehat{\mathbf{T}}_{\mu}^{\widehat{x}}$. This means that

$$\widehat{x}(k + zp + k' + 1) = \mu = \widehat{\sigma}^{k'+1}(\widehat{x})(k + zp) = \widehat{\sigma}^{k''}(\widehat{x})(k + zp)$$

i.e. $k + zp \in \widehat{\mathbf{T}}_{\mu}^{\widehat{\sigma}^{k''}(\widehat{x})}$.

(3.2)$\Longrightarrow$(3.3) Let $p \geq 1, \mu \in \widehat{Or}(\widehat{x})$ arbitrary. From (3.2) we have the existence of $k'' \in \mathbf{N}$ such that

$$\forall k \in \widehat{\mathbf{T}}_{\mu}^{\widehat{\sigma}^{k''}(\widehat{x})}, \{k + zp | z \in \mathbf{Z}\} \cap \mathbf{N}_{\_} \subset \widehat{\mathbf{T}}_{\mu}^{\widehat{\sigma}^{k''}(\widehat{x})}. \tag{3.17}$$

We define $k' = k'' - 1$ and we have two possibilities.

Case $\forall k \geq k', \widehat{x}(k) \neq \mu$

Then $\widehat{\mathbf{T}}_{\mu}^{\widehat{\sigma}^{k''}(\widehat{x})} = \varnothing$ (as we have $\mu \in \widehat{Or}(\widehat{x}) \setminus \widehat{Or}(\widehat{\sigma}^{k''}(\widehat{x}))$) and (3.17) is trivially fulfilled, as well as the statement

$$\begin{cases} \forall k \geq k', \widehat{x}(k) = \mu \Longrightarrow \\ \Longrightarrow (\widehat{x}(k) = \widehat{x}(k + p) \text{ and } k - p \geq k' \Longrightarrow \widehat{x}(k) = \widehat{x}(k - p)). \end{cases}$$

Case $\exists k \geq k', \widehat{x}(k) = \mu$,

when $\mu \in \widehat{Or}(\widehat{\sigma}^{k''}(\widehat{x}))$. We take $k \geq k'$ arbitrary, such that $\widehat{x}(k) = \mu$. Then $k - k'' = k - k' - 1 \geq -1$ and

$$\widehat{\sigma}^{k''}(\widehat{x})(k - k'') = \widehat{x}(k - k'' + k'') = \widehat{x}(k) = \mu,$$

thus $k - k'' \in \widehat{\mathbf{T}}_{\mu}^{\widehat{\sigma}^{k''}(\widehat{x})}$. Furthermore

$$k - k'' + p \in \{k - k'' + zp | z \in \mathbf{Z}\} \cap \mathbf{N}_{\_} \overset{(3.17)}{\subset} \widehat{\mathbf{T}}_{\mu}^{\widehat{\sigma}^{k''}(\widehat{x})},$$

meaning that

$$\widehat{\sigma}^{k''}(\widehat{x})(k - k'' + p) = \mu = \widehat{x}(k - k'' + p + k'') = \widehat{x}(k + p).$$

If $k - p \geq k'$, then $k - p - k'' = k - p - k' - 1 \geq -1$, thus

$$k - k'' - p \in \{k - k'' + zp | z \in \mathbf{Z}\} \cap \mathbf{N}_{\_} \overset{(3.17)}{\subset} \widehat{\mathbf{T}}_{\mu}^{\widehat{\sigma}^{k''}(\widehat{x})}$$

and finally

$$\widehat{\sigma}^{k''}(\widehat{x})(k-k''-p)=\mu=\widehat{x}(k-k''-p+k'')=\widehat{x}(k-p).$$

(3.3)$\Longrightarrow$(3.4) We take $p\geq 1,\mu\in\widehat{Or}(\widehat{x})$ arbitrarily and we infer from (3.3) that $k'\in\mathbf{N}_{\_}$ exists with

$$\forall k\geq k',\widehat{x}(k)=\mu\Longrightarrow\widehat{x}(k)=\widehat{x}(k+p), \tag{3.18}$$

$$\forall k\geq k',(\widehat{x}(k)=\mu \text{ and } k-p\geq k')\Longrightarrow\widehat{x}(k)=\widehat{x}(k-p). \tag{3.19}$$

We define $k''=k'+1$ and there are two possibilities.

Case $\forall k\in\mathbf{N}_{\_},\widehat{\sigma}^{k''}(\widehat{x})(k)\neq\mu$

This corresponds to the situation when $\mu\in\widehat{Or}(\widehat{x})\setminus\widehat{Or}(\widehat{\sigma}^{k''}(\widehat{x}))$. The statement

$$\begin{cases}\forall k\in\mathbf{N}_{\_},\widehat{\sigma}^{k''}(\widehat{x})(k)=\mu\Longrightarrow\\ \Longrightarrow(\widehat{\sigma}^{k''}(\widehat{x})(k)=\widehat{\sigma}^{k''}(\widehat{x})(k+p) \text{ and}\\ \text{and } k-p\geq -1\Longrightarrow\widehat{\sigma}^{k''}(\widehat{x})(k)=\widehat{\sigma}^{k''}(\widehat{x})(k-p))\end{cases}$$

is true in a trivial manner.

Case $\exists k\in\mathbf{N}_{\_},\widehat{\sigma}^{k''}(\widehat{x})(k)=\mu,$

i.e. $\mu\in\widehat{Or}(\widehat{\sigma}^{k''}(\widehat{x}))$. We take $k\in\mathbf{N}_{\_}$ arbitrarily with $\widehat{\sigma}^{k''}(\widehat{x})(k)=\mu$, thus $\widehat{x}(k+k'')=\mu=\widehat{x}(k+k'+1)$. We have $k+k'+1\geq k'$ and then

$$\widehat{\sigma}^{k''}(\widehat{x})(k)=\widehat{x}(k+k'')=\widehat{x}(k+k'+1)\overset{(3.18)}{=}\widehat{x}(k+k'+1+p)=$$

$$=\widehat{x}(k+k''+p)=\widehat{\sigma}^{k''}(\widehat{x})(k+p).$$

If in addition $k-p\geq -1$, as $k-p+k'+1\geq k'$, we can write that

$$\widehat{\sigma}^{k''}(\widehat{x})(k)=\widehat{x}(k+k'')=\widehat{x}(k+k'+1)\overset{(3.19)}{=}\widehat{x}(k-p+k'+1)=$$

$$=\widehat{x}(k-p+k'')=\widehat{\sigma}^{k''}(\widehat{x})(k-p).$$

(3.4)$\Longrightarrow$(1.1) We write (3.4) for $p=1$ and for an arbitrary $\mu\in\widehat{\omega}(\widehat{x})$ (we have $\widehat{\omega}(\widehat{x})\neq\emptyset$). Some $k''\in\mathbf{N}$ exists then with

$$\begin{cases}\forall k\in\mathbf{N}_{\_},\widehat{x}(k+k'')=\mu\Longrightarrow(\widehat{x}(k+k'')=\widehat{x}(k+k''+1) \text{ and}\\ \text{and } k\geq 0\Longrightarrow\widehat{x}(k+k'')=\widehat{x}(k+k''-1))\end{cases} \tag{3.20}$$

and, whichever k'' might be, some $k\in\mathbf{N}_{\_}$ exists such that $\widehat{x}(k+k'')=\mu$ (from the hypothesis that $\mu\in\widehat{\omega}(\widehat{x})$). We get from (3.20) that

$$\mu=\widehat{x}(k''-1)=\widehat{x}(k'')=\widehat{x}(k''+1)=...$$

i.e. (1.1) holds with $k'=k''-1$.

b) (1.6)$\Longrightarrow$(3.5) We have the existence of $\mu\in\mathbf{B}^n$ and $t_1\in\mathbf{R}$ with $\{x(t)|t\geq t_1\}=\omega(x)=\{\mu\}$ and let $T>0,\mu'\in Or(x)$ arbitrary[1]. Let $t'\in I^x$ arbitrary. We take $t_1'\geq\max\{t',t_1\}$ arbitrarily also and we have two possibilities.

Case $\mu'\neq\mu$

[1]The fact that we shall take $t'\in I^x$ arbitrary shows that we prove at this moment a statement that is stronger than (3.5).

This corresponds to the situation when $\mu' \in Or(x) \setminus \omega(x)$. Then $\mathbf{T}^x_{\mu'} \cap [t'_1, \infty) = \varnothing$ and the statement

$$\forall t \in \mathbf{T}^x_{\mu'} \cap [t'_1, \infty), \{t + zT | z \in \mathbf{Z}\} \cap [t'_1, \infty) \subset \mathbf{T}^x_{\mu'}$$

takes place trivially.

Case $\mu' = \mu$,

when $\mu' \in \omega(x)$. We have $\mathbf{T}^x_{\mu} \cap [t'_1, \infty) = [t'_1, \infty)$ thus let $t \in \mathbf{T}^x_{\mu} \cap [t'_1, \infty)$ and $z \in \mathbf{Z}$ with the property that $t + zT \geq t'_1$. As $t + zT \geq t_1$, we have $t + zT \in \mathbf{T}^x_{\mu}$.

(3.5)$\Longrightarrow$(3.6) Obvious.

(3.6)$\Longrightarrow$(3.7) Let $T > 0, \mu \in Or(x)$ arbitrary. From (3.6) we have the existence of $t'_1 \in \mathbf{R}$ with

$$\forall t \in \mathbf{T}^x_{\mu} \cap [t'_1, \infty), \{t + zT | z \in \mathbf{Z}\} \cap [t'_1, \infty) \subset \mathbf{T}^x_{\mu}. \tag{3.21}$$

We take $t'' > t'_1$ arbitrary and let $\varepsilon > 0$ having the property that

$$\forall t \in (t'' - \varepsilon, t''), x(t) = x(t'' - 0).$$

We take $t' \in (t'' - \varepsilon, t'') \cap [t'_1, \infty)$ arbitrary and we notice that

$$\sigma^{t''}(x)(t) = \begin{cases} x(t), t > t'' - \varepsilon, \\ x(t'' - 0), t < t''. \end{cases} \tag{3.22}$$

Obviously $t' \in (-\infty, t'') \subset I^{\sigma^{t''}(x)}$ and we have also

$$\forall t \in \mathbf{T}^x_{\mu} \cap [t', \infty), \{t + zT | z \in \mathbf{Z}\} \cap [t', \infty) \subset \mathbf{T}^x_{\mu}, \tag{3.23}$$

from Lemma 3, page 224, (3.21) and taking into account the fact that $t' \geq t'_1$.

The truth of

$$\forall t \in \mathbf{T}^{\sigma^{t''}(x)}_{\mu} \cap [t', \infty), \{t + zT | z \in \mathbf{Z}\} \cap [t', \infty) \subset \mathbf{T}^{\sigma^{t''}(x)}_{\mu} \tag{3.24}$$

results from (3.23) and from the fact that $\forall t \geq t', \sigma^{t''}(x)(t) = x(t)$, see (3.22).

(3.7)$\Longrightarrow$(3.8) Obvious.

(3.8)$\Longrightarrow$(3.9) We take arbitrarily $T > 0, \mu \in Or(x)$. From (3.8) we have the existence of $t'' \in \mathbf{R}$ and $t''' \in \mathbf{R}$ such that

$$\forall t \in \mathbf{T}^{\sigma^{t''}(x)}_{\mu} \cap [t''', \infty), \{t + zT | z \in \mathbf{Z}\} \cap [t''', \infty) \subset \mathbf{T}^{\sigma^{t''}(x)}_{\mu}. \tag{3.25}$$

Let $t' \in I^x$ arbitrary[2] and we take $t'_1 \geq \max\{t', t'', t'''\}$ arbitrarily also. From $t'_1 \geq t'''$, from (3.25) and from Lemma 3 we infer

$$\forall t \in \mathbf{T}^{\sigma^{t''}(x)}_{\mu} \cap [t'_1, \infty), \{t + zT | z \in \mathbf{Z}\} \cap [t'_1, \infty) \subset \mathbf{T}^{\sigma^{t''}(x)}_{\mu}. \tag{3.26}$$

Let $t \geq t'_1$ arbitrary and we have two possibilities.

Case $\forall t \geq t'_1, x(t) \neq \mu$

Then the implication

$$\forall t \geq t'_1, x(t) = \mu \Longrightarrow (x(t) = x(t+T) \text{ and } t - T \geq t'_1 \Longrightarrow x(t) = x(t-T)) \tag{3.27}$$

[2]The statement that we prove is stronger than (3.9).

is trivially true.

Case $\exists t \geq t_1', x(t) = \mu$

We take $t \geq t_1'$ arbitrarily such that $x(t) = \mu$. Because $t \geq t''$, we have $\sigma^{t''}(x)(t) = x(t) = \mu$, thus $t \in \mathbf{T}_\mu^{\sigma^{t''}(x)} \cap [t_1', \infty)$. We have

$$t + T \in \{t + zT | z \in \mathbf{Z}\} \cap [t_1', \infty) \overset{(3.26)}{\subset} \mathbf{T}_\mu^{\sigma^{t''}(x)},$$

i.e. $\sigma^{t''}(x)(t+T) = \mu$. On the other hand $\sigma^{t''}(x)(t+T) = x(t+T)$, thus $x(t+T) = \mu = x(t)$.

We suppose now that we have in addition $t - T \geq t_1'$. In a similar way with the previous situation,

$$t - T \in \{t + zT | z \in \mathbf{Z}\} \cap [t_1', \infty) \overset{(3.26)}{\subset} \mathbf{T}_\mu^{\sigma^{t''}(x)},$$

i.e. $\sigma^{t''}(x)(t-T) = \mu$. As $\sigma^{t''}(x)(t-T) = x(t-T)$, we have obtained that $x(t-T) = \mu = x(t)$.

(3.9)$\Longrightarrow$(3.10) Obvious.

(3.10)$\Longrightarrow$(3.11) Let $T > 0, \mu \in Or(x)$ arbitrary. From (3.10) we have the existence of $t_1' \in \mathbf{R}$ such that the property

$$\forall t \geq t_1', x(t) = \mu \Longrightarrow (x(t) = x(t+T) \text{ and } t - T \geq t_1' \Longrightarrow x(t) = x(t-T)) \tag{3.28}$$

holds. We take $t'' > t_1'$ arbitrary. Some $\varepsilon > 0$ exists with $\forall t \in (t'' - \varepsilon, t''), x(t) = x(t'' - 0)$. Obviously,

$$\sigma^{t''}(x)(t) = \begin{cases} x(t), t > t'' - \varepsilon, \\ x(t'' - 0), t < t''. \end{cases} \tag{3.29}$$

We take $t' \in (t'' - \varepsilon, t'') \cap [t_1', \infty)$ arbitrary and we have $t' \in (-\infty, t'') \subset I^{\sigma^{t''}(x)}$. On the other hand $t' \geq t_1'$, (3.28) and Lemma 3, page 224 imply the truth of

$$\forall t \geq t', x(t) = \mu \Longrightarrow (x(t) = x(t+T) \text{ and } t - T \geq t' \Longrightarrow x(t) = x(t-T)). \tag{3.30}$$

As $\forall t \geq t', \sigma^{t''}(x)(t) = x(t)$, (3.30) implies that

$$\begin{cases} \forall t \geq t', \sigma^{t''}(x)(t) = \mu \Longrightarrow (\sigma^{t''}(x)(t) = \sigma^{t''}(x)(t+T) \text{ and} \\ \quad \text{and } t - T \geq t' \Longrightarrow \sigma^{t''}(x)(t) = \sigma^{t''}(x)(t-T)) \end{cases}$$

is true.

(3.11)$\Longrightarrow$(3.12) Obvious.

(3.12)$\Longrightarrow$(1.6) We suppose against all reason that $\mu, \mu' \in \omega(x)$ exist, $\mu \neq \mu'$. We write (3.12) for an arbitrary $T > 0$, thus $t_1', t_2', t_1'', t_2'' \in \mathbf{R}$ exist such that

$$\begin{cases} \forall t \geq t_1', \sigma^{t_1''}(x)(t) = \mu \Longrightarrow (\sigma^{t_1''}(x)(t) = \sigma^{t_1''}(x)(t+T) \text{ and} \\ \quad \text{and } t - T \geq t_1' \Longrightarrow \sigma^{t_1''}(x)(t) = \sigma^{t_1''}(x)(t-T)), \end{cases} \tag{3.31}$$

$$\begin{cases} \forall t \geq t_2', \sigma^{t_2''}(x)(t) = \mu' \Longrightarrow (\sigma^{t_2''}(x)(t) = \sigma^{t_2''}(x)(t+T) \text{ and} \\ \quad \text{and } t - T \geq t_2' \Longrightarrow \sigma^{t_2''}(x)(t) = \sigma^{t_2''}(x)(t-T)). \end{cases} \tag{3.32}$$

Let $t_3' \geq \max\{t_1', t_2', t_1'', t_2''\}$. From (3.31), (3.32) and Lemma 3, page 224 we get

$$\forall t \geq t_3', x(t) = \mu \Longrightarrow x(t) = x(t+T), \tag{3.33}$$

$$\forall t \geq t_3', x(t) = \mu' \Longrightarrow x(t) = x(t+T). \tag{3.34}$$

As $\mu, \mu' \in \omega(x)$, some $t_1 \geq t_3'$ exists such that $x(t_1) = \mu$ and some $t_2 \geq t_3'$ also exists such that $x(t_2) = \mu'$ and, from (3.33), (3.34) we infer

$$\mu = x(t_1) = x(t_1 + T) = x(t_1 + 2T) = ... \tag{3.35}$$

$$\mu' = x(t_2) = x(t_2 + T) = x(t_2 + 2T) = ... \tag{3.36}$$

We suppose without loss that $t_1 < t_2$. We write (3.12) for $T' = t_2 - t_1$, thus $t_4', t_4'' \in \mathbf{R}$ exist with

$$\left\{ \begin{array}{l} \forall t \geq t_4', \sigma^{t_4''}(x)(t) = \mu \Longrightarrow (\sigma^{t_4''}(x)(t) = \sigma^{t_4''}(x)(t+T') \text{ and} \\ \quad \text{and } t - T' \geq t_4' \Longrightarrow \sigma^{t_4''}(x)(t) = \sigma^{t_4''}(x)(t-T')). \end{array} \right. \tag{3.37}$$

For $t_5' \geq \max\{t_4', t_4''\}$ we infer from (3.37) that

$$\forall t \geq t_5', x(t) = \mu \Longrightarrow x(t) = x(t+T'). \tag{3.38}$$

Let now $k \in \mathbf{N}$ having the property that $t_1 + kT \geq t_5'$. As $t_2 + kT - T' = t_1 + kT$, we have

$$\mu \overset{(3.35)}{=} x(t_1 + kT) = x(t_2 + kT - T') \overset{(3.38)}{=} x(t_2 + kT - T' + T') = x(t_2 + kT) \overset{(3.36)}{=} \mu',$$

contradiction. We have obtained that $\omega(x)$ has a single point μ, thus (1.6) is true. □

4. The Third Group of Eventual Constancy Properties

Remark 45. *The third group of eventual constancy properties involves eventual periodicity properties of some point μ of the orbit. These properties result one by one from the properties of the second group, by the replacement of $\forall \mu \in \widehat{Or}(\widehat{x}), \forall \mu \in Or(x)$ with $\exists \mu \in \widehat{Or}(\widehat{x}), \exists \mu \in Or(x)$. We notice that we have avoided each time the trivialities by requests of the kind $\widehat{\mathbf{T}}_\mu^{\widehat{x}} \cap \{k', k'+1, k'+2, ...\} \neq \varnothing, \mathbf{T}_\mu^x \cap [t_1', \infty) \neq \varnothing$. The possibility of replacing the universal quantifier with the existential quantifier when passing from Theorem 17, page 27 to Theorem 18 is given by the fact that the final value, if it exists, is unique.*

Theorem 18. *Let the signals $\widehat{x} \in \widehat{S}^{(n)}, x \in S^{(n)}$.*

a) The following statements are equivalent with the eventual constancy of $\widehat{x}$:

$$\left\{ \begin{array}{c} \forall p \geq 1, \exists \mu \in \widehat{Or}(\widehat{x}), \exists k' \in \mathbf{N}_{\_}, \\ \widehat{\mathbf{T}}_\mu^{\widehat{x}} \cap \{k', k'+1, k'+2, ...\} \neq \varnothing \text{ and } \forall k \in \widehat{\mathbf{T}}_\mu^{\widehat{x}} \cap \{k', k'+1, k'+2, ...\}, \\ \{k + zp | z \in \mathbf{Z}\} \cap \{k', k'+1, k'+2, ...\} \subset \widehat{\mathbf{T}}_\mu^{\widehat{x}}, \end{array} \right. \tag{4.1}$$

$$\left\{ \begin{array}{l} \forall p \geq 1, \exists \mu \in \widehat{Or}(\widehat{x}), \exists k'' \in \mathbf{N}, \widehat{\mathbf{T}}_\mu^{\widehat{\sigma}^{k''}(\widehat{x})} \neq \varnothing \text{ and } \forall k \in \widehat{\mathbf{T}}_\mu^{\widehat{\sigma}^{k''}(\widehat{x})}, \\ \quad \{k + zp | z \in \mathbf{Z}\} \cap \mathbf{N}_{\_} \subset \widehat{\mathbf{T}}_\mu^{\widehat{\sigma}^{k''}(\widehat{x})}, \end{array} \right. \tag{4.2}$$

$$\left\{ \begin{array}{c} \forall p \geq 1, \exists \mu \in \widehat{Or}(\widehat{x}), \exists k' \in \mathbf{N}_{\_}, \exists k_1 \geq k', \widehat{x}(k_1) = \mu \text{ and} \\ \text{and } \forall k \geq k', \widehat{x}(k) = \mu \Longrightarrow \\ \Longrightarrow (\widehat{x}(k) = \widehat{x}(k+p) \text{ and } k - p \geq k' \Longrightarrow \widehat{x}(k) = \widehat{x}(k-p)), \end{array} \right. \tag{4.3}$$

$$\begin{cases} \forall p \geq 1, \exists \mu \in \widehat{Or}(\widehat{x}), \exists k'' \in \mathbf{N}, \exists k_1 \in \mathbf{N}_{\_}, \widehat{\sigma}^{k''}(\widehat{x})(k_1) = \mu \textit{ and} \\ \textit{and } \forall k \in \mathbf{N}_{\_}, \widehat{\sigma}^{k''}(\widehat{x})(k) = \mu \Longrightarrow (\widehat{\sigma}^{k''}(\widehat{x})(k) = \widehat{\sigma}^{k''}(\widehat{x})(k+p) \textit{ and} \\ \textit{and } k - p \geq -1 \Longrightarrow \widehat{\sigma}^{k''}(\widehat{x})(k) = \widehat{\sigma}^{k''}(\widehat{x})(k-p)). \end{cases} \tag{4.4}$$

b) The following statements are equivalent with the eventual constancy of x:

$$\begin{cases} \forall T > 0, \exists \mu \in Or(x), \exists t' \in I^x, \exists t_1' \geq t', \mathbf{T}_\mu^x \cap [t_1', \infty) \neq \varnothing \textit{ and} \\ \textit{and } \forall t \in \mathbf{T}_\mu^x \cap [t_1', \infty), \{t + zT | z \in \mathbf{Z}\} \cap [t_1', \infty) \subset \mathbf{T}_\mu^x, \end{cases} \tag{4.5}$$

$$\begin{cases} \forall T > 0, \exists \mu \in Or(x), \exists t_1' \in \mathbf{R}, \mathbf{T}_\mu^x \cap [t_1', \infty) \neq \varnothing \textit{ and} \\ \textit{and } \forall t \in \mathbf{T}_\mu^x \cap [t_1', \infty), \{t + zT | z \in \mathbf{Z}\} \cap [t_1', \infty) \subset \mathbf{T}_\mu^x, \end{cases} \tag{4.6}$$

$$\begin{cases} \forall T > 0, \exists \mu \in Or(x), \exists t'' \in \mathbf{R}, \exists t' \in I^{\sigma^{t''}(x)}, \mathbf{T}_\mu^{\sigma^{t''}(x)} \cap [t', \infty) \neq \varnothing \textit{ and} \\ \textit{and } \forall t \in \mathbf{T}_\mu^{\sigma^{t''}(x)} \cap [t', \infty), \{t + zT | z \in \mathbf{Z}\} \cap [t', \infty) \subset \mathbf{T}_\mu^{\sigma^{t''}(x)}, \end{cases} \tag{4.7}$$

$$\begin{cases} \forall T > 0, \exists \mu \in Or(x), \exists t'' \in \mathbf{R}, \exists t' \in \mathbf{R}, \\ \mathbf{T}_\mu^{\sigma^{t''}(x)} \cap [t', \infty) \neq \varnothing \textit{ and} \\ \textit{and } \forall t \in \mathbf{T}_\mu^{\sigma^{t''}(x)} \cap [t', \infty), \{t + zT | z \in \mathbf{Z}\} \cap [t', \infty) \subset \mathbf{T}_\mu^{\sigma^{t''}(x)}, \end{cases} \tag{4.8}$$

$$\begin{cases} \forall T > 0, \exists \mu \in Or(x), \exists t' \in I^x, \\ \exists t_1' \geq t', \exists t_2' \geq t_1', x(t_2') = \mu \textit{ and } \forall t \geq t_1', x(t) = \mu \Longrightarrow \\ \Longrightarrow (x(t) = x(t+T) \textit{ and } t - T \geq t_1' \Longrightarrow x(t) = x(t-T)), \end{cases} \tag{4.9}$$

$$\begin{cases} \forall T > 0, \exists \mu \in Or(x), \exists t_1' \in \mathbf{R}, \exists t_2' \geq t_1', x(t_2') = \mu \textit{ and} \\ \textit{and } \forall t \geq t_1', x(t) = \mu \Longrightarrow \\ \Longrightarrow (x(t) = x(t+T) \textit{ and } t - T \geq t_1' \Longrightarrow x(t) = x(t-T)), \end{cases} \tag{4.10}$$

$$\begin{cases} \forall T > 0, \exists \mu \in Or(x), \exists t'' \in \mathbf{R}, \exists t' \in I^{\sigma^{t''}(x)}, \exists t''' \geq t', \\ \sigma^{t''}(x)(t''') = \mu \textit{ and } \forall t \geq t', \sigma^{t''}(x)(t) = \mu \Longrightarrow \\ \Longrightarrow (\sigma^{t''}(x)(t) = \sigma^{t''}(x)(t+T) \textit{ and} \\ \textit{and } t - T \geq t' \Longrightarrow \sigma^{t''}(x)(t) = \sigma^{t''}(x)(t-T)), \end{cases} \tag{4.11}$$

$$\begin{cases} \forall T > 0, \exists \mu \in Or(x), \exists t'' \in \mathbf{R}, \exists t' \in \mathbf{R}, \exists t''' \geq t', \sigma^{t''}(x)(t''') = \mu \textit{ and} \\ \textit{and } \forall t \geq t', \sigma^{t''}(x)(t) = \mu \Longrightarrow (\sigma^{t''}(x)(t) = \sigma^{t''}(x)(t+T) \textit{ and} \\ \textit{and } t - T \geq t' \Longrightarrow \sigma^{t''}(x)(t) = \sigma^{t''}(x)(t-T)). \end{cases} \tag{4.12}$$

Proof. a) (1.1)$\Longrightarrow$(4.1) Let $p \geq 1$ arbitrary. From (1.1) we have the existence of $\mu \in \mathbf{B}^n$ and $k' \in \mathbf{N}_{\_}$ with the property

$$\forall k \geq k', \widehat{x}(k) = \mu. \tag{4.13}$$

We have that $\widehat{\mathbf{T}}_\mu^{\widehat{x}} \cap \{k', k'+1, k'+2, ...\} \neq \varnothing$ and let $k \in \widehat{\mathbf{T}}_\mu^{\widehat{x}}, z \in \mathbf{Z}$ arbitrary such that $k \geq k', k + zp \geq k'$. We get from (4.13) that $k + zp \in \widehat{\mathbf{T}}_\mu^{\widehat{x}}$.

(4.1)$\Longrightarrow$(4.2) Let $p \geq 1$ arbitrary. Hypothesis (4.1) shows the existence of $\mu \in \widehat{Or}(\widehat{x})$ and $k' \in \mathbf{N}_{\_}$ such that

$$\widehat{\mathbf{T}}_\mu^{\widehat{x}} \cap \{k', k'+1, k'+2, ...\} \neq \varnothing, \tag{4.14}$$

$$\begin{cases} \forall k \in \widehat{\mathbf{T}}_\mu^{\widehat{x}} \cap \{k', k'+1, k'+2, ...\}, \\ \{k + zp | z \in \mathbf{Z}\} \cap \{k', k'+1, k'+2, ...\} \subset \widehat{\mathbf{T}}_\mu^{\widehat{x}}. \end{cases} \tag{4.15}$$

We put $k'' = k' + 1$. The existence from (4.14) of some $k \in \widehat{\mathbf{T}}_{\mu}^{\widehat{x}} \cap \{k', k'+1, k'+2, ...\}$ means that $\widehat{x}(k) = \mu$ and $k \geq k'$, thus $k \geq k'' - 1$; the number $k_1 = k - k''$ is ≥ -1 and it fulfills $\widehat{\sigma}^{k''}(\widehat{x})(k_1) = \widehat{x}(k_1 + k'') = \widehat{x}(k) = \mu$, thus $k_1 \in \widehat{\mathbf{T}}_{\mu}^{\widehat{\sigma}^{k''}(\widehat{x})}$ and $\widehat{\mathbf{T}}_{\mu}^{\widehat{\sigma}^{k''}(\widehat{x})} \neq \varnothing$. Let now $k \in \widehat{\mathbf{T}}_{\mu}^{\widehat{\sigma}^{k''}(\widehat{x})}$ and $z \in \mathbf{Z}$ arbitrary such that $k + zp \geq -1$. We have

$$\mu = \widehat{\sigma}^{k''}(\widehat{x})(k) = \widehat{x}(k + k''),$$

where $k \geq -1$ means that $k + k'' = k + k' + 1 \geq k'$ and on the other hand $k + k'' + zp \geq k'' - 1 = k'$, thus we can apply (4.15). We have:

$$\widehat{\sigma}^{k''}(\widehat{x})(k + zp) = \widehat{x}(k + k'' + zp) \overset{(4.15)}{=} \widehat{x}(k + k'') = \mu,$$

in other words $k + zp \in \widehat{\mathbf{T}}_{\mu}^{\widehat{\sigma}^{k''}(\widehat{x})}$.

(4.2)$\Longrightarrow$(4.3) Let $p \geq 1$. Property (4.2) states the existence of $\mu \in \widehat{Or}(\widehat{x})$ and $k'' \in \mathbf{N}$ such that

$$\widehat{\mathbf{T}}_{\mu}^{\widehat{\sigma}^{k''}(\widehat{x})} \neq \varnothing, \tag{4.16}$$

$$\forall k \in \widehat{\mathbf{T}}_{\mu}^{\widehat{\sigma}^{k''}(\widehat{x})}, \{k + zp | z \in \mathbf{Z}\} \cap \mathbf{N}_{\_} \subset \widehat{\mathbf{T}}_{\mu}^{\widehat{\sigma}^{k''}(\widehat{x})}. \tag{4.17}$$

We define $k' = k'' - 1$. Inequality (4.16) shows the existence of $k \in \widehat{\mathbf{T}}_{\mu}^{\widehat{\sigma}^{k''}(\widehat{x})}$, thus $\widehat{x}(k + k'') = \mu$. With the notation $k_1 = k + k''$ we have $k_1 = k + k' + 1 \geq k'$ (because $k \geq -1$).

Let now $k \geq k'$ arbitrary such that $\widehat{x}(k) = \mu$. The number $k - k'' = k - k' - 1 \geq -1$ satisfies $\widehat{\sigma}^{k''}(\widehat{x})(k - k'') = \widehat{x}(k) = \mu$, thus $k - k'' \in \widehat{\mathbf{T}}_{\mu}^{\widehat{\sigma}^{k''}(\widehat{x})}$. We infer

$$k - k'' + p \in \{k - k'' + zp | z \in \mathbf{Z}\} \cap \mathbf{N}_{\_} \overset{(4.17)}{\subset} \widehat{\mathbf{T}}_{\mu}^{\widehat{\sigma}^{k''}(\widehat{x})},$$

thus $\mu = \widehat{\sigma}^{k''}(\widehat{x})(k - k'' + p) = \widehat{x}(k + p)$. Moreover, if $k - p \geq k'$, then $k - k'' - p = k - k' - 1 - p \geq -1$ and

$$k - k'' - p \in \{k - k'' + zp | z \in \mathbf{Z}\} \cap \mathbf{N}_{\_} \overset{(4.17)}{\subset} \widehat{\mathbf{T}}_{\mu}^{\widehat{\sigma}^{k''}(\widehat{x})},$$

thus $\mu = \widehat{\sigma}^{k''}(\widehat{x})(k - k'' - p) = \widehat{x}(k - p)$.

(4.3)$\Longrightarrow$(4.4) Let $p \geq 1$ arbitrary. From (4.3) we infer the existence of $\mu \in \widehat{Or}(\widehat{x})$ and $k' \in \mathbf{N}_{\_}$ such that

$$\exists k_1 \geq k', \widehat{x}(k_1) = \mu, \tag{4.18}$$

$$\left\{ \begin{array}{c} \forall k \geq k', \widehat{x}(k) = \mu \Longrightarrow \\ \Longrightarrow (\widehat{x}(k) = \widehat{x}(k+p) \text{ and } k - p \geq k' \Longrightarrow \widehat{x}(k) = \widehat{x}(k-p)). \end{array} \right. \tag{4.19}$$

We define $k'' = k' + 1$ and let $k_1 \geq k'$ such that $\widehat{x}(k_1) = \mu$. The number $k_1' = k_1 - k''$ belongs to $\mathbf{N}_{\_}$ and fulfills $\widehat{\sigma}^{k''}(\widehat{x})(k_1') = \widehat{x}(k_1' + k'') = \widehat{x}(k_1) = \mu$, in other words

$$\exists k_1' \in \mathbf{N}_{\_}, \widehat{\sigma}^{k''}(\widehat{x})(k_1') = \mu. \tag{4.20}$$

Let now $k \in \mathbf{N}_{\_}$ arbitrary, with the property that $\widehat{\sigma}^{k''}(\widehat{x})(k) = \mu$, thus $\widehat{x}(k + k'') = \mu$. In this situation we have $k + k'' = k + k' + 1 \geq k'$ and we can apply (4.19), resulting

$$\widehat{\sigma}^{k''}(\widehat{x})(k) = \widehat{x}(k + k'') \overset{(4.19)}{=} \widehat{x}(k + k'' + p) = \widehat{\sigma}^{k''}(\widehat{x})(k + p).$$

In the case when in addition $k-p\geq -1$, we have $k+k''-p\geq k''-1=k'$, thus we can apply (4.19) again, with the result

$$\widehat{\sigma}^{k''}(\widehat{x})(k)=\widehat{x}(k+k'')\overset{(4.19)}{=}\widehat{x}(k+k''-p)=\widehat{\sigma}^{k''}(\widehat{x})(k-p).$$

(4.4)$\Longrightarrow$(1.1) We put $p=1$ in (4.4); then $\mu\in\widehat{Or}(\widehat{x})$ and $k''\in\mathbf{N}$ exist such that

$$\exists k_1\in\mathbf{N}_{\_},\widehat{\sigma}^{k''}(\widehat{x})(k_1)=\mu, \tag{4.21}$$

$$\left\{\begin{array}{c}\forall k\in\mathbf{N}_{\_},\widehat{\sigma}^{k''}(\widehat{x})(k)=\mu\Longrightarrow(\widehat{\sigma}^{k''}(\widehat{x})(k)=\widehat{\sigma}^{k''}(\widehat{x})(k+1)\text{ and}\\ \text{and }k\geq 0\Longrightarrow\widehat{\sigma}^{k''}(\widehat{x})(k)=\widehat{\sigma}^{k''}(\widehat{x})(k-1)),\end{array}\right. \tag{4.22}$$

thus

$$\exists k_1\in\mathbf{N}_{\_},\widehat{x}(k_1+k'')=\mu, \tag{4.23}$$

$$\left\{\begin{array}{c}\forall k\in\mathbf{N}_{\_},\widehat{x}(k+k'')=\mu\Longrightarrow(\widehat{x}(k+k'')=\widehat{x}(k+k''+1)\text{ and}\\ \text{and }k\geq 0\Longrightarrow\widehat{x}(k+k'')=\widehat{x}(k+k''-1)).\end{array}\right. \tag{4.24}$$

We define $k'=k''-1$. With the notation $k_2=k+k''$, where $k\in\mathbf{N}_{\_}$, we get $k_2=k+k'+1\geq k'$ and (4.23), (4.24) become

$$\exists k_2\geq k',\widehat{x}(k_2)=\mu, \tag{4.25}$$

$$\left\{\begin{array}{c}\forall k_2\geq k',\widehat{x}(k_2)=\mu\Longrightarrow(\widehat{x}(k_2)=\widehat{x}(k_2+1)\text{ and}\\ \text{and }k_2\geq k'+1\Longrightarrow\widehat{x}(k_2)=\widehat{x}(k_2-1)).\end{array}\right. \tag{4.26}$$

From (4.25), (4.26) we infer

$$\mu=\widehat{x}(k')=\widehat{x}(k'+1)=\widehat{x}(k'+2)=...$$

thus (1.1) holds.

b) (1.4)$\Longrightarrow$(4.5) Let $T>0$ arbitrary. From (1.4) some $\mu\in\mathbf{B}^n$ and $t_1'\in\mathbf{R}$ exist such that

$$\forall t\geq t_1',x(t)=\mu. \tag{4.27}$$

Let $t'\in I^x$ that we can choose, without restricting the generality, $t'\leq t_1'$. From (4.27) we have $\mathbf{T}_\mu^x\cap[t_1',\infty)=[t_1',\infty)\neq\varnothing$ and, on the other hand, let $t\in\mathbf{T}_\mu^x\cap[t_1',\infty),z\in\mathbf{Z}$ be arbitrary with $t+zT\geq t_1'$. Then, from (4.27), $x(t+zT)=\mu$, i.e. $t+zT\in\mathbf{T}_\mu^x$. Conclusion (4.5) results.

(4.5)$\Longrightarrow$(4.6) Obvious.

(4.6)$\Longrightarrow$(4.7) Let $T>0$ arbitrary. From (4.6) we have the existence of $\mu\in Or(x)$ and $t_1'\in\mathbf{R}$ such that

$$\mathbf{T}_\mu^x\cap[t_1',\infty)\neq\varnothing, \tag{4.28}$$

$$\forall t\in\mathbf{T}_\mu^x\cap[t_1',\infty),\{t+zT|z\in\mathbf{Z}\}\cap[t_1',\infty)\subset\mathbf{T}_\mu^x. \tag{4.29}$$

Because some $t\in\mathbf{T}_\mu^x\cap[t_1',\infty)$ exists (from (4.28)) and then $\{t,t+T,t+2T,...\}\subset\mathbf{T}_\mu^x$ (from (4.29)) we infer $\mu\in\omega(x)$.

We take $t''>t_1'$ arbitrary. Some $\varepsilon>0$ exists with $\forall\xi\in(t''-\varepsilon,t''),x(\xi)=x(t''-0)$. We take $t'\in(\max\{t_1',t''-\varepsilon\},t'')$ arbitrarily and we have

$$\sigma^{t''}(x)(t)=\left\{\begin{array}{c}x(t),t\geq t'\\ x(t''-0),t<t''.\end{array}\right.$$

We infer the truth of $t' \in (-\infty, t'') \subset I^{\sigma^{t''}(x)}$. The fact that $\mathbf{T}_{\mu}^{\sigma^{t''}(x)} \cap [t', \infty) \neq \varnothing$ results from the remark that $\mu \in \omega(x)$. Let us take now some $t \in \mathbf{T}_{\mu}^{\sigma^{t''}(x)} \cap [t', \infty)$ and $z \in \mathbf{Z}$ arbitrarily such that $t + zT \geq t'$. Obviously $\mathbf{T}_{\mu}^{\sigma^{t''}(x)} \cap [t', \infty) = \mathbf{T}_{\mu}^{x} \cap [t', \infty)$. As far as $t \in \mathbf{T}_{\mu}^{x} \cap [t_1', \infty), t + zT \geq t_1'$, we can apply (4.29) and we have that $t + zT \in \mathbf{T}_{\mu}^{x}$, i.e. $x(t + zT) = \mu$. As $t + zT \geq t'$, and consequently $x(t + zT) = \sigma^{t''}(x)(t + zT)$, we conclude that $t + zT \in \mathbf{T}_{\mu}^{\sigma^{t''}(x)}$.

(4.7)$\Longrightarrow$(4.8) Obvious.

(4.8)$\Longrightarrow$(4.9) Let $T > 0$. From (4.8) we have the existence of $\mu \in Or(x), t'' \in \mathbf{R}$ and $t''' \in \mathbf{R}$ such that

$$\mathbf{T}_{\mu}^{\sigma^{t''}(x)} \cap [t''', \infty) \neq \varnothing, \tag{4.30}$$

$$\forall t \in \mathbf{T}_{\mu}^{\sigma^{t''}(x)} \cap [t''', \infty), \{t + zT | z \in \mathbf{Z}\} \cap [t''', \infty) \subset \mathbf{T}_{\mu}^{\sigma^{t''}(x)}. \tag{4.31}$$

From (4.30) some $t \in \mathbf{T}_{\mu}^{\sigma^{t''}(x)} \cap [t''', \infty)$ exists and from (4.31) $\{t, t+T, t+2T, ...\} \subset \mathbf{T}_{\mu}^{\sigma^{t''}(x)}$, thus $\mu \in \omega(x)$.

We define $t_1' = \max\{t'', t'''\}$. $t' \in \mathbf{R}$ is chosen without loss $\leq t_1'$ such that $t' \in I^x$. Since $\mu \in \omega(x)$ we get $\exists t_2' \geq t_1', x(t_2') = \mu$.

Let now $t \geq t_1'$ such that $x(t) = \mu$, thus $t \in \mathbf{T}_{\mu}^{\sigma^{t''}(x)} \cap [t''', \infty)$. We have

$$t + T \in \{t + zT | z \in \mathbf{Z}\} \cap [t''', \infty) \overset{(4.31)}{\subset} \mathbf{T}_{\mu}^{\sigma^{t''}(x)}$$

and, because $t + T \geq t_1'$, we infer $\mu = \sigma^{t''}(x)(t + T) = x(t + T)$. Similarly, if $t - T \geq t_1'$, we can apply (4.31) again and we get $\mu = \sigma^{t''}(x)(t - T) = x(t - T)$.

(4.9)$\Longrightarrow$(4.10) Obvious.

(4.10)$\Longrightarrow$(4.11) Let $T > 0$ arbitrary. Then $\mu \in Or(x), t_1' \in \mathbf{R}$ exist such that

$$\exists t_2' \geq t_1', x(t_2') = \mu, \tag{4.32}$$

$$\left\{ \begin{array}{c} \forall t \geq t_1', x(t) = \mu \Longrightarrow \\ \Longrightarrow (x(t) = x(t + T) \text{ and } t - T \geq t_1' \Longrightarrow x(t) = x(t - T)). \end{array} \right. \tag{4.33}$$

From (4.32), (4.33) we infer $\{t_2', t_2' + T, t_2' + 2T, ...\} \subset \mathbf{T}_{\mu}^{x}$, thus $\mu \in \omega(x)$.

Let $\varepsilon > 0$ with $\forall \xi \in [t_1', t_1' + \varepsilon), x(\xi) = x(t_1')$ and we take $t' = t'' \in (t_1', t_1' + \varepsilon)$ arbitrarily. We have

$$\sigma^{t''}(x)(t) = \left\{ \begin{array}{l} x(t), t \geq t', \\ x(t_1'), t \leq t', \end{array} \right. \tag{4.34}$$

wherefrom $t' \in I^{\sigma^{t''}(x)}$. As $\mu \in \omega(x)$, we obtain the existence of $t''' > t'$ with $\sigma^{t''}(x)(t''') = x(t''') = \mu$. The truth of

$$\left\{ \begin{array}{l} \forall t \geq t', \sigma^{t''}(x)(t) = \mu \Longrightarrow (\sigma^{t''}(x)(t) = \sigma^{t''}(x)(t + T) \text{ and} \\ \qquad \text{and } t - T \geq t' \Longrightarrow \sigma^{t''}(x)(t) = \sigma^{t''}(x)(t - T)) \end{array} \right. \tag{4.35}$$

results from (4.33), $t' \geq t_1'$, (4.34) and Lemma 3, page 224.

(4.11)$\Longrightarrow$(4.12) Obvious.

(4.12)$\Longrightarrow$(1.5) Let $T > 0$ arbitrary. Then $\mu \in Or(x)$, $t'' \in \mathbf{R}$ and $t' \in \mathbf{R}$ exist such that

$$\exists t''' \geq t', \sigma^{t''}(x)(t''') = \mu, \tag{4.36}$$

$$\begin{cases} \forall t \geq t', \sigma^{t''}(x)(t) = \mu \Longrightarrow (\sigma^{t''}(x)(t) = \sigma^{t''}(x)(t+T) \text{ and} \\ \text{and } t - T \geq t' \Longrightarrow \sigma^{t''}(x)(t) = \sigma^{t''}(x)(t-T)). \end{cases} \tag{4.37}$$

From (4.37), Lemma 3, page 224 and from the fact that $\forall t \geq \max\{t', t''\}, \sigma^{t''}(x)(t) = x(t)$, we have

$$\begin{cases} \forall t \geq \max\{t', t''\}, x(t) = \mu \Longrightarrow \\ \Longrightarrow (x(t) = x(t+T) \text{ and } t - T \geq \max\{t', t''\} \Longrightarrow x(t) = x(t-T)). \end{cases} \tag{4.38}$$

On the other hand from (4.36), (4.37) we have the existence of $t''' \geq t'$ with

$$\{t''', t''' + T, t''' + 2T, ...\} \subset \mathbf{T}_{\mu}^{\sigma^{t''}(x)},$$

meaning that $\mu \in \omega(\sigma^{t''}(x)) = \omega(x)$ (Theorem 5, page 9). If $\omega(x) = \{\mu\}$ then the implication is proved, so let us suppose against all reason that this is not true. Some $t_1, t_2 \in \mathbf{R}$ exist with the property $\max\{t', t''\} < t_1 < t_2$, $[t_1, t_2) \subset \mathbf{T}_{\mu}^{x}$, $x(t_1 - 0) \neq \mu$, $x(t_2) \neq \mu$. This shows from Lemma 4, page 224 that

$$[t_1, t_2) \cup [t_1 + T, t_2 + T) \cup [t_1 + 2T, t_2 + 2T) \cup ... \subset \mathbf{T}_{\mu}^{x} \tag{4.39}$$

and from Lemma 5, page 224 that $\forall k \in \mathbf{N}$,

$$x(t_1 + kT - 0) \neq \mu, \tag{4.40}$$

$$x(t_2 + kT) \neq \mu. \tag{4.41}$$

Let us write now (4.12) for $T' \in (0, t_2 - t_1)$. There exist $\mu' \in Or(x)$, $t_1'' \in \mathbf{R}$ and $t_1' \in \mathbf{R}$ with

$$\exists t_1''' \geq t_1', \sigma^{t_1''}(x)(t_1''') = \mu', \tag{4.42}$$

$$\begin{cases} \forall t \geq t_1', \sigma^{t_1''}(x)(t) = \mu' \Longrightarrow (\sigma^{t_1''}(x)(t) = \sigma^{t_1''}(x)(t+T') \text{ and} \\ \text{and } t - T' \geq t_1' \Longrightarrow \sigma^{t_1''}(x)(t) = \sigma^{t_1''}(x)(t-T')). \end{cases} \tag{4.43}$$

We infer like before the existence of $t_3, t_4 \in \mathbf{R}$ having the property that $\max\{t_1', t_1''\} < t_3 < t_4$, $[t_3, t_4) \subset \mathbf{T}_{\mu'}^{x}$, $x(t_3 - 0) \neq \mu'$, $x(t_4) \neq \mu'$ and

$$[t_3, t_4) \cup [t_3 + T', t_4 + T') \cup [t_3 + 2T', t_4 + 2T') \cup ... \subset \mathbf{T}_{\mu'}^{x} \tag{4.44}$$

and for any $k \in \mathbf{N}$ we have

$$x(t_3 + kT' - 0) \neq \mu', \tag{4.45}$$

$$x(t_4 + kT') \neq \mu'. \tag{4.46}$$

From the fact that $T' < t_2 - t_1$ and from Lemma 6, page 226 we have that

$$\varnothing \neq ([t_1, t_2) \cup [t_1 + T, t_2 + T) \cup [t_1 + 2T, t_2 + 2T) \cup ...) \cap$$

$$\cap([t_3, t_4) \cup [t_3 + T', t_4 + T') \cup [t_3 + 2T', t_4 + 2T') \cup ...) \subset \mathbf{T}_{\mu}^{x} \cap \mathbf{T}_{\mu'}^{x}$$

wherefrom $\mu = \mu'$.

Let now $k_2, k_3 \in \mathbf{N}$ with the property that $[t_1 + k_2T, t_2 + k_2T) \cap [t_3 + k_3T', t_4 + k_3T') \neq \varnothing$. We have the following non-exclusive cases, that cover all the possibilities.

Case $t_1 + k_2T = t_3 + k_3T'$
As $T' < t_2 - t_1$, we have $t_3 + (k_3+1)T' \in (t_1 + k_2T, t_2 + k_2T)$, contradiction with (4.45).
Case $t_1 + k_2T \in (t_3 + k_3T', t_4 + k_3T')$,
contradiction with (4.40).
Case $t_2 + k_2T \in (t_3 + k_3T', t_4 + k_3T')$,
contradiction with (4.41).
Case $t_3 + k_3T' \in (t_1 + k_2T, t_2 + k_2T)$,
contradiction with (4.45).
Case $t_4 + k_3T' \in (t_1 + k_2T, t_2 + k_2T)$,
contradiction with (4.46).

The fact that we have obtained in all these cases a contradiction shows the falsity of (4.39), with $x(t_1 - 0) \neq \mu$, $x(t_2) \neq \mu$. These should be replaced by an inclusion of the form $[t_1, \infty) \subset \mathbf{T}^x_\mu$. The truth of (1.5) is proved. □

5. The Third Group of Eventual Constancy Properties, Version

Remark 46. *These properties are a version of the properties of the third group from the previous section. To be noticed that the universal quantifier* $\forall \mu \in \widehat{Or}(\widehat{x}), \forall \mu \in Or(x)$ *in Theorem 17, page 27 can be replaced by the existential quantifier in two different ways; the first possibility expressed at Theorem 18, page 33 is:* $\exists \mu \in \widehat{Or}(\widehat{x}), ..., \widehat{\mathbf{T}}^{\widehat{x}}_\mu \cap \{k', k'+1, k'+2, ...\} \neq \varnothing, \exists \mu \in Or(x), ..., \mathbf{T}^x_\mu \cap [t'_1, \infty) \neq \varnothing$ *and the second possibility expressed at Theorem 19 to follow is:* $\exists \mu \in \widehat{\omega}(\widehat{x}), \exists \mu \in \omega(x)$, *when the previous non-triviality conditions* $\widehat{\mathbf{T}}^{\widehat{x}}_\mu \cap \{k', k'+1, k'+2, ...\} \neq \varnothing, \mathbf{T}^x_\mu \cap [t'_1, \infty) \neq \varnothing$ *are fulfilled see also Lemma 1, page 223.*

Theorem 19. *Let the signals* $\widehat{x} \in \widehat{S}^{(n)}, x \in S^{(n)}$.

a) The following statements are equivalent with the eventual constancy of $\widehat{x}$:

$$\left\{ \begin{array}{c} \forall p \geq 1, \exists \mu \in \widehat{\omega}(\widehat{x}), \exists k' \in \mathbf{N}_{\_}, \forall k \in \widehat{\mathbf{T}}^{\widehat{x}}_\mu \cap \{k', k'+1, k'+2, ...\}, \\ \{k + zp | z \in \mathbf{Z}\} \cap \{k', k'+1, k'+2, ...\} \subset \widehat{\mathbf{T}}^{\widehat{x}}_\mu, \end{array} \right. \tag{5.1}$$

$$\left\{ \begin{array}{c} \forall p \geq 1, \exists \mu \in \widehat{\omega}(\widehat{x}), \exists k'' \in \mathbf{N}, \forall k \in \widehat{\mathbf{T}}^{\widehat{\sigma}^{k''}(\widehat{x})}_\mu, \\ \{k + zp | z \in \mathbf{Z}\} \cap \mathbf{N}_{\_} \subset \widehat{\mathbf{T}}^{\widehat{\sigma}^{k''}(\widehat{x})}_\mu, \end{array} \right. \tag{5.2}$$

$$\left\{ \begin{array}{c} \forall p \geq 1, \exists \mu \in \widehat{\omega}(\widehat{x}), \exists k' \in \mathbf{N}_{\_}, \forall k \geq k', \widehat{x}(k) = \mu \Longrightarrow \\ \Longrightarrow (\widehat{x}(k) = \widehat{x}(k+p) \textit{ and } k - p \geq k' \Longrightarrow \widehat{x}(k) = \widehat{x}(k-p)), \end{array} \right. \tag{5.3}$$

$$\left\{ \begin{array}{c} \forall p \geq 1, \exists \mu \in \widehat{\omega}(\widehat{x}), \exists k'' \in \mathbf{N}, \forall k \in \mathbf{N}_{\_}, \widehat{\sigma}^{k''}(\widehat{x})(k) = \mu \Longrightarrow \\ \Longrightarrow (\widehat{\sigma}^{k''}(\widehat{x})(k) = \widehat{\sigma}^{k''}(\widehat{x})(k+p) \textit{ and} \\ \textit{and } k - p \geq -1 \Longrightarrow \widehat{\sigma}^{k''}(\widehat{x})(k) = \widehat{\sigma}^{k''}(\widehat{x})(k-p)). \end{array} \right. \tag{5.4}$$

b) The following statements are equivalent with the eventual constancy of x:

$$\left\{ \begin{array}{c} \forall T > 0, \exists \mu \in \omega(x), \exists t' \in I^x, \\ \exists t'_1 \geq t', \forall t \in \mathbf{T}^x_\mu \cap [t'_1, \infty), \{t + zT | z \in \mathbf{Z}\} \cap [t'_1, \infty) \subset \mathbf{T}^x_\mu, \end{array} \right. \tag{5.5}$$

$$\left\{ \begin{array}{c} \forall T > 0, \exists \mu \in \omega(x), \exists t'_1 \in \mathbf{R}, \\ \forall t \in \mathbf{T}^x_\mu \cap [t'_1, \infty), \{t + zT | z \in \mathbf{Z}\} \cap [t'_1, \infty) \subset \mathbf{T}^x_\mu, \end{array} \right. \tag{5.6}$$

$$\begin{cases} \forall T>0, \exists \mu \in \omega(x), \exists t'' \in \mathbf{R}, \exists t' \in I^{\sigma^{t''}(x)}, \\ \forall t \in \mathbf{T}_{\mu}^{\sigma^{t''}(x)} \cap [t',\infty), \{t+zT|z\in \mathbf{Z}\} \cap [t',\infty) \subset \mathbf{T}_{\mu}^{\sigma^{t''}(x)}, \end{cases} \tag{5.7}$$

$$\begin{cases} \forall T>0, \exists \mu \in \omega(x), \exists t'' \in \mathbf{R}, \exists t' \in \mathbf{R}, \\ \forall t \in \mathbf{T}_{\mu}^{\sigma^{t''}(x)} \cap [t',\infty), \{t+zT|z\in \mathbf{Z}\} \cap [t',\infty) \subset \mathbf{T}_{\mu}^{\sigma^{t''}(x)}, \end{cases} \tag{5.8}$$

$$\begin{cases} \forall T>0, \exists \mu \in \omega(x), \exists t' \in I^x, \exists t_1' \geq t', \forall t \geq t_1', x(t)=\mu \Longrightarrow \\ \Longrightarrow (x(t)=x(t+T) \; and \; t-T \geq t_1' \Longrightarrow x(t)=x(t-T)), \end{cases} \tag{5.9}$$

$$\begin{cases} \forall T>0, \exists \mu \in \omega(x), \exists t_1' \in \mathbf{R}, \forall t \geq t_1', x(t)=\mu \Longrightarrow \\ \Longrightarrow (x(t)=x(t+T) \; and \; t-T \geq t_1' \Longrightarrow x(t)=x(t-T)), \end{cases} \tag{5.10}$$

$$\begin{cases} \forall T>0, \exists \mu \in \omega(x), \exists t'' \in \mathbf{R}, \exists t' \in I^{\sigma^{t''}(x)}, \\ \forall t \geq t', \sigma^{t''}(x)(t)=\mu \Longrightarrow (\sigma^{t''}(x)(t)=\sigma^{t''}(x)(t+T) \; and \\ and \; t-T \geq t' \Longrightarrow \sigma^{t''}(x)(t)=\sigma^{t''}(x)(t-T)), \end{cases} \tag{5.11}$$

$$\begin{cases} \forall T>0, \exists \mu \in \omega(x), \exists t'' \in \mathbf{R}, \exists t' \in \mathbf{R}, \forall t \geq t', \sigma^{t''}(x)(t)=\mu \Longrightarrow \\ \Longrightarrow (\sigma^{t''}(x)(t)=\sigma^{t''}(x)(t+T) \; and \\ and \; t-T \geq t' \Longrightarrow \sigma^{t''}(x)(t)=\sigma^{t''}(x)(t-T)). \end{cases} \tag{5.12}$$

6. The Fourth Group of Eventual Constancy Properties

Remark 47. *This group of eventual constancy properties involves the eventual periodicity of the signals.*

Theorem 20. *Let the signals $\widehat{x} \in \widehat{S}^{(n)}, x \in S^{(n)}$.*

a) The following statements are equivalent with the eventual constancy of $\widehat{x}$:

$$\forall p \geq 1, \exists k' \in \mathbf{N}_{\_}, \forall k \geq k', \widehat{x}(k)=\widehat{x}(k+p), \tag{6.1}$$

$$\forall p \geq 1, \exists k'' \in \mathbf{N}, \forall k \in \mathbf{N}_{\_}, \widehat{\sigma}^{k''}(\widehat{x})(k)=\widehat{\sigma}^{k''}(\widehat{x})(k+p). \tag{6.2}$$

b) The following statements are equivalent with the eventual constancy of x:

$$\forall T>0, \exists t' \in I^x, \exists t_1' \geq t', \forall t \geq t_1', x(t)=x(t+T), \tag{6.3}$$

$$\forall T>0, \exists t_1' \in \mathbf{R}, \forall t \geq t_1', x(t)=x(t+T), \tag{6.4}$$

$$\forall T>0, \exists t'' \in \mathbf{R}, \exists t' \in I^{\sigma^{t''}(x)}, \forall t \geq t', \sigma^{t''}(x)(t)=\sigma^{t''}(x)(t+T), \tag{6.5}$$

$$\forall T>0, \exists t'' \in \mathbf{R}, \exists t' \in \mathbf{R}, \forall t \geq t', \sigma^{t''}(x)(t)=\sigma^{t''}(x)(t+T). \tag{6.6}$$

Proof. a) $(1.1)_{page\,25} \Longrightarrow$(6.1) Let $p \geq 1$ arbitrary. We have from (1.1) the existence of $\mu \in \mathbf{B}^n$ and $k' \in \mathbf{N}_{\_}$ such that

$$\forall k \geq k', \widehat{x}(k)=\mu.$$

Then for any $k \geq k'$ we have $\widehat{x}(k+p)=\mu$, thus (6.1) holds.

(6.1)$\Longrightarrow$(6.2) Let $p \geq 1$. From (6.1), some $k' \in \mathbf{N}_{\_}$ exists with

$$\forall k \geq k', \widehat{x}(k)=\widehat{x}(k+p). \tag{6.7}$$

We define $k'' = k'+1$ and let $k \in \mathbf{N}_{\_}$ arbitrary. As $k+k'' = k+k'+1 \geq k'$, we can write that

$$\widehat{\sigma}^{k''}(\widehat{x})(k) = \widehat{x}(k+k'') = \widehat{x}(k+k'+1) \overset{(6.7)}{=} \widehat{x}(k+k'+1+p) =$$

$$= \widehat{x}(k+k''+p) = \widehat{\sigma}^{k''}(\widehat{x})(k+p).$$

(6.2)$\Longrightarrow$(1.1) We write (6.2) for $p=1$ under the form: $k'' \in \mathbf{N}$ exists with

$$\forall k \in \mathbf{N}_{\_}, \widehat{\sigma}^{k''}(\widehat{x})(k) = \widehat{\sigma}^{k''}(\widehat{x})(k+1), \tag{6.8}$$

i.e. $\widehat{\sigma}^{k''}(\widehat{x})$ is constant. We denote with μ the value of this constant, for which we have from (6.8):

$$\forall k \in \mathbf{N}_{\_}, \widehat{x}(k+k'') = \mu. \tag{6.9}$$

We denote $k' = k''-1, k' \in \mathbf{N}_{\_}$. As $k+k'' = k+k'+1 \geq k'$, (6.9) shows that

$$\forall k \geq k', \widehat{x}(k) = \mu.$$

b) (1.4)$_{page\,25}$ $\Longrightarrow$(6.3) Let $T>0$ arbitrary. Some $\mu \in \mathbf{B}^n$ and some $t_1' \in \mathbf{R}$ exist from (1.4) with

$$\forall t \geq t_1', x(t) = \mu. \tag{6.10}$$

There also exists an initial time instant $t' \in I^x$ that can be chosen without loss $\leq t_1'$.

We fix in (6.10) an arbitrary $t \geq t_1'$. We have $x(t+T) = \mu$, thus (6.3) holds.

(6.3)$\Longrightarrow$(6.4) Obvious.

(6.4)$\Longrightarrow$(6.5) Let $T>0$ arbitrary. Some $t_1' \in \mathbf{R}$ exists from (6.4) such that

$$\forall t \geq t_1', x(t) = x(t+T). \tag{6.11}$$

We take $t'' > t_1'$ arbitrary. Some $\varepsilon > 0$ exists then with $\forall t \in (t''-\varepsilon, t''), x(t) = x(t''-0)$. We also take $t' \in (t''-\varepsilon, t'') \cap [t_1', \infty)$ arbitrarily and we infer that we have

$$\sigma^{t''}(x)(t) = \begin{cases} x(t), t > t''-\varepsilon, \\ x(t''-0), t < t''. \end{cases} \tag{6.12}$$

The fact that $\forall t \leq t', \sigma^{t''}(x)(t) = x(t''-0)$ is obvious, wherefrom $t' \in I^{\sigma^{t''}(x)}$. For any $t \geq t'$ we have

$$\sigma^{t''}(x)(t) \overset{(6.12)}{=} x(t) \overset{(6.11)}{=} x(t+T) \overset{(6.12)}{=} \sigma^{t''}(x)(t+T).$$

(6.5)$\Longrightarrow$(6.6) Obvious.

(6.6)$\Longrightarrow$(1.6)$_{page\,25}$ We suppose against all reason that (1.6) is false, meaning that $\mu', \mu'' \in \omega(x)$ exist, with $\mu' \neq \mu''$. Let $T>0$ be arbitrary. From (6.6) we have the existence of $t'' \in \mathbf{R}, t' \in \mathbf{R}$ such that

$$\forall t \geq t', \sigma^{t''}(x)(t) = \sigma^{t''}(x)(t+T), \tag{6.13}$$

wherefrom

$$\forall t \geq \max\{t', t''\}, x(t) = x(t+T). \tag{6.14}$$

Then $t_0 \geq \max\{t', t''\}$ and $t_1 \geq \max\{t', t''\}$ exist such that $x(t_0) = \mu', x(t_1) = \mu''$ thus, from (6.14),

$$\forall k \in \mathbf{N}, x(t_0 + kT) = \mu', \tag{6.15}$$

$$\forall k \in \mathbf{N}, x(t_1 + kT) = \mu''. \tag{6.16}$$

Obviously $t_0 \neq t_1$ and, in order to make a choice, we suppose that $t_0 < t_1$.

We write now (6.6) for $T' = t_1 - t_0$ and we get the existence of $t_1'' \in \mathbf{R}, t_1' \in \mathbf{R}$ with

$$\forall t \geq t_1', \sigma^{t_1''}(x)(t) = \sigma^{t_1''}(x)(t + t_1 - t_0). \tag{6.17}$$

Let $k_1 \in \mathbf{N}$ satisfying $t_0 + k_1 T \geq \max\{t_1', t_1''\}$. We have $t_1 + k_1 T > t_0 + k_1 T \geq \max\{t_1', t_1''\}$, wherefrom:

$$\mu' \overset{(6.15)}{=} x(t_0 + k_1 T) = \sigma^{t_1''}(x)(t_0 + k_1 T) \overset{(6.17)}{=} \sigma^{t_1''}(x)(t_0 + k_1 T + t_1 - t_0)$$

$$= \sigma^{t_1''}(x)(t_1 + k_1 T) = x(t_1 + k_1 T) \overset{(6.16)}{=} \mu'',$$

representing a contradiction with our supposition that $\mu' \neq \mu''$. We infer the truth of (1.6). □

7. Discrete Time vs Real Time

Theorem 21. *We suppose that $(t_k) \in Seq$ exists with*

$$x(t) = \widehat{x}(-1) \cdot \chi_{(-\infty, t_0)}(t) \oplus \widehat{x}(0) \cdot \chi_{[t_0, t_1)}(t) \oplus \ldots \oplus \widehat{x}(k) \cdot \chi_{[t_k, t_{k+1})}(t) \oplus \ldots \tag{7.1}$$

Then the eventual constancies of $\widehat{x}$ and x are equivalent.

Proof. If (7.1) is true, then $(1.1)_{page\,25}$ and $(1.4)_{page\,25}$ are obviously equivalent, with $k' = \begin{cases} -1, \; if\; t' < t_0, \\ k, \; if\; t' \in [t_k, t_{k+1}), k \geq 0 \end{cases}$ and $t' = \begin{cases} t_{k'}, if\; k' \geq 0, \\ t_0 - \varepsilon, if\; k' = -1, \end{cases}$ where $\varepsilon > 0$ is arbitrary. □

8. Discussion

Remark 48. *The statements of Theorem 15, page 25,..., Theorem 20, page 40, are structured in discrete time - real time analogue properties and we notice that to a discrete time statement there correspond either one or (in Theorems 17, 18, 19, 20) two real time statements. This is principially based on the fact that we may omit in these requirements to refer to the initial time.*

Remark 49. *To be compared Theorem 21 with Theorem 3, page 6.*

Remark 50. *Properties (1.1),..., (1.3) and (1.4),..., (1.6) do not involve periodicity. The other properties that are equivalent with eventual constancy are divided into two groups:*

- (3.1),...,(3.4) and (3.5),...,(3.12); (4.1),...,(4.4) and (4.5),...,(4.12); (5.1),..., (5.4) and (5.5),...,(5.12) are of eventual periodicity of the points, and

- (6.1), (6.2) and (6.3),..., (6.6) are of eventual periodicity of the signals.

Remark 51. *The previous groups of periodicity properties have in common the fact that the eventual periodicity of a signal exists if all the points of the orbit are eventually periodic, with the same period and with the same limit of periodicity.*

Remark 52. *The statements (3.2), (3.4) and (3.7), (3.11); (4.2), (4.4) and (4.7), (4.11); (5.2), (5.4) and (5.7), (5.11); (6.2) and (6.5) are of periodicity of* $\widehat{\sigma}^{k''}(\widehat{x}), \sigma^{t''}(x)$. *The eventual periodicity of* $\widehat{x}, x$ *results from the fact that* $\widehat{\sigma}^{k''}(\widehat{x})$, $\sigma^{t''}(x)$ *ignore the first values of* $\widehat{x}, x$.

Remark 53. *We ask, in order that the eventual periodicity be equivalent with the eventual constancy, that it should take place with any period* $p \geq 1, T > 0$.

Remark 54. *In (1.1),..., (1.3) and (1.4),..., (1.6) the existence of a unique* μ *is asked and we have* $\mu = \lim\limits_{k \to \infty} \widehat{x}(k), \mu = \lim\limits_{t \to \infty} x(t)$.

Remark 55. *The eventually constant signals* $\widehat{x}, x$ *fulfill* $\widehat{\omega}(\widehat{x}) = \{\mu\}, \omega(x) = \{\mu\}$ *like the constant signals, but* $\widehat{Or}(\widehat{x}), Or(x)$ *contain also other points than* μ, *in general. The points of* $\widehat{Or}(\widehat{x}) \setminus \widehat{\omega}(\widehat{x})$, $Or(x) \setminus \omega(x)$ *are some 'first values' of these signals.*

Chapter 4

Constant Signals

The chapter presents properties that are equivalent with the constancy of the signals and that are related, most of them, with periodicity. The key aspect is that periodicity must hold with any period in order to be equivalent with constancy.

We have gathered these properties in four groups, in order to analyze them better and make them be better understood. Section 1 presents the first group of constancy properties, gathering these properties that are not related with periodicity. Sections 2, respectively 3 present the groups of constancy properties of the signals involving periodicity and eventual periodicity properties of all the points of the orbit, respectively of some point of the orbit. The fourth group of constancy properties, involving the periodicity and the eventual periodicity of the signals, is introduced in Section 4. Section 5 relates the constancy of the discrete time and of the real time signals. The last section contains the interpretation of the constancy properties.

1. The First Group of Constancy Properties

Remark 56. *The first group of constancy properties of the signals contains these properties that are not related with periodicity. These properties are inspired one by one by the properties of eventual constancy from Theorem 15, page 25.*

Theorem 22. *We consider the signals $\widehat{x} \in \widehat{S}^{(n)}, x \in S^{(n)}$.*

a) The following requirements stating the constancy of $\widehat{x}$ are equivalent

$$\exists \mu \in \mathbf{B}^n, \forall k \in \mathbf{N}_{\_}, \widehat{x}(k) = \mu, \tag{1.1}$$

$$\exists \mu \in \mathbf{B}^n, \widehat{\mathbf{T}}_{\mu}^{\widehat{x}} = \mathbf{N}_{\_}, \tag{1.2}$$

$$\exists \mu \in \mathbf{B}^n, \widehat{Or}(\widehat{x}) = \{\mu\}. \tag{1.3}$$

b) The following requirements stating the constancy of x are also equivalent

$$\exists \mu \in \mathbf{B}^n, \forall t \in \mathbf{R}, x(t) = \mu, \tag{1.4}$$

$$\exists \mu \in \mathbf{B}^n, \mathbf{T}_{\mu}^{x} = \mathbf{R}, \tag{1.5}$$

$$\exists \mu \in \mathbf{B}^n, Or(x) = \{\mu\}. \tag{1.6}$$

Proof. a) (1.1)$\Longrightarrow$(1.2) If $\mu \in \mathbf{B}^n$ exists such that $\forall k \in \mathbf{N}_{\text{-}}, \widehat{x}(k) = \mu$, then $\{k | k \in \mathbf{N}_{\text{-}}, \widehat{x}(k) = \mu\} = \mathbf{N}_{\text{-}}$.

(1.2)$\Longrightarrow$(1.3) If $\mu \in \mathbf{B}^n$ exists such that $\{k | k \in \mathbf{N}_{\text{-}}, \widehat{x}(k) = \mu\} = \mathbf{N}_{\text{-}}$, then $\{\widehat{x}(k) | k \in \mathbf{N}_{\text{-}}\} = \{\mu\}$.

(1.3)$\Longrightarrow$(1.1) The existence of $\mu \in \mathbf{B}^n$ such that $\{\widehat{x}(k) | k \in \mathbf{N}_{\text{-}}\} = \{\mu\}$ implies $\forall k \in \mathbf{N}_{\text{-}}, \widehat{x}(k) = \mu$.

b) (1.4)$\Longrightarrow$(1.5) If $\mu \in \mathbf{B}^n$ exists such that $\forall t \in \mathbf{R}, x(t) = \mu$, then $\{t | t \in \mathbf{R}, x(t) = \mu\} = \mathbf{R}$.

(1.5)$\Longrightarrow$(1.6) If $\mu \in \mathbf{B}^n$ exists such that $\{t | t \in \mathbf{R}, x(t) = \mu\} = \mathbf{R}$, then $\{x(t) | t \in \mathbf{R}\} = \{\mu\}$.

(1.6)$\Longrightarrow$(1.4) The existence of μ with $\{x(t) | t \in \mathbf{R}\} = \{\mu\}$ implies that $\forall t \in \mathbf{R}, x(t) = \mu$ is true. □

Remark 57. *Note that if $\widehat{x}, x$ is constant and $f : \mathbf{B}^n \to \mathbf{B}^m$ is some function, then $\widehat{f}(\widehat{x})$ and $f(x)$ are constant and we have*

$$\widehat{Or}(\widehat{f}(\widehat{x})) \overset{Theorem\ 14,\ page\ 14}{=} f(\widehat{Or}(\widehat{x})) \overset{(1.3)}{=} \{f(\mu)\},$$

$$Or(f(x)) \overset{Theorem\ 14,\ page\ 14}{=} f(Or(x)) \overset{(1.6)}{=} \{f(\mu)\}.$$

2. The Second Group of Constancy Properties

Remark 58. *This group of constancy properties of the signals involves periodicity and eventual periodicity properties of all the points μ of the orbit, i.e. in (2.1),...,(2.6), (2.7),...,(2.12) to follow we ask $\forall \mu \in \widehat{Or}(\widehat{x}), \forall \mu \in Or(x)$.*

Remark 59. *In order to understand better the way that these properties were written, to be noticed the existence of the couples and triples:*

- (2.1)-(2.7), (2.2)-(2.8),...,(2.6)-(2.12);

- (2.1)-(2.2)-(2.3), (2.4)-(2.5)-(2.6) and (2.7)-(2.8)-(2.9), (2.10)-(2.11)-(2.12);

- (2.1)-(2.4),(2.2)-(2.5),(2.3)-(2.6) and (2.7)-(2.10), (2.8)-(2.11), (2.9)-(2.12).

Remark 60. *These properties are inspired by the equivalent properties of Theorem 17, page 27. Note that:*

- (2.1) and (2.2) (the last contains $\forall k' \in \mathbf{N}_{\text{-}}$) are inspired by $(3.1)_{page\,27}$ (containing $\exists k' \in \mathbf{N}_{\text{-}}$);

- (2.3) (containing $\forall k'' \in \mathbf{N}$) is inspired by $(3.2)_{page\,27}$ (containing $\exists k'' \in \mathbf{N}$);

- (2.4) and (2.5) (the last contains $\forall k' \in \mathbf{N}_{\text{-}}$) are inspired by $(3.3)_{page\,27}$ (containing $\exists k' \in \mathbf{N}_{\text{-}}$);

- (2.6) (containing $\forall k'' \in \mathbf{N}$) is inspired by $(3.4)_{page\,27}$ (containing $\exists k'' \in \mathbf{N}$);

- (2.7) and (2.8) (the last contains $\forall t_1' \geq t'$) are inspired by $(3.5)_{page\,28}$ and $(3.6)_{page\,28}$ (containing $\exists t_1' \geq t'$ and $\exists t_1' \in \mathbf{R}$);

- (2.9) (containing $\forall t'' \in \mathbf{R}$) is inspired by $(3.7)_{page\,28}$ and $(3.8)_{page\,28}$ (containing $\exists t'' \in \mathbf{R}$);

- (2.10) and (2.11) (the last contains $\forall t_1' \geq t'$) are inspired by $(3.9)_{page\,28}$ and $(3.10)_{page\,28}$ (containing $\exists t_1' \geq t'$ and $\exists t_1' \in \mathbf{R}$);

- (2.12) (containing $\forall t'' \in \mathbf{R}$) is inspired by $(3.11)_{page\,28}$ and $(3.12)_{page\,28}$ (containing $\exists t'' \in \mathbf{R}$).

Theorem 23. *a) Any of the following properties is equivalent with the constancy of $\widehat{x} \in \widehat{S}^{(n)}$:*

$$\forall p \geq 1, \forall \mu \in \widehat{Or}(\widehat{x}), \forall k \in \widehat{\mathbf{T}}_{\mu}^{\widehat{x}}, \{k+zp | z \in \mathbf{Z}\} \cap \mathbf{N}_{\_} \subset \widehat{\mathbf{T}}_{\mu}^{\widehat{x}}, \tag{2.1}$$

$$\begin{cases} \forall p \geq 1, \forall \mu \in \widehat{Or}(\widehat{x}), \forall k' \in \mathbf{N}_{\_}, \forall k \in \widehat{\mathbf{T}}_{\mu}^{\widehat{x}} \cap \{k', k'+1, k'+2, ...\}, \\ \{k+zp | z \in \mathbf{Z}\} \cap \{k', k'+1, k'+2, ...\} \subset \widehat{\mathbf{T}}_{\mu}^{\widehat{x}}, \end{cases} \tag{2.2}$$

$$\begin{cases} \forall p \geq 1, \forall \mu \in \widehat{Or}(\widehat{x}), \forall k'' \in \mathbf{N}, \forall k \in \widehat{\mathbf{T}}_{\mu}^{\widehat{\sigma}^{k''}(\widehat{x})}, \\ \{k+zp | z \in \mathbf{Z}\} \cap \mathbf{N}_{\_} \subset \widehat{\mathbf{T}}_{\mu}^{\widehat{\sigma}^{k''}(\widehat{x})}, \end{cases} \tag{2.3}$$

$$\begin{cases} \forall p \geq 1, \forall \mu \in \widehat{Or}(\widehat{x}), \forall k \in \mathbf{N}_{\_}, \widehat{x}(k) = \mu \Longrightarrow \\ \Longrightarrow (\widehat{x}(k) = \widehat{x}(k+p) \text{ and } k-p \geq -1 \Longrightarrow \widehat{x}(k) = \widehat{x}(k-p)), \end{cases} \tag{2.4}$$

$$\begin{cases} \forall p \geq 1, \forall \mu \in \widehat{Or}(\widehat{x}), \forall k' \in \mathbf{N}_{\_}, \forall k \geq k', \widehat{x}(k) = \mu \Longrightarrow \\ \Longrightarrow (\widehat{x}(k) = \widehat{x}(k+p) \text{ and } k-p \geq k' \Longrightarrow \widehat{x}(k) = \widehat{x}(k-p)), \end{cases} \tag{2.5}$$

$$\begin{cases} \forall p \geq 1, \forall \mu \in \widehat{Or}(\widehat{x}), \forall k'' \in \mathbf{N}, \forall k \in \mathbf{N}_{\_}, \widehat{\sigma}^{k''}(\widehat{x})(k) = \mu \Longrightarrow \\ \Longrightarrow (\widehat{\sigma}^{k''}(\widehat{x})(k) = \widehat{\sigma}^{k''}(\widehat{x})(k+p) \text{ and} \\ \text{and } k-p \geq -1 \Longrightarrow \widehat{\sigma}^{k''}(\widehat{x})(k) = \widehat{\sigma}^{k''}(\widehat{x})(k-p)). \end{cases} \tag{2.6}$$

b) Any of the following properties is equivalent with the constancy of $x \in S^{(n)}$:

$$\begin{cases} \forall T > 0, \forall \mu \in Or(x), \exists t' \in I^x, \\ \forall t \in \mathbf{T}_{\mu}^{x} \cap [t', \infty), \{t+zT | z \in \mathbf{Z}\} \cap [t', \infty) \subset \mathbf{T}_{\mu}^{x}, \end{cases} \tag{2.7}$$

$$\begin{cases} \forall T > 0, \forall \mu \in Or(x), \exists t' \in I^x, \\ \forall t_1' \geq t', \forall t \in \mathbf{T}_{\mu}^{x} \cap [t_1', \infty), \{t+zT | z \in \mathbf{Z}\} \cap [t_1', \infty) \subset \mathbf{T}_{\mu}^{x}, \end{cases} \tag{2.8}$$

$$\begin{cases} \forall T > 0, \forall \mu \in Or(x), \forall t'' \in \mathbf{R}, \exists t' \in I^{\sigma^{t''}(x)}, \\ \forall t \in \mathbf{T}_{\mu}^{\sigma^{t''}(x)} \cap [t', \infty), \{t+zT | z \in \mathbf{Z}\} \cap [t', \infty) \subset \mathbf{T}_{\mu}^{\sigma^{t''}(x)}, \end{cases} \tag{2.9}$$

$$\begin{cases} \forall T > 0, \forall \mu \in Or(x), \exists t' \in I^x, \forall t \geq t', \\ x(t) = \mu \Longrightarrow (x(t) = x(t+T) \text{ and } t-T \geq t' \Longrightarrow x(t) = x(t-T)), \end{cases} \tag{2.10}$$

$$\begin{cases} \forall T > 0, \forall \mu \in Or(x), \exists t' \in I^x, \forall t_1' \geq t', \forall t \geq t_1', x(t) = \mu \Longrightarrow \\ \Longrightarrow (x(t) = x(t+T) \text{ and } t-T \geq t_1' \Longrightarrow x(t) = x(t-T)), \end{cases} \tag{2.11}$$

$$\begin{cases} \forall T > 0, \forall \mu \in Or(x), \forall t'' \in \mathbf{R}, \exists t' \in I^{\sigma^{t''}(x)}, \\ \forall t \geq t', \sigma^{t''}(x)(t) = \mu \Longrightarrow (\sigma^{t''}(x)(t) = \sigma^{t''}(x)(t+T) \text{ and} \\ \text{and } t-T \geq t' \Longrightarrow \sigma^{t''}(x)(t) = \sigma^{t''}(x)(t-T)). \end{cases} \tag{2.12}$$

Proof. a) (1.3)$\Longrightarrow$(2.1) We suppose that $\mu \in \mathbf{B}^n$ exists with $\{\widehat{x}(k) | k \in \mathbf{N}_{\_}\} = \{\mu\}$ and let $p \geq 1$, $k \in \widehat{\mathbf{T}}_{\mu}^{\widehat{x}}$, $z \in \mathbf{Z}$ arbitrary such that $k+zp \geq -1$. Obviously $\widehat{x}(k+zp) = \mu$, thus $k+zp \in \widehat{\mathbf{T}}_{\mu}^{\widehat{x}}$.

(2.1)$\Longrightarrow$(2.2) We take $p \geq 1, \mu \in \widehat{Or}(\widehat{x}), k' \in \mathbf{N}_{\_}$ arbitrary. If $\widehat{\mathbf{T}}_{\mu}^{\widehat{x}} \cap \{k', k'+1, k'+2, ...\} = \varnothing$, then the statement

$$\forall k \in \widehat{\mathbf{T}}_{\mu}^{\widehat{x}} \cap \{k', k'+1, k'+2, ...\}, \{k+zp | z \in \mathbf{Z}\} \cap \{k', k'+1, k'+2, ...\} \subset \widehat{\mathbf{T}}_{\mu}^{\widehat{x}}$$

holds trivially, thus we can suppose from now that $\widehat{\mathbf{T}}_{\mu}^{\widehat{x}} \cap \{k', k'+1, k'+2, ...\} \neq \varnothing$ and let $k \in \widehat{\mathbf{T}}_{\mu}^{\widehat{x}}$, $z \in \mathbf{Z}$ arbitrary, fixed, such that $k \geq k'$ and $k+zp \geq k'$. As $k+zp \geq -1$, we have from (2.1) that $k+zp \in \widehat{\mathbf{T}}_{\mu}^{\widehat{x}}$.

(2.2)$\Longrightarrow$(2.3) Let $p \geq 1$, $\mu \in \widehat{Or}(\widehat{x})$, $k'' \in \mathbf{N}$ arbitrary. If $\widehat{\mathbf{T}}_{\mu}^{\widehat{\sigma}^{k''}(\widehat{x})} = \varnothing$, the statement

$$\forall k \in \widehat{\mathbf{T}}_{\mu}^{\widehat{\sigma}^{k''}(\widehat{x})}, \{k+zp | z \in \mathbf{Z}\} \cap \mathbf{N}_{\text{-}} \subset \widehat{\mathbf{T}}_{\mu}^{\widehat{\sigma}^{k''}(\widehat{x})}$$

is trivially fulfilled, and we can suppose from now that $\widehat{\mathbf{T}}_{\mu}^{\widehat{\sigma}^{k''}(\widehat{x})} \neq \varnothing$. Let $k \in \widehat{\mathbf{T}}_{\mu}^{\widehat{\sigma}^{k''}(\widehat{x})}$, $z \in \mathbf{Z}$ arbitrary such that $k+zp \geq -1$. We have $\widehat{x}(k+k'') = \mu$ and, if we denote $k' = k''-1$, then $\widehat{x}(k+k'+1) = \mu$, where $k' \in \mathbf{N}_{\text{-}}$. Of course that $k+k'+1 \geq k'$, thus $k+k'+1 \in \widehat{\mathbf{T}}_{\mu}^{\widehat{x}} \cap \{k', k'+1, k'+2, ...\}$ and, on the other hand, $k+k'+1+zp \geq k'+1-1$, thus we can apply (2.2), wherefrom $\widehat{x}(k+k'+1+zp) = \mu$. It has resulted that $\widehat{\sigma}^{k''}(\widehat{x})(k+zp) = \widehat{x}(k+k''+zp) = \mu$, in other words $k+zp \in \widehat{\mathbf{T}}_{\mu}^{\widehat{\sigma}^{k''}(\widehat{x})}$.

(2.3)$\Longrightarrow$(2.4) Let $p \geq 1, \mu \in \widehat{Or}(\widehat{x})$ and $k \in \mathbf{N}_{\text{-}}$ such that $\widehat{x}(k) = \mu$. Hypothesis (2.3) written for $k'' = 0$ gives $\{k+zp | z \in \mathbf{Z}\} \cap \mathbf{N}_{\text{-}} \subset \widehat{\mathbf{T}}_{\mu}^{\widehat{x}}$, thus

$$k+p \in \{k+zp | z \in \mathbf{Z}\} \cap \mathbf{N}_{\text{-}} \overset{(2.3)}{\subset} \widehat{\mathbf{T}}_{\mu}^{\widehat{x}},$$

wherefrom $\widehat{x}(k+p) = \mu = \widehat{x}(k)$.

If in addition $k-p \geq -1$, then

$$k-p \in \{k+zp | z \in \mathbf{Z}\} \cap \mathbf{N}_{\text{-}} \overset{(2.3)}{\subset} \widehat{\mathbf{T}}_{\mu}^{\widehat{x}},$$

wherefrom $\widehat{x}(k-p) = \mu = \widehat{x}(k)$.

(2.4)$\Longrightarrow$(2.5) We take $p \geq 1, \mu \in \widehat{Or}(\widehat{x})$, $k' \in \mathbf{N}_{\text{-}}$ arbitrary. If $\forall k \geq k', \widehat{x}(k) \neq \mu$, then the statement

$$\forall k \geq k', \widehat{x}(k) = \mu \Longrightarrow (\widehat{x}(k) = \widehat{x}(k+p) \text{ and } k-p \geq k' \Longrightarrow \widehat{x}(k) = \widehat{x}(k-p))$$

is trivially fulfilled, thus we can take $k \geq k'$ arbitrarily with $\widehat{x}(k) = \mu$. From (2.4) we have that $\widehat{x}(k) = \widehat{x}(k+p)$. In the case that in addition $k-p \geq k'$, as $k-p \geq -1$, we can apply (2.4) again in order to infer that $\widehat{x}(k) = \widehat{x}(k-p)$.

(2.5)$\Longrightarrow$(2.6) Let $p \geq 1$, $\mu \in \widehat{Or}(\widehat{x})$, $k'' \in \mathbf{N}$ arbitrary. If $\forall k \in \mathbf{N}_{\text{-}}, \widehat{\sigma}^{k''}(\widehat{x})(k) \neq \mu$ then the statement

$$\begin{cases} \forall k \in \mathbf{N}_{\text{-}}, \widehat{\sigma}^{k''}(\widehat{x})(k) = \mu \Longrightarrow (\widehat{\sigma}^{k''}(\widehat{x})(k) = \widehat{\sigma}^{k''}(\widehat{x})(k+p) \text{ and} \\ \quad \text{and } k-p \geq -1 \Longrightarrow \widehat{\sigma}^{k''}(\widehat{x})(k) = \widehat{\sigma}^{k''}(\widehat{x})(k-p)) \end{cases}$$

is trivially true, thus we take $k \in \mathbf{N}_{\text{-}}$ arbitrary such that $\widehat{\sigma}^{k''}(\widehat{x})(k) = \widehat{x}(k+k'') = \mu$. We denote $k' = k''-1$ and we see that $\widehat{x}(k+k'+1) = \mu$, where $k+k'+1 \geq k'$. We can apply (2.5) and we infer that

$$\widehat{\sigma}^{k''}(\widehat{x})(k) = \widehat{x}(k+k'') = \widehat{x}(k+k'+1) \overset{(2.5)}{=} \widehat{x}(k+k'+1+p) =$$

$$= \widehat{x}(k+k''+p) = \widehat{\sigma}^{k''}(\widehat{x})(k+p).$$

We suppose now that in addition $k-p\geq -1$, thus $k+k'+1-p\geq k'$ and we can apply again (2.5) in order to obtain

$$\widehat{\sigma}^{k''}(\widehat{x})(k)=\widehat{x}(k+k'')=\widehat{x}(k+k'+1)\overset{(2.5)}{=}\widehat{x}(k+k'+1-p)=$$

$$=\widehat{x}(k+k''-p)=\widehat{\sigma}^{k''}(\widehat{x})(k-p).$$

(2.6)$\Longrightarrow$(1.1) The statement (2.6) written for $p=1$ and $k''=0$ becomes:

$$\begin{cases} \forall\mu\in\widehat{Or}(\widehat{x}),\forall k\in\mathbf{N}_{\_},\widehat{x}(k)=\mu\Longrightarrow \\ \Longrightarrow(\widehat{x}(k)=\widehat{x}(k+1)\text{ and }k\geq 0\Longrightarrow\widehat{x}(k)=\widehat{x}(k-1)). \end{cases} \tag{2.13}$$

Let $\mu\in\widehat{Or}(\widehat{x})$ arbitrary, thus $k_1\in\mathbf{N}_{\_}$ exists with $\widehat{x}(k_1)=\mu$. From (2.13) we infer:

$$\widehat{x}(k_1)=\widehat{x}(k_1-1)=\widehat{x}(k_1-2)=...=\widehat{x}(-1),$$

$$\widehat{x}(k_1)=\widehat{x}(k_1+1)=\widehat{x}(k_1+2)=...$$

We have obtained that (1.1) holds.

b) (1.6)$\Longrightarrow$(2.7) The hypothesis states the existence of $\mu\in\mathbf{B}^n$ such that $\{x(t)|t\in\mathbf{R}\}=\{\mu\}$. Let us take $T>0$ and $t'\in I^x$ arbitrarily. Let furthermore $t\in\mathbf{T}_{\mu}^{x}\cap[t',\infty)(=[t',\infty))$ and $z\in\mathbf{Z}$ having the property that $t+zT\geq t'$. We have $x(t+zT)=\mu$, thus $t+zT\in\mathbf{T}_{\mu}^{x}$. These imply the truth of (2.7).

(2.7)$\Longrightarrow$(2.8) Let $T>0$ and $\mu\in Or(x)$ arbitrary. From (2.7) we have the existence of $t'\in I^x$ with the property

$$\forall t\in\mathbf{T}_{\mu}^{x}\cap[t',\infty),\{t+zT|z\in\mathbf{Z}\}\cap[t',\infty)\subset\mathbf{T}_{\mu}^{x}. \tag{2.14}$$

Let us take $t_1'\geq t'$ arbitrary. If $\mathbf{T}_{\mu}^{x}\cap[t_1',\infty)=\varnothing$, then the statement

$$\forall t\in\mathbf{T}_{\mu}^{x}\cap[t_1',\infty),\{t+zT|z\in\mathbf{Z}\}\cap[t_1',\infty)\subset\mathbf{T}_{\mu}^{x}$$

is trivially true, so we suppose $\mathbf{T}_{\mu}^{x}\cap[t_1',\infty)\neq\varnothing$ and let $t\in\mathbf{T}_{\mu}^{x}\cap[t_1',\infty),z\in\mathbf{Z}$ arbitrary such that $t+zT\geq t_1'$. We have $t\in\mathbf{T}_{\mu}^{x}\cap[t',\infty),t+zT\geq t'$ and we can apply (2.14). We infer $t+zT\in\mathbf{T}_{\mu}^{x}$.

(2.8)$\Longrightarrow$(2.9) Let $T>0,\mu\in Or(x)$ arbitrary. Hypothesis (2.8) shows the existence of $t'''\in I^x$ with

$$\forall t\in\mathbf{T}_{\mu}^{x}\cap[t''',\infty),\{t+zT|z\in\mathbf{Z}\}\cap[t''',\infty)\subset\mathbf{T}_{\mu}^{x} \tag{2.15}$$

true. When writing (2.15) we have taken in (2.8) $t_1'=t'''(=t')$.

Let $t''\in\mathbf{R}$ arbitrary. We have the following possibilities.

Case $t''\leq t'''$

Then, from $t'''\in I^x$, $\sigma^{t''}(x)=x$ and from (2.15) we have the truth of (2.9) with $t'=t'''$.

Case $t''>t'''$

Some $\varepsilon>0$ exists with the property that $\forall t\in(t''-\varepsilon,t''),x(t)=x(t''-0)$. We take $t'\in(t''-\varepsilon,t'')\cap(t''',t'')$ arbitrarily. The fact that

$$(-\infty,t']\subset\mathbf{T}_{x(t''-0)}^{\sigma^{t''}(x)}$$

is obvious, thus $t' \in I^{\sigma^{t''}(x)}$. If $\mathbf{T}_{\mu}^{\sigma^{t''}(x)} \cap [t',\infty) = \varnothing$, then the property

$$\forall t \in \mathbf{T}_{\mu}^{\sigma^{t''}(x)} \cap [t',\infty), \{t+zT|z \in \mathbf{Z}\} \cap [t',\infty) \subset \mathbf{T}_{\mu}^{\sigma^{t''}(x)} \tag{2.16}$$

is true, thus we can suppose that $\mathbf{T}_{\mu}^{\sigma^{t''}(x)} \cap [t',\infty) \neq \varnothing$ and let $t \in \mathbf{T}_{\mu}^{\sigma^{t''}(x)} \cap [t',\infty)$ arbitrary, fixed. We notice that $\forall \xi \geq t', \sigma^{t''}(x)(\xi) = x(\xi)$, thus $\mathbf{T}_{\mu}^{\sigma^{t''}(x)} \cap [t',\infty) = \mathbf{T}_{\mu}^{x} \cap [t',\infty)$. We take $z \in \mathbf{Z}$ arbitrary with $t+zT \geq t'$. Because in this situation $t \in \mathbf{T}_{\mu}^{x} \cap [t''',\infty)$ and $t+zT \geq t'''$, we can apply (2.15) and we infer $t+zT \in \mathbf{T}_{\mu}^{x}$, i.e. $x(t+zT) = \mu = \sigma^{t''}(x)(t+zT)$ and finally $t+zT \in \mathbf{T}_{\mu}^{\sigma^{t''}(x)}$.

(2.9)$\Longrightarrow$(2.10) Let $T > 0, \mu \in Or(x)$ arbitrary, fixed. The existence of $x(-\infty+0)$ shows that in (2.9) we can choose $t'' \in \mathbf{R}$ sufficiently small such that $\sigma^{t''}(x) = x$ (see Theorem 4 a), page 7). For that choice of t'', (2.9) shows the existence of $t' \in I^{\sigma^{t''}(x)} = I^x$ with

$$\forall t \in \mathbf{T}_{\mu}^{x} \cap [t',\infty), \{t+zT|z \in \mathbf{Z}\} \cap [t',\infty) \subset \mathbf{T}_{\mu}^{x} \tag{2.17}$$

true. If $\forall t \geq t', x(t) \neq \mu$, then the statement

$$\forall t \geq t', x(t) = \mu \Longrightarrow (x(t) = x(t+T) \text{ and } t-T \geq t' \Longrightarrow x(t) = x(t-T))$$

is trivially true, so we can suppose that $\mathbf{T}_{\mu}^{x} \cap [t',\infty) \neq \varnothing$. Let $t \geq t'$ arbitrary such that $x(t) = \mu$, in other words $t \in \mathbf{T}_{\mu}^{x} \cap [t',\infty)$. As far as

$$t+T \in \{t+zT|z \in \mathbf{Z}\} \cap [t',\infty) \overset{(2.17)}{\subset} \mathbf{T}_{\mu}^{x},$$

we conclude that $t+T \in \mathbf{T}_{\mu}^{x}$, i.e. $x(t+T) = \mu = x(t)$. If in addition $t-T \geq t'$, then

$$t-T \in \{t+zT|z \in \mathbf{Z}\} \cap [t',\infty) \overset{(2.17)}{\subset} \mathbf{T}_{\mu}^{x},$$

wherefrom $t-T \in \mathbf{T}_{\mu}^{x}$, i.e. $x(t-T) = \mu = x(t)$.

(2.10)$\Longrightarrow$(2.11) We take $T > 0$ and $\mu \in Or(x)$ arbitrary, for which the truth of (2.10) shows the existence of $t' \in I^x$ such that

$$\forall t \geq t', x(t) = \mu \Longrightarrow (x(t) = x(t+T) \text{ and } t-T \geq t' \Longrightarrow x(t) = x(t-T)). \tag{2.18}$$

Let now $t'_1 \geq t'$ arbitrary. If $\forall t \geq t'_1, x(t) \neq \mu$, the statement

$$\forall t \geq t'_1, x(t) = \mu \Longrightarrow (x(t) = x(t+T) \text{ and } t-T \geq t'_1 \Longrightarrow x(t) = x(t-T))$$

is trivially true, so we suppose that we can take $t \geq t'_1$ arbitrarily with $x(t) = \mu$. As $t \geq t'$, we conclude from (2.18) that

$$x(t) = x(t+T) \text{ and } t-T \geq t' \Longrightarrow x(t) = x(t-T) \tag{2.19}$$

holds. If $t-T \geq t'_1$, then $t-T \geq t'$ and from (2.19) we have that $x(t) = x(t-T)$.

(2.11)$\Longrightarrow$(2.12) Let $T > 0, \mu \in Or(x)$ arbitrary, fixed. From (2.11) we get the existence of $t''' \in I^x$ such that, in the special case when $t_1' \geq t'''$ holds as equality, we have

$$\forall t \geq t''', x(t) = \mu \Longrightarrow (x(t) = x(t+T) \text{ and } t - T \geq t''' \Longrightarrow x(t) = x(t-T)). \qquad (2.20)$$

Let $t'' \in \mathbf{R}$ arbitrary, fixed. We have the following possibilities.

Case $t'' \leq t'''$

We have $\sigma^{t''}(x) = x$ and, from (2.20) we infer the truth of (2.12), with $t' = t'''$.

Case $t'' > t'''$

Some $\varepsilon > 0$ exists with the property that $\forall t \in (t'' - \varepsilon, t''), x(t) = x(t'' - 0)$. We take $t' \in (t'' - \varepsilon, t'') \cap (t''', t'')$ arbitrary and, from (2.20) and Lemma 3, page 224, we obtain

$$\forall t \geq t', x(t) = \mu \Longrightarrow (x(t) = x(t+T) \text{ and } t - T \geq t' \Longrightarrow x(t) = x(t-T)). \qquad (2.21)$$

We notice that $(-\infty, t'] \subset \mathbf{T}^{\sigma^{t''}(x)}_{x(t''-0)}$, thus $t' \in I^{\sigma^{t''}(x)}$. If $\forall t \geq t', \sigma^{t''}(x)(t) \neq \mu$, then

$$\begin{cases} \forall t \geq t', \sigma^{t''}(x)(t) = \mu \Longrightarrow (\sigma^{t''}(x)(t) = \sigma^{t''}(x)(t+T) \text{ and} \\ \quad \text{and } t - T \geq t' \Longrightarrow \sigma^{t''}(x)(t) = \sigma^{t''}(x)(t-T)) \end{cases}$$

is trivially true, so let $t \geq t'$ arbitrary with $\sigma^{t''}(x)(t) = \mu$. As for $t \geq t', \sigma^{t''}(x)(t) = x(t)$ (irrespective of the fact that $t < t''$ or $t \geq t''$), we can apply (2.21).

(2.12)$\Longrightarrow$(1.6) We suppose against all reason that $\mu, \mu' \in Or(x)$ exist, $\mu \neq \mu'$, meaning the existence of $t_1, t_2 \in \mathbf{R}$ with $x(t_1) = \mu, x(t_2) = \mu'$. We can suppose without loss that $t_1 > t_2$. We write (2.12) for the period $T' = t_1 - t_2 > 0$, for μ' and for $t'' \in \mathbf{R}$ sufficiently small such that $\sigma^{t''}(x) = x$. Some $t' \in I^x$ exists with

$$\forall t \geq t', x(t) = \mu' \Longrightarrow (x(t) = x(t+T') \text{ and } t - T' \geq t' \Longrightarrow x(t) = x(t-T')). \qquad (2.22)$$

We use the fact that $Or(x) = \{x(t) | t \geq t'\}$, thus we can suppose that $t' \leq t_2 < t_1$ and we have

$$\mu' = x(t_2) \overset{(2.22)}{=} x(t_2 + T') = x(t_2 + t_1 - t_2) = x(t_1) = \mu,$$

contradiction with the supposition that $\mu \neq \mu'$. Statement (1.6) holds. □

3. The Third Group of Constancy Properties

Remark 61. *The third group of constancy properties involves periodicity and eventual periodicity properties of some point μ of the orbit. The constancy properties result from (2.1),...,(2.6) and (2.7),...,(2.12) of the second group of properties, by the replacement of $\forall \mu \in \widehat{Or}(\widehat{x}), \forall \mu \in Or(x)$ with $\exists \mu \in \widehat{Or}(\widehat{x}), \exists \mu \in Or(x)$. The proofs of the implications are similar, most of them, with the proofs of Theorem 23, page 47.*

Remark 62. *The properties are also inspired by the eventual constancy properties of Theorem 18, page 33. Note that:*

- (3.1), (3.2) (the last contains $\forall k' \in \mathbf{N}_{\_}$) are inspired by (4.1)$_{page\,33}$ (containing $\exists k' \in \mathbf{N}_{\_}, \widehat{\mathbf{T}}^{\widehat{x}}_{\mu} \cap \{k', k'+1, k'+2, ...\} \neq \emptyset$);

- *(3.3) (containing* $\forall k'' \in \mathbf{N}$*) is inspired by (4.2)*$_{page\ 33}$ *(containing* $\exists k'' \in \mathbf{N}, \widehat{\mathbf{T}}_{\mu}^{\widehat{\sigma}^{k''}(\widehat{x})} \neq \varnothing$*);*

- *(3.4), (3.5) (the last contains* $\forall k' \in \mathbf{N}_{\text{-}}$*) are inspired by (4.3)*$_{page\ 33}$ *(containing* $\exists k' \in \mathbf{N}_{\text{-}}, \exists k_1 \geq k', \widehat{x}(k_1) = \mu$*);*

- *(3.6) (containing* $\forall k'' \in \mathbf{N}$*) is inspired by (4.4)*$_{page\ 34}$ *(containing* $\exists k'' \in \mathbf{N}, \exists k_1 \in \mathbf{N}_{\text{-}}, \widehat{\sigma}^{k''}(\widehat{x})(k_1) = \mu$*);*

- *(3.7) and (3.8) (the last contains* $\forall t_1' \geq t'$*) are inspired by (4.5)*$_{page\ 34}$ *(containing* $\exists t_1' \geq t', \mathbf{T}_{\mu}^{x} \cap [t_1', \infty) \neq \varnothing$*) and (4.6)*$_{page\ 34}$ *(containing* $\exists t_1' \in \mathbf{R}, \mathbf{T}_{\mu}^{x} \cap [t_1', \infty) \neq \varnothing$*);*

- *(3.9) (containing* $\forall t'' \in \mathbf{R}$*) is inspired by (4.7)*$_{page\ 34}$ *and (4.8)*$_{page\ 34}$ *(containing* $\exists t'' \in \mathbf{R} ... \mathbf{T}_{\mu}^{\sigma^{t''}(x)} \cap [t', \infty) \neq \varnothing$*);*

- *(3.10) and (3.11) (the last contains* $\forall t_1' \geq t'$*) are inspired by (4.9)*$_{page\ 34}$ *(containing* $\exists t_1' \geq t', \exists t_2' \geq t_1', x(t_2') = \mu$*) and (4.10)*$_{page\ 34}$ *(containing* $\exists t_1' \in \mathbf{R}, \exists t_2' \geq t_1', x(t_2') = \mu$*);*

- *(3.12) (containing* $\forall t'' \in \mathbf{R}$*) is inspired by (4.11)*$_{page\ 34}$ *and (4.12)*$_{page\ 34}$ *(containing* $\exists t'' \in \mathbf{R} ... \exists t''' \geq t', \sigma^{t''}(x)(t''') = \mu$*).*

Remark 63. *We refer also to Theorem 19, page 39 that makes use of the eventual periodicity of some points of the omega limit set. Here are the differences:*

- *(3.1), (3.2) (the last has* $\exists \mu \in \widehat{Or}(\widehat{x}), \forall k' \in \mathbf{N}_{\text{-}}$*) are inspired by (5.1)*$_{page\ 39}$ *(containing* $\exists \mu \in \widehat{\omega}(\widehat{x}), \exists k' \in \mathbf{N}_{\text{-}}$*);*

- *(3.3) (containing* $\exists \mu \in \widehat{Or}(\widehat{x}), \forall k'' \in \mathbf{N}$*) is inspired by (5.2)*$_{page\ 39}$ *(containing* $\exists \mu \in \widehat{\omega}(\widehat{x}), \exists k'' \in \mathbf{N}$*);*

- *(3.4), (3.5) (the last has* $\exists \mu \in \widehat{Or}(\widehat{x}), \forall k' \in \mathbf{N}_{\text{-}}$*) are inspired by (5.3)*$_{page\ 39}$ *(containing* $\exists \mu \in \widehat{\omega}(\widehat{x}), \exists k' \in \mathbf{N}_{\text{-}}$*);*

- *(3.6) (containing* $\exists \mu \in \widehat{Or}(\widehat{x}), \forall k'' \in \mathbf{N}$*) is inspired by (5.4)*$_{page\ 39}$ *(containing* $\exists \mu \in \widehat{\omega}(\widehat{x}), \exists k'' \in \mathbf{N}$*);*

- *(3.7) and (3.8) (the last contains* $\exists \mu \in Or(x), ..., \forall t_1' \geq t'$*) are inspired by (5.5)*$_{page\ 39}$ *(containing* $\exists \mu \in \omega(x), ..., \exists t_1' \geq t'$*) and (5.6)*$_{page\ 39}$ *(containing* $\exists \mu \in \omega(x), \exists t_1' \in \mathbf{R}$*);*

- *(3.9) (with* $\exists \mu \in Or(x), \forall t'' \in \mathbf{R}$*) is inspired by (5.7)*$_{page\ 40}$ *and (5.8)*$_{page\ 40}$ *(containing* $\exists \mu \in \omega(x), \exists t'' \in \mathbf{R}$*);*

- *(3.10) and (3.11) (the last contains* $\exists \mu \in Or(x), ..., \forall t_1' \geq t'$*) are inspired by (5.9)*$_{page\ 40}$ *(containing* $\exists \mu \in \omega(x), ..., \exists t_1' \geq t'$*) and (5.10)*$_{page\ 40}$ *(containing* $\exists \mu \in \omega(x), \exists t_1' \in \mathbf{R}$*);*

- *(3.12) (containing* $\exists \mu \in Or(x), \forall t'' \in \mathbf{R}$*) is inspired by (5.11)*$_{page\ 40}$ *and (5.12)*$_{page\ 40}$ *(containing* $\exists \mu \in \omega(x), \exists t'' \in \mathbf{R}$*).*

Theorem 24. *Let the signals* $\widehat{x} \in \widehat{S}^{(n)}, x \in S^{(n)}$.

a) The following properties are equivalent with the constancy of $\widehat{x} \in \widehat{S}^{(n)}$*:*

$$\forall p \geq 1, \exists \mu \in \widehat{Or}(\widehat{x}), \forall k \in \widehat{\mathbf{T}}_{\mu}^{\widehat{x}}, \{k + zp | z \in \mathbf{Z}\} \cap \mathbf{N}_{\text{-}} \subset \widehat{\mathbf{T}}_{\mu}^{\widehat{x}}, \tag{3.1}$$

$$\begin{cases} \forall p \geq 1, \exists \mu \in \widehat{Or}(\widehat{x}), \forall k' \in \mathbf{N}_{\text{-}}, \forall k \in \widehat{\mathbf{T}}_{\mu}^{\widehat{x}} \cap \{k', k'+1, k'+2, ...\}, \\ \{k + zp | z \in \mathbf{Z}\} \cap \{k', k'+1, k'+2, ...\} \subset \widehat{\mathbf{T}}_{\mu}^{\widehat{x}}, \end{cases} \tag{3.2}$$

$$\begin{cases} \forall p \geq 1, \exists \mu \in \widehat{Or}(\widehat{x}), \forall k'' \in \mathbf{N}, \forall k \in \widehat{\mathbf{T}}_{\mu}^{\widehat{\sigma}^{k''}(\widehat{x})}, \\ \{k + zp | z \in \mathbf{Z}\} \cap \mathbf{N}_{\text{-}} \subset \widehat{\mathbf{T}}_{\mu}^{\widehat{\sigma}^{k''}(\widehat{x})}, \end{cases} \tag{3.3}$$

$$\begin{cases} \forall p \geq 1, \exists \mu \in \widehat{Or}(\widehat{x}), \forall k \in \mathbf{N}_{\text{-}}, \widehat{x}(k) = \mu \Longrightarrow \\ \Longrightarrow (\widehat{x}(k) = \widehat{x}(k+p) \text{ and } k - p \geq -1 \Longrightarrow \widehat{x}(k) = \widehat{x}(k-p)), \end{cases} \tag{3.4}$$

$$\left\{\begin{array}{c}\forall p\geq 1,\exists\mu\in\widehat{Or}(\widehat{x}),\forall k'\in\mathbf{N}_{\_},\forall k\geq k',\widehat{x}(k)=\mu\Longrightarrow\\ \Longrightarrow(\widehat{x}(k)=\widehat{x}(k+p)\ and\ k-p\geq k'\Longrightarrow\widehat{x}(k)=\widehat{x}(k-p)),\end{array}\right. \tag{3.5}$$

$$\left\{\begin{array}{c}\forall p\geq 1,\exists\mu\in\widehat{Or}(\widehat{x}),\forall k''\in\mathbf{N},\forall k\in\mathbf{N}_{\_},\widehat{\sigma}^{k''}(\widehat{x})(k)=\mu\Longrightarrow\\ \Longrightarrow(\widehat{\sigma}^{k''}(\widehat{x})(k)=\widehat{\sigma}^{k''}(\widehat{x})(k+p)\ and\\ and\ k-p\geq -1\Longrightarrow\widehat{\sigma}^{k''}(\widehat{x})(k)=\widehat{\sigma}^{k''}(\widehat{x})(k-p)).\end{array}\right. \tag{3.6}$$

b) The following properties are equivalent with the constancy of $x\in S^{(n)}$*:*

$$\left\{\begin{array}{c}\forall T>0,\exists\mu\in Or(x),\exists t'\in I^{x},\\ \forall t\in\mathbf{T}_{\mu}^{x}\cap[t',\infty),\{t+zT|z\in\mathbf{Z}\}\cap[t',\infty)\subset\mathbf{T}_{\mu}^{x},\end{array}\right. \tag{3.7}$$

$$\left\{\begin{array}{c}\forall T>0,\exists\mu\in Or(x),\exists t'\in I^{x},\\ \forall t_1'\geq t',\forall t\in\mathbf{T}_{\mu}^{x}\cap[t_1',\infty),\{t+zT|z\in\mathbf{Z}\}\cap[t_1',\infty)\subset\mathbf{T}_{\mu}^{x},\end{array}\right. \tag{3.8}$$

$$\left\{\begin{array}{c}\forall T>0,\exists\mu\in Or(x),\forall t''\in\mathbf{R},\exists t'\in I^{\sigma^{t''}(x)},\\ \forall t\in\mathbf{T}_{\mu}^{\sigma^{t''}(x)}\cap[t',\infty),\{t+zT|z\in\mathbf{Z}\}\cap[t',\infty)\subset\mathbf{T}_{\mu}^{\sigma^{t''}(x)},\end{array}\right. \tag{3.9}$$

$$\left\{\begin{array}{c}\forall T>0,\exists\mu\in Or(x),\exists t'\in I^{x},\forall t\geq t',\\ x(t)=\mu\Longrightarrow(x(t)=x(t+T)\ and\ t-T\geq t'\Longrightarrow x(t)=x(t-T)),\end{array}\right. \tag{3.10}$$

$$\left\{\begin{array}{c}\forall T>0,\exists\mu\in Or(x),\exists t'\in I^{x},\forall t_1'\geq t',\forall t\geq t_1',x(t)=\mu\Longrightarrow\\ \Longrightarrow(x(t)=x(t+T)\ and\ t-T\geq t_1'\Longrightarrow x(t)=x(t-T)),\end{array}\right. \tag{3.11}$$

$$\left\{\begin{array}{c}\forall T>0,\exists\mu\in Or(x),\forall t''\in\mathbf{R},\exists t'\in I^{\sigma^{t''}(x)},\\ \forall t\geq t',\sigma^{t''}(x)(t)=\mu\Longrightarrow(\sigma^{t''}(x)(t)=\sigma^{t''}(x)(t+T)\ and\\ and\ t-T\geq t'\Longrightarrow\sigma^{t''}(x)(t)=\sigma^{t''}(x)(t-T)).\end{array}\right. \tag{3.12}$$

Proof. a) (1.1)$\Longrightarrow$(3.1) Let us prove first that (1.1) implies

$$\exists\mu\in\widehat{Or}(\widehat{x}),\forall p\geq 1,\forall k\in\widehat{\mathbf{T}}_{\mu}^{\widehat{x}},\{k+zp|z\in\mathbf{Z}\}\cap\mathbf{N}_{\_}\subset\widehat{\mathbf{T}}_{\mu}^{\widehat{x}}. \tag{3.13}$$

Indeed, the hypothesis states the existence of $\mu\in\widehat{Or}(\widehat{x})$ with $\forall k\in\mathbf{N}_{\_},\widehat{x}(k)=\mu$ and let $p\geq 1,k\in\widehat{\mathbf{T}}_{\mu}^{\widehat{x}},z\in\mathbf{Z}$ arbitrary such that $k+zp\geq -1$. Then $\widehat{x}(k+zp)=\mu$, thus $k+zp\in\widehat{\mathbf{T}}_{\mu}^{\widehat{x}}$ and (3.13) holds. From (3.13) we obviously get (3.1).

(3.1)$\Longrightarrow$(3.2) We take $p\geq 1$ arbitrarily. The truth of (3.1) shows the existence of $\mu\in\widehat{Or}(\widehat{x})$ with

$$\forall k\in\widehat{\mathbf{T}}_{\mu}^{\widehat{x}},\{k+zp|z\in\mathbf{Z}\}\cap\mathbf{N}_{\_}\subset\widehat{\mathbf{T}}_{\mu}^{\widehat{x}}. \tag{3.14}$$

Let now $k'\in\mathbf{N}_{\_}$ arbitrary. If $\widehat{\mathbf{T}}_{\mu}^{\widehat{x}}\cap\{k',k'+1,k'+2,...\}=\varnothing$, then

$$\forall k\in\widehat{\mathbf{T}}_{\mu}^{\widehat{x}}\cap\{k',k'+1,k'+2,...\},\{k+zp|z\in\mathbf{Z}\}\cap\{k',k'+1,k'+2,...\}\subset\widehat{\mathbf{T}}_{\mu}^{\widehat{x}}$$

is trivially true, so we suppose $\widehat{\mathbf{T}}_{\mu}^{\widehat{x}}\cap\{k',k'+1,k'+2,...\}\neq\varnothing$ and let $k\in\widehat{\mathbf{T}}_{\mu}^{\widehat{x}}$, $z\in\mathbf{Z}$ arbitrary, fixed, such that $k\geq k'$ and $k+zp\geq k'$. As $k+zp\geq -1$, we have from (3.14) that $k+zp\in\widehat{\mathbf{T}}_{\mu}^{\widehat{x}}$.

(3.2)$\Longrightarrow$(3.3) Let an arbitrary $p\geq 1$. We have from (3.2) the existence of $\mu\in\widehat{Or}(\widehat{x})$ such that

$$\left\{\begin{array}{c}\forall k'\in\mathbf{N}_{\_},\forall k\in\widehat{\mathbf{T}}_{\mu}^{\widehat{x}}\cap\{k',k'+1,k'+2,...\},\\ \{k+zp|z\in\mathbf{Z}\}\cap\{k',k'+1,k'+2,...\}\subset\widehat{\mathbf{T}}_{\mu}^{\widehat{x}}\end{array}\right. \tag{3.15}$$

holds and we take $k'' \in \mathbf{N}$ arbitrary. If $\widehat{\mathbf{T}}_{\mu}^{\widehat{\sigma}^{k''}(\widehat{x})} = \varnothing$, then

$$\forall k \in \widehat{\mathbf{T}}_{\mu}^{\widehat{\sigma}^{k''}(\widehat{x})}, \{k+zp|z \in \mathbf{Z}\} \cap \mathbf{N}_{\_} \subset \widehat{\mathbf{T}}_{\mu}^{\widehat{\sigma}^{k''}(\widehat{x})}$$

is trivially true, thus we can suppose $\widehat{\mathbf{T}}_{\mu}^{\widehat{\sigma}^{k''}(\widehat{x})} \neq \varnothing$ and let $k \in \widehat{\mathbf{T}}_{\mu}^{\widehat{\sigma}^{k''}(\widehat{x})}$, $z \in \mathbf{Z}$ arbitrary such that $k+zp \geq -1$. We have $\widehat{x}(k+k'') = \mu$ and, if we denote $k' = k''-1$, then $\widehat{x}(k+k'+1) = \mu$, where $k' \in \mathbf{N}_{\_}$. Of course that $k+k'+1 \geq k'$, thus $k+k'+1 \in \widehat{\mathbf{T}}_{\mu}^{\widehat{x}} \cap \{k', k'+1, k'+2, ...\}$ and, on the other hand, $k+k'+1+zp \geq k'+1-1$, resulting that we can apply (3.15), wherefrom $\widehat{x}(k+k'+1+zp) = \mu$. It has resulted that $\widehat{\sigma}^{k''}(\widehat{x})(k+zp) = \widehat{x}(k+k''+zp) = \mu$, in other words $k+zp \in \widehat{\mathbf{T}}_{\mu}^{\widehat{\sigma}^{k''}(\widehat{x})}$.

(3.3)$\Longrightarrow$(3.4) Let $p \geq 1$ arbitrary. Hypothesis (3.3) shows the existence of $\mu \in \widehat{Or}(\widehat{x})$ with

$$\forall k \in \widehat{\mathbf{T}}_{\mu}^{\widehat{x}}, \{k+zp|z \in \mathbf{Z}\} \cap \mathbf{N}_{\_} \subset \widehat{\mathbf{T}}_{\mu}^{\widehat{x}} \tag{3.16}$$

true (for $k'' = 0$ and $\widehat{\sigma}^{k''}(\widehat{x}) = \widehat{x}$) and let $k \in \mathbf{N}_{\_}$ such that $\widehat{x}(k) = \mu$. We obtain

$$k+p \in \{k+zp|z \in \mathbf{Z}\} \cap \mathbf{N}_{\_} \overset{(3.16)}{\subset} \widehat{\mathbf{T}}_{\mu}^{\widehat{x}},$$

wherefrom $\widehat{x}(k+p) = \mu = \widehat{x}(k)$.

If in addition $k-p \geq -1$, then

$$k-p \in \{k+zp|z \in \mathbf{Z}\} \cap \mathbf{N}_{\_} \overset{(3.16)}{\subset} \widehat{\mathbf{T}}_{\mu}^{\widehat{x}},$$

wherefrom $\widehat{x}(k-p) = \mu = \widehat{x}(k)$.

(3.4)$\Longrightarrow$(3.5) We take an arbitrary $p \geq 1$ and we have from (3.4) the existence of $\mu \in \widehat{Or}(\widehat{x})$, with

$$\begin{cases} \forall k \in \mathbf{N}_{\_}, \widehat{x}(k) = \mu \Longrightarrow \\ \Longrightarrow (\widehat{x}(k) = \widehat{x}(k+p) \text{ and } k-p \geq -1 \Longrightarrow \widehat{x}(k) = \widehat{x}(k-p)) \end{cases} \tag{3.17}$$

fulfilled. We take $k' \in \mathbf{N}_{\_}$ arbitrary. If $\forall k \geq k', \widehat{x}(k) \neq \mu$ then

$$\forall k \geq k', \widehat{x}(k) = \mu \Longrightarrow (\widehat{x}(k) = \widehat{x}(k+p) \text{ and } k-p \geq k' \Longrightarrow \widehat{x}(k) = \widehat{x}(k-p))$$

is trivially true, thus we can take $k \geq k'$ arbitrarily such that $\widehat{x}(k) = \mu$. From (3.17) we have that $\widehat{x}(k) = \widehat{x}(k+p)$. In case that $k-p \geq k'$, as $k-p \geq -1$, we can apply (3.17) once again in order to infer that $\widehat{x}(k) = \widehat{x}(k-p)$.

(3.5)$\Longrightarrow$(3.6) For an arbitrary $p \geq 1$, the hypothesis states the existence of $\mu \in \widehat{Or}(\widehat{x})$ with

$$\begin{cases} \forall k' \in \mathbf{N}_{\_}, \forall k \geq k', \widehat{x}(k) = \mu \Longrightarrow \\ \Longrightarrow (\widehat{x}(k) = \widehat{x}(k+p) \text{ and } k-p \geq k' \Longrightarrow \widehat{x}(k) = \widehat{x}(k-p)) \end{cases} \tag{3.18}$$

true. We take $k'' \in \mathbf{N}$ arbitrary. If $\forall k \in \mathbf{N}_{\_}, \widehat{\sigma}^{k''}(\widehat{x})(k) \neq \mu$, then

$$\begin{cases} \forall k \in \mathbf{N}_{\_}, \widehat{\sigma}^{k''}(\widehat{x})(k) = \mu \Longrightarrow (\widehat{\sigma}^{k''}(\widehat{x})(k) = \widehat{\sigma}^{k''}(\widehat{x})(k+p) \text{ and} \\ \text{and } k-p \geq -1 \Longrightarrow \widehat{\sigma}^{k''}(\widehat{x})(k) = \widehat{\sigma}^{k''}(\widehat{x})(k-p)) \end{cases}$$

is trivially true, thus let $k \in \mathbf{N}\_$ arbitrary with $\widehat{\sigma}^{k''}(\widehat{x})(k) = \widehat{x}(k+k'') = \mu$. We denote $k' = k'' - 1$ and we see that $\widehat{x}(k+k'+1) = \mu$, where $k+k'+1 \geq k'$. We can apply (3.18) and we infer that

$$\widehat{\sigma}^{k''}(\widehat{x})(k) = \widehat{x}(k+k'') = \widehat{x}(k+k'+1) \overset{(3.18)}{=} \widehat{x}(k+k'+1+p) =$$

$$= \widehat{x}(k+k''+p) = \widehat{\sigma}^{k''}(\widehat{x})(k+p).$$

We suppose now that in addition we have $k - p \geq -1$, thus $k+k'+1-p \geq k'$ and we can apply again (3.18) in order to obtain

$$\widehat{\sigma}^{k''}(\widehat{x})(k) = \widehat{x}(k+k'') = \widehat{x}(k+k'+1) \overset{(3.18)}{=} \widehat{x}(k+k'+1-p) =$$

$$= \widehat{x}(k+k''-p) = \widehat{\sigma}^{k''}(\widehat{x})(k-p).$$

(3.6)$\Longrightarrow$(1.1) The hypothesis written for $p = 1$ shows the existence of $\mu \in \widehat{Or}(\widehat{x})$ such that, in the special case when $k'' = 0$,

$$\begin{cases} \forall k \in \mathbf{N}\_, \widehat{x}(k) = \mu \Longrightarrow \\ \Longrightarrow (\widehat{x}(k) = \widehat{x}(k+1) \text{ and } k \geq 0 \Longrightarrow \widehat{x}(k) = \widehat{x}(k-1)) \end{cases} \tag{3.19}$$

is fulfilled. Some $k_1 \in \mathbf{N}\_$ exists with $\widehat{x}(k_1) = \mu$ and from (3.19) we get:

$$\widehat{x}(k_1) = \widehat{x}(k_1 - 1) = \widehat{x}(k_1 - 2) = \ldots = \widehat{x}(-1),$$

$$\widehat{x}(k_1) = \widehat{x}(k_1 + 1) = \widehat{x}(k_1 + 2) = \ldots$$

i.e. (1.1) holds.

b) (1.5)$\Longrightarrow$(3.7) Let $T > 0$ arbitrary. The hypothesis states the existence of $\mu \in \mathbf{B}^n$ such that $\mathbf{T}^x_\mu = \{t | t \in \mathbf{R}, x(t) = \mu\} = \mathbf{R}$, in particular $\mu = x(-\infty + 0)$. We take $t' \in I^x$ arbitrarily and let $t \in \mathbf{T}^x_\mu \cap [t', \infty) = [t', \infty), z \in \mathbf{Z}$ with $t + zT \geq t'$. We conclude that $t + zT \in \mathbf{T}^x_\mu$. These imply the truth of (3.7).

(3.7)$\Longrightarrow$(3.8) Let $T > 0$ arbitrary. The hypothesis states the existence of $\mu \in Or(x), t' \in I^x$ with the property

$$\forall t \in \mathbf{T}^x_\mu \cap [t', \infty), \{t + zT | z \in \mathbf{Z}\} \cap [t', \infty) \subset \mathbf{T}^x_\mu. \tag{3.20}$$

Let us take $t'_1 \geq t'$ arbitrary. If $\mathbf{T}^x_\mu \cap [t'_1, \infty) = \varnothing$, then the statement

$$\forall t \in \mathbf{T}^x_\mu \cap [t'_1, \infty), \{t + zT | z \in \mathbf{Z}\} \cap [t'_1, \infty) \subset \mathbf{T}^x_\mu$$

is trivially true, thus we can suppose from now that $\mathbf{T}^x_\mu \cap [t'_1, \infty) \neq \varnothing$. We take $t \in \mathbf{T}^x_\mu \cap [t'_1, \infty), z \in \mathbf{Z}$ arbitrarily such that $t + zT \geq t'_1$. We have $t \in \mathbf{T}^x_\mu \cap [t', \infty), t + zT \geq t'$ and we can apply (3.20). We infer $t + zT \in \mathbf{T}^x_\mu$.

(3.8)$\Longrightarrow$(3.9) Let $T > 0$ arbitrary. From (3.8) we get the existence of $\mu \in Or(x)$ and $t''' \in I^x$ with

$$\forall t \in \mathbf{T}^x_\mu \cap [t''', \infty), \{t + zT | z \in \mathbf{Z}\} \cap [t''', \infty) \subset \mathbf{T}^x_\mu \tag{3.21}$$

true. When writing (3.21) we have taken in (3.8) $t'_1 = t'''(= t')$. As $Or(x) = \{x(t) | t \geq t'''\}$, we have $\mathbf{T}^x_\mu \cap [t''', \infty) \neq \varnothing$ and (3.21) shows that $\mu \in \omega(x)$.

Let $t'' \in \mathbf{R}$ arbitrary. We have the following possibilities.

Case $t'' \leq t'''$

Then, since $t'' \in I^x$, we get $\sigma^{t''}(x) = x$ thus from (3.21) we have the truth of (3.9) with $t' = t'''$.

Case $t'' > t'''$

Some $\varepsilon > 0$ exists with the property that $\forall t \in (t'' - \varepsilon, t''), x(t) = x(t'' - 0)$. We take $t' \in (t'' - \varepsilon, t'') \cap (t''', t'')$ arbitrarily and we have

$$\sigma^{t''}(x)(t) = \begin{cases} x(t), t \geq t', \\ x(t'' - 0), t < t''. \end{cases}$$

The fact that $t' \in (-\infty, t'') \subset I^{\sigma^{t''}(x)}$ is obvious. As far as $\mu \in \omega(x)$, we have $\mathbf{T}_\mu^{\sigma^{t''}(x)} \cap [t', \infty) \neq \varnothing$. Let $t \in \mathbf{T}_\mu^{\sigma^{t''}(x)} \cap [t', \infty)$ arbitrary, fixed and we notice that $\mathbf{T}_\mu^{\sigma^{t''}(x)} \cap [t', \infty) = \mathbf{T}_\mu^x \cap [t', \infty)$. We take also $z \in \mathbf{Z}$ arbitrary with $t + zT \geq t'$. Because in this situation $t \in \mathbf{T}_\mu^x \cap [t''', \infty)$ and $t + zT \geq t'''$, we can apply (3.21) and we infer $t + zT \in \mathbf{T}_\mu^x$, i.e. $x(t + zT) = \mu = \sigma^{t''}(x)(t + zT)$ and finally $t + zT \in \mathbf{T}_\mu^{\sigma^{t''}(x)}$.

(3.9)$\Longrightarrow$(3.10) Let $T > 0$ arbitrary, fixed. From (3.9), some $\mu \in Or(x)$ exists such that

$$\begin{cases} \forall t'' \in \mathbf{R}, \exists t' \in I^{\sigma^{t''}(x)}, \\ \forall t \in \mathbf{T}_\mu^{\sigma^{t''}(x)} \cap [t', \infty), \{t + zT | z \in \mathbf{Z}\} \cap [t', \infty) \subset \mathbf{T}_\mu^{\sigma^{t''}(x)}. \end{cases} \tag{3.22}$$

The existence of $x(-\infty + 0)$ shows that in (3.22) we can choose $t'' \in \mathbf{R}$ sufficiently small such that $\sigma^{t''}(x) = x$. For this choice of t'', (3.22) shows the existence of $t' \in I^x$ with

$$\forall t \in \mathbf{T}_\mu^x \cap [t', \infty), \{t + zT | z \in \mathbf{Z}\} \cap [t', \infty) \subset \mathbf{T}_\mu^x \tag{3.23}$$

true. We have $Or(x) = \{x(t) | t \geq t'\}$, wherefrom we get $\mathbf{T}_\mu^x \cap [t', \infty) \neq \varnothing$. Let $t \geq t'$ arbitrary with $x(t) = \mu$, in other words $t \in \mathbf{T}_\mu^x \cap [t', \infty)$. As far as

$$t + T \in \{t + zT | z \in \mathbf{Z}\} \cap [t', \infty) \overset{(3.23)}{\subset} \mathbf{T}_\mu^x,$$

we conclude that $t + T \in \mathbf{T}_\mu^x$, i.e. $x(t + T) = \mu = x(t)$. If in addition $t - T \geq t'$, then

$$t - T \in \{t + zT | z \in \mathbf{Z}\} \cap [t', \infty) \overset{(3.23)}{\subset} \mathbf{T}_\mu^x,$$

wherefrom $t - T \in \mathbf{T}_\mu^x$, i.e. $x(t - T) = \mu = x(t)$.

(3.10)$\Longrightarrow$(3.11) We take $T > 0$ arbitrarily, for which the truth of (3.10) shows the existence of $\mu \in Or(x)$ and $t' \in I^x$ such that

$$\forall t \geq t', x(t) = \mu \Longrightarrow (x(t) = x(t + T) \text{ and } t - T \geq t' \Longrightarrow x(t) = x(t - T)). \tag{3.24}$$

Let now $t_1' \geq t'$ arbitrary. As $\mu \in Or(x)$ we get the existence of $t \geq t'$ with $x(t) = \mu$, wherefrom taking into account (3.24) we infer $\mu \in \omega(x)$. Let $t \geq t_1'$ arbitrarily such that $x(t) = \mu$ (we can take such t's because $\mu \in \omega(x)$ and $\mathbf{T}_\mu^x$ is superiorly unbounded). As $t \geq t'$, we conclude from (3.24) that $x(t) = x(t + T)$ holds. If in addition $t - T \geq t_1'$, then $t - T \geq t'$ and from (3.24) we have that $x(t) = x(t - T)$.

(3.11)$\Longrightarrow$(3.12) Let $T>0$ arbitrary, fixed. Hypothesis (3.11) shows the existence of $\mu\in Or(x)$ and $t'''\in I^x$ such that, in the special case when $t_1'\geq t'''$ is true as equality, we have

$$\forall t\geq t''',x(t)=\mu\Longrightarrow (x(t)=x(t+T) \text{ and } t-T\geq t'''\Longrightarrow x(t)=x(t-T)) \tag{3.25}$$

and in particular we notice that $\mu\in\omega(x)$ and $\mathbf{T}_\mu^x$ is superiorly unbounded.

Let $t''\in\mathbf{R}$ arbitrary, fixed. We have the following possibilities.

Case $t''\leq t'''$

We have $\sigma^{t''}(x)=x$ and, from (3.25) we have the truth of (3.12), with $t'=t'''$.

Case $t''>t'''$

Some $\varepsilon>0$ exists with the property that $\forall t\in(t''-\varepsilon,t''),x(t)=x(t''-0)$. We take $t'\in(t''-\varepsilon,t'')\cap(t''',t'')$ arbitrary. We notice that

$$\sigma^{t''}(x)(t)=\begin{cases} x(t),t\geq t', \\ x(t''-0),t<t'', \end{cases}$$

$t'\in(-\infty,t'')\subset I^{\sigma^{t''}(x)}$ hold and let now $t\geq t'$ arbitrary with $\sigma^{t''}(x)(t)=\mu$. Such a choice of t is possible since $\mathbf{T}_\mu^x$ is superiorly unbounded. We infer from (3.25) and Lemma 3, page 224 that

$$\forall t\geq t',x(t)=\mu\Longrightarrow (x(t)=x(t+T) \text{ and } t-T\geq t'\Longrightarrow x(t)=x(t-T)). \tag{3.26}$$

As for $t\geq t',\sigma^{t''}(x)(t)=x(t)$, we can apply (3.26) in order to conclude the truth of (3.12).

(3.12)$\Longrightarrow$(1.4) We suppose against all reason that x is not constant, thus $t_0\in\mathbf{R}$ exists with

$$\forall t<t_0,x(t)=x(-\infty+0), \tag{3.27}$$

$$x(t_0)\neq x(-\infty+0). \tag{3.28}$$

Let $T>0$ arbitrary. Some $\mu\in Or(x)$ exists from the hypothesis (3.12) such that

$$\left\{\begin{array}{c} \forall t''\in\mathbf{R},\exists t'\in I^{\sigma^{t''}(x)}, \\ \forall t\geq t',\sigma^{t''}(x)(t)=\mu\Longrightarrow(\sigma^{t''}(x)(t)=\sigma^{t''}(x)(t+T) \text{ and} \\ \text{and } t-T\geq t'\Longrightarrow\sigma^{t''}(x)(t)=\sigma^{t''}(x)(t-T)). \end{array}\right. \tag{3.29}$$

We take in (3.29) $t''\leq t_0$, for which we have $\sigma^{t''}(x)=x$ and from (3.27), (3.28), (3.29), $t'<t_0$ exists with $t'\in I^x$,

$$\forall t\geq t',x(t)=\mu\Longrightarrow (x(t)=x(t+T) \text{ and } t-T\geq t'\Longrightarrow x(t)=x(t-T)). \tag{3.30}$$

As $\mu\in Or(x)=\{x(t)|t\geq t'\}$, some $t'''\geq t'$ exists with $x(t''')=\mu$ and, from (3.30), $\mu\in\omega(x)$. In both situations: $t'''\in[t',t_0)$ and $t'''\geq t_0$, we have the existence of $t_1,t_2\in\mathbf{R}$ with the properties $t'\leq t_1<t_2$, $[t_1,t_2)\subset\mathbf{T}_\mu^x$ and at least one of $x(t_1-0)\neq\mu,x(t_2)\neq\mu$ is true. We note from Lemma 4, page 224 that in this situation the inclusion

$$[t_1,t_2)\cup[t_1+T,t_2+T)\cup[t_1+2T,t_2+2T)\cup\ldots\subset\mathbf{T}_\mu^x \tag{3.31}$$

holds. Moreover, Lemma 5, page 224 shows that $\forall k\in\mathbf{N}$, one of $x(t_1+kT-0)\neq\mu,x(t_2+kT)\neq\mu$ is also fulfilled.

Let now $T' \in (0, t_2 - t_1)$. From the hypothesis (3.12), $\mu' \in Or(x)$ exists such that

$$\begin{cases} \forall t'' \in \mathbf{R}, \exists t' \in I^{\sigma^{t''}(x)}, \\ \forall t \geq t', \sigma^{t''}(x)(t) = \mu' \Longrightarrow (\sigma^{t''}(x)(t) = \sigma^{t''}(x)(t+T') \text{ and} \\ \text{and } t - T' \geq t' \Longrightarrow \sigma^{t''}(x)(t) = \sigma^{t''}(x)(t-T')). \end{cases} \quad (3.32)$$

For $t'' \leq t_0$, as $\sigma^{t''}(x) = x$, (3.27), (3.28), (3.32) imply the existence of $t_0' < t_0$ such that $t_0' \in I^x$,

$$\begin{cases} \forall t \geq t_0', x(t) = \mu' \Longrightarrow \\ \Longrightarrow (x(t) = x(t+T') \text{ and } t - T' \geq t_0' \Longrightarrow x(t) = x(t-T')). \end{cases} \quad (3.33)$$

But $\mu' \in Or(x) = \{x(t) | t \geq t_0'\}$ and, like before, $t_1', t_2' \in \mathbf{R}$ exist such that $t_0' \leq t_1' < t_2'$, $[t_1', t_2') \subset \mathbf{T}^x_{\mu'}$ and

$$[t_1', t_2') \cup [t_1' + T', t_2' + T') \cup [t_1' + 2T', t_2' + 2T') \cup \ldots \subset \mathbf{T}^x_{\mu'}. \quad (3.34)$$

The fact that $T' < t_2 - t_1$ implies however from Lemma 6, page 226 that

$$\varnothing \neq ([t_1, t_2) \cup [t_1 + T, t_2 + T) \cup [t_1 + 2T, t_2 + 2T) \cup \ldots) \cap$$

$$\cap([t_1', t_2') \cup [t_1' + T', t_2' + T') \cup [t_1' + 2T', t_2' + 2T') \cup \ldots) \subset \mathbf{T}^x_{\mu} \cap \mathbf{T}^x_{\mu'},$$

thus $\mu = \mu'$. As we have already mentioned, two possibilities exist.

Case $x(t_1 - 0) \neq \mu$.

Let $k \in \mathbf{N}$ with $t_1 + kT > t_0'$. Some $\varepsilon > 0$ exists with $\forall \xi \in (t_1 + kT - \varepsilon, t_1 + kT), x(\xi) = x(t_1 + kT - 0)$ and $t_1 + kT - \varepsilon \geq t_0'$. But then $t \in (t_1 + kT - \varepsilon, t_1 + kT)$ exists such that $t + T' \in [t_1 + kT, t_2 + kT)$[1] and we have

$$\mu \overset{Lemma\ 5}{\neq} x(t_1 + kT - 0) = x(t),$$

$$\mu \overset{(3.31)}{=} x(t + T') \overset{(3.33) \text{ with } \mu = \mu'}{=} x(t),$$

contradiction.

Case $x(t_2) \neq \mu$.

Let $k \in \mathbf{N}$ such that $t_1 + kT > t_0'$ and $t \in [t_1 + kT, t_2 + kT)$ such that $t + T' = t_2 + kT$[2]. We have

$$\mu \overset{(3.31)}{=} x(t) \overset{(3.33) \text{ with } \mu = \mu'}{=} x(t + T') = x(t_2 + kT) \overset{(3.30)}{=} x(t_2) \overset{Lemma\ 5}{\neq} \mu,$$

contradiction.

We have obtained that x is constant. □

[1] Proving that $max\{t_1 + kT - \varepsilon, t_1 + kT - T'\} < \min\{t_1 + kT, t_2 + kT - T'\}$ is easy and we take $t \in (max\{t_1 + kT - \varepsilon, t_1 + kT - T'\}, \min\{t_1 + kT, t_2 + kT - T'\})$ arbitrarily.

[2] Such a t exists since $t_1 + kT \leq t_2 + kT - T' < t_2 + kT$ holds.

4. The Fourth Group of Constancy Properties

Remark 64. *The constancy properties to follow have their origin in the eventual constancy properties from Theorem 20, page 40 and they use the periodicity and the eventual periodicity of the signals. We see that:*

- (4.1) and (4.2) (the last contains $\forall k' \in \mathbf{N}_{\_}$) have their origin in $(6.1)_{page\,40}$ (containing $\exists k' \in \mathbf{N}_{\_}$);

- (4.3) (containing $\forall k'' \in \mathbf{N}$) has its origin in $(6.2)_{page\,40}$ (containing $\exists k'' \in \mathbf{N}$);

- (4.4) and (4.5) (the last contains $\forall t_1' \geq t'$) have their origin in $(6.3)_{page\,40}$ (containing $\exists t_1' \geq t'$) and $(6.4)_{page\,40}$ (containing $\exists t_1' \in \mathbf{R}$);

- (4.6) (containing $\forall t'' \in \mathbf{R}$) has its origin in $(6.5)_{page\,40}$ and $(6.6)_{page\,40}$ (containing both $\exists t'' \in \mathbf{R}$).

Theorem 25. *a) The following properties are equivalent with the constancy of the signal $\widehat{x} \in \widehat{S}^{(n)}$:*

$$\forall p \geq 1, \forall k \in \mathbf{N}_{\_}, \widehat{x}(k) = \widehat{x}(k+p), \tag{4.1}$$

$$\forall p \geq 1, \forall k' \in \mathbf{N}_{\_}, \forall k \geq k', \widehat{x}(k) = \widehat{x}(k+p), \tag{4.2}$$

$$\forall p \geq 1, \forall k'' \in \mathbf{N}, \forall k \in \mathbf{N}_{\_}, \widehat{\sigma}^{k''}(\widehat{x})(k) = \widehat{\sigma}^{k''}(\widehat{x})(k+p). \tag{4.3}$$

b) The following properties are equivalent with the constancy of $x \in S^{(n)}$:

$$\forall T > 0, \exists t' \in I^x, \forall t \geq t', x(t) = x(t+T), \tag{4.4}$$

$$\forall T > 0, \exists t' \in I^x, \forall t_1' \geq t', \forall t \geq t_1', x(t) = x(t+T), \tag{4.5}$$

$$\forall T > 0, \forall t'' \in \mathbf{R}, \exists t' \in I^{\sigma^{t''}(x)}, \forall t \geq t', \sigma^{t''}(x)(t) = \sigma^{t''}(x)(t+T). \tag{4.6}$$

Proof. a) (1.1)$\Longrightarrow$(4.1) We suppose that $\mu \in \mathbf{B}^n$ exists with $\forall k \in \mathbf{N}_{\_}, \widehat{x}(k) = \mu$ and let $p \geq 1, k \in \mathbf{N}_{\_}$ arbitrary. We have

$$\widehat{x}(k) = \mu = \widehat{x}(k+p),$$

making (4.1) true.

(4.1)$\Longrightarrow$(4.2) Let $p \geq 1, k' \in \mathbf{N}_{\_}, k \geq k'$ arbitrary. From (4.1) we infer that $\widehat{x}(k) = \widehat{x}(k+p)$.

(4.2)$\Longrightarrow$(4.3) We take $p \geq 1, k'' \in \mathbf{N}, k \in \mathbf{N}_{\_}$ arbitrarily. We denote $k' = k'' - 1$ and we notice that $k + k'' = k + k' + 1 \geq k'$, thus we can apply (4.2) and we obtain

$$\widehat{\sigma}^{k''}(\widehat{x})(k) = \widehat{x}(k+k'') = \widehat{x}(k+k'+1) \overset{(4.2)}{=} \widehat{x}(k+k'+1+p) =$$

$$= \widehat{x}(k+k''+p) = \widehat{\sigma}^{k''}(\widehat{x})(k+p).$$

(4.3)$\Longrightarrow$(1.1) We write (4.3) for $p = 1$ and $k'' = 0$, when $\widehat{\sigma}^{k''}(\widehat{x})(k) = \widehat{x}(k)$, with $k = -1, 0, 1, ...$ and we get

$$\widehat{x}(-1) = \widehat{x}(0) = \widehat{x}(1) = ...$$

We denote with μ the common value of $\widehat{x}(-1), \widehat{x}(0), \widehat{x}(1), ...$ We infer that (1.1) holds.

b) (1.4)$\Longrightarrow$(4.4) If $\mu \in \mathbf{B}^n$ exists with $\forall t \in \mathbf{R}, x(t) = \mu$, then for arbitrary $T > 0$ and $t' \in \mathbf{R}$,

$$\forall t \leq t', x(t) = \mu,$$

$$\forall t \geq t', x(t) = x(t+T) = \mu$$

hold. We have that $t' \in I^x$ thus (4.4) is true.

(4.4)$\Longrightarrow$(4.5) Let $T > 0$ arbitrary. Hypothesis (4.4) shows that $t' \in I^x$ exists such that

$$\forall t \geq t', x(t) = x(t+T). \tag{4.7}$$

We take $t_1' \geq t'$ and $t \geq t_1'$ arbitrarily. From the fact that $t \geq t'$, the statement (4.7) gives $x(t) = x(t+T)$, i.e. (4.5) is true.

(4.5)$\Longrightarrow$(4.6) Let $T > 0$ arbitrary. Property (4.5) gives the existence of $t''' \in I^x$ such that, in the special case when $t_1' \geq t'''$ holds as equality, the statement

$$\forall t \geq t''', x(t) = x(t+T) \tag{4.8}$$

is fulfilled. We suppose that an arbitrary $t'' \in \mathbf{R}$ is given and we have the following possibilities.

Case $t'' \leq t'''$

From $\forall t \leq t''', x(t) = x(-\infty+0)$ we infer that $\sigma^{t''}(x) = x$ and, taking into account (4.8) also, we get that (4.6) is true with $t' = t'''$.

Case $t'' > t'''$

Some $\varepsilon > 0$ exists with the property that $\forall t \in (t''-\varepsilon, t''), x(t) = x(t''-0)$. We take arbitrarily a $t' \in (t''-\varepsilon, t'') \cap (t''', t'')$ and we get that $\forall t \leq t', \sigma^{t''}(x)(t) = x(t''-0)$ is true. We notice that for any $t \geq t'$ we have $\sigma^{t''}(x)(t) = x(t)$, irrespective of the fact that $t < t''$ or $t \geq t''$ and let us fix an arbitrary $t \geq t'$. We have

$$\sigma^{t''}(x)(t) = x(t) \overset{(4.8)}{=} x(t+T) = \sigma^{t''}(x)(t+T).$$

(4.6)$\Longrightarrow$(1.4) Let us suppose against all reason that (1.4) is false, meaning that $t_0 < t_1$ exist with the property

$$\forall t < t_0, x(t) = x(-\infty+0), \tag{4.9}$$

$$\forall t \in [t_0, t_1), x(t) \neq x(-\infty+0). \tag{4.10}$$

We write (4.6) for $T \in (0, t_1 - t_0)$ and t'' sufficiently small such that $\sigma^{t''}(x) = x$ and we obtain the existence of $t' \in I^x$ with

$$\forall t \geq t', x(t) = x(t+T). \tag{4.11}$$

From (4.9), (4.10) we infer $t' < t_0$.

Let now $t \in [t', t_0) \cap [t_0 - T, t_0)$ fixed. We have $t + T \in [t_0, t_0 + T) \subset [t_0, t_1)$, thus

$$x(-\infty+0) = x(t) \overset{(4.11)}{=} x(t+T) \overset{(4.10)}{\neq} x(-\infty+0),$$

contradiction. We conclude that (1.4) holds. □

5. Discrete Time vs Real Time

Theorem 26. *Let us suppose that the sequence* $(t_k) \in Seq$ *exists such that*

$$x(t) = \widehat{x}(-1) \cdot \chi_{(-\infty,t_0)}(t) \oplus \widehat{x}(0) \cdot \chi_{[t_0,t_1)}(t) \oplus \ldots \oplus \widehat{x}(k) \cdot \chi_{[t_k,t_{k+1})}(t) \oplus \ldots \tag{5.1}$$

Then the constancy of $\widehat{x}$ *is equivalent with the constancy of* x.

Proof. Obvious, but let us take a look at Theorem 6 a) also, page 11, stating that the hypothesis implies $\widehat{Or}(\widehat{x}) = Or(x)$. We infer then the equivalence between $(1.3)_{page\,45}$ and $(1.6)_{page\,45}$:

$$\exists \mu \in \mathbf{B}^n, \widehat{Or}(\widehat{x}) = \{\mu\} \Longleftrightarrow \exists \mu \in \mathbf{B}^n, Or(x) = \{\mu\}.$$

□

6. Discussion

Remark 65. *The statements from Theorems 22, 23, 24, 25 are present in discrete time - real time couples: (1.1)-(1.4), (1.2)-(1.5),... This continues the previous style of organizing the exposure, corresponding to our intuition that strong analogies work between the discrete time and the real time properties of the signals. Theorem 26 gives the relation between the discrete time and the real time constancy of the signals.*

Remark 66. *The three groups 2,3,4 of properties of periodicity have in common the periodicity of the signal, which is present under the form: all the points of its orbit are periodic, with the same period.*

Remark 67. *The key request in all these periodicity properties in order to be equivalent with constancy is that they hold for any period* $p \geq 1, T > 0$.

Remark 68. *In (1.1),..., (1.3) the existence of a unique* $\mu \in \mathbf{B}^n$ *is stated. Similarly, in (1.4),..., (1.6)* $\exists \mu \in \mathbf{B}^n$ *must be understood as* $\exists! \mu \in \mathbf{B}^n$.

Remark 69. *The statement*

$$\forall k \in \widehat{\mathbf{T}}^{\widehat{x}}_{\mu}, \{k + zp | z \in \mathbf{Z}\} \cap \mathbf{N}_{\_} \subset \widehat{\mathbf{T}}^{\widehat{x}}_{\mu}$$

from (2.1) is one of periodicity of μ *with the period* p *and the statement*

$$\exists t' \in I^x, \forall t \in \mathbf{T}^x_{\mu} \cap [t', \infty), \{t + zT | z \in \mathbf{Z}\} \cap [t', \infty) \subset \mathbf{T}^x_{\mu}$$

from (2.7) is one of periodicity of μ *with the period* T. *Both these requirements are related with an initial time=limit of periodicity, which is* -1 *and* t'. *Their demand is that if* $\widehat{x}(k) = \mu, x(t) = \mu$, *then right translations along the time axis are allowed giving the same value* μ *of* $\widehat{x}, x : \widehat{x}(k + zp) = \mu, z \geq 0, x(t + zT) = \mu, z \geq 0$ *and left translations along the time axis are also allowed as long as the argument still exceeds the limit of periodicity- and they give the same value* μ *of* $\widehat{x}, x : \widehat{x}(k + zp) = \mu, z < 0, x(t + zT) = \mu, z < 0$. *And this should happen for all the periods* $p \geq 1, T > 0$ *and all the points of the orbit* $\mu \in \widehat{Or}(\widehat{x}), \mu \in Or(x)$.

Remark 70. *The properties*

$$\begin{cases} \forall k' \in \mathbf{N}_{\_}, \forall k \in \widehat{\mathbf{T}}_{\mu}^{\widehat{x}} \cap \{k', k'+1, k'+2, ...\}, \\ \{k+zp | z \in \mathbf{Z}\} \cap \{k', k'+1, k'+2, ...\} \subset \widehat{\mathbf{T}}_{\mu}^{\widehat{x}}, \end{cases}$$

$$\exists t' \in I^x, \forall t_1' \geq t', \forall t \in \mathbf{T}_{\mu}^x \cap [t_1', \infty), \{t+zT | z \in \mathbf{Z}\} \cap [t_1', \infty) \subset \mathbf{T}_{\mu}^x$$

from (2.2), (2.8) are of eventual periodicity of μ *with the period* p, T. *Here the periodicity of* μ *starts not from the very beginning* $-1, t'$ *like previously, but from a time instant* $k' \in \mathbf{N}_{\_}, t_1' \geq t'$. *In order to have periodicity, we ask that such properties hold for any* k', t_1'. *And in order to rediscover constancy, they should hold for any* p, T *and any* $\mu \in \widehat{Or}(\widehat{x}), \mu \in Or(x)$.

Remark 71. *Let us fix in (2.3)* $p \geq 1, \mu \in \widehat{Or}(\widehat{x})$ *and* $k'' \in \mathbf{N}$. *In general we have* $\widehat{Or}(\widehat{\sigma}^{k''}(\widehat{x})) \subset \widehat{Or}(\widehat{x})$ *(we have shown this at Theorem 5 c), page 9) and the points* $\mu \in \widehat{Or}(\widehat{x}) \setminus \widehat{Or}(\widehat{\sigma}^{k''}(\widehat{x}))$ *may satisfy* $\widehat{\mathbf{T}}_{\mu}^{\widehat{\sigma}^{k''}(\widehat{x})} = \varnothing$, *when*

$$\forall k \in \widehat{\mathbf{T}}_{\mu}^{\widehat{\sigma}^{k''}(\widehat{x})}, \{k+zp | z \in \mathbf{Z}\} \cap \mathbf{N}_{\_} \subset \widehat{\mathbf{T}}_{\mu}^{\widehat{\sigma}^{k''}(\widehat{x})}$$

is trivially fulfilled. This is not the case if the previous property takes place for any $k'' \in \mathbf{N}$, *including the case* $k'' = 0$, *when* $\widehat{Or}(\widehat{\sigma}^{k''}(\widehat{x})) = \widehat{Or}(\widehat{x})$. *This discussion is in principle the same for (2.9).*

Remark 72. *The requests*

$$\forall k \in \widehat{\mathbf{T}}_{\mu}^{\widehat{\sigma}^{k''}(\widehat{x})}, \{k+zp | z \in \mathbf{Z}\} \cap \mathbf{N}_{\_} \subset \widehat{\mathbf{T}}_{\mu}^{\widehat{\sigma}^{k''}(\widehat{x})},$$

$$\begin{cases} \forall k \in \mathbf{N}_{\_}, \widehat{\sigma}^{k''}(\widehat{x})(k) = \mu \Longrightarrow (\widehat{\sigma}^{k''}(\widehat{x})(k) = \widehat{\sigma}^{k''}(\widehat{x})(k+p) \text{ and} \\ \quad \text{and } k-p \geq -1 \Longrightarrow \widehat{\sigma}^{k''}(\widehat{x})(k) = \widehat{\sigma}^{k''}(\widehat{x})(k-p)) \end{cases}$$

derived from (2.3), (2.6) and the requests

$$\exists t' \in I^{\sigma^{t''}(x)}, \forall t \in \mathbf{T}_{\mu}^{\sigma^{t''}(x)} \cap [t', \infty), \{t+zT | z \in \mathbf{Z}\} \cap [t', \infty) \subset \mathbf{T}_{\mu}^{\sigma^{t''}(x)},$$

$$\begin{cases} \exists t' \in I^{\sigma^{t''}(x)}, \forall t \geq t', \sigma^{t''}(x)(t) = \mu \Longrightarrow (\sigma^{t''}(x)(t) = \sigma^{t''}(x)(t+T) \text{ and} \\ \quad \text{and } t-T \geq t' \Longrightarrow \sigma^{t''}(x)(t) = \sigma^{t''}(x)(t-T)) \end{cases}$$

derived from (2.9), (2.12) are of eventual periodicity of μ *with the period* p, T. *In such requests, the fact that periodicity might not start from the very beginning is indicated by working with the signals* $\widehat{\sigma}^{k''}(\widehat{x}), \sigma^{t''}(x)$ *that have forgotten their first values. Note that all these properties are of periodicity of* μ *-related to* $\widehat{\sigma}^{k''}(\widehat{x}), \sigma^{t''}(x)$ *instead of* $\widehat{x}, x$*- and that they must hold, for constancy, for all* p, T, *all* $\mu \in \widehat{Or}(\widehat{x}), \mu \in Or(x)$ *and all* $k'' \in \mathbf{N}, t'' \in \mathbf{R}$.

Remark 73. *The statements*

$$\begin{cases} \forall k \in \mathbf{N}_{\_}, \widehat{x}(k) = \mu \Longrightarrow \\ \Longrightarrow (\widehat{x}(k) = \widehat{x}(k+p) \text{ and } k-p \geq -1 \Longrightarrow \widehat{x}(k) = \widehat{x}(k-p)), \end{cases}$$

$$\exists t' \in I^x, \forall t \geq t', x(t) = \mu \Longrightarrow (x(t) = x(t+T) \text{ and } t-T \geq t' \Longrightarrow x(t) = x(t-T))$$

from (2.4), (2.10) refer also to the periodicity of μ *with the period* p, T. *The difference from the previous property consists in the fact that the translations along the time axis are with one period only, and the general case is rediscovered by iterating these translations. We must have periodicity with any period* p, T, *of all the points* $\mu \in \widehat{Or}(\widehat{x}), \mu \in Or(x)$ *for constancy.*

Remark 74. *The case of (2.5) is similar with that of (2.3) (see Remark 71). Points* $\mu \in \widehat{Or}(\widehat{x})$ *might exist for which* $\widehat{x}(k) = \mu$ *is false if* $k \geq k'$ *and then*

$$\begin{cases} \forall k \geq k', \widehat{x}(k) = \mu \Longrightarrow \\ \Longrightarrow (\widehat{x}(k) = \widehat{x}(k+p) \text{ and } k - p \geq k' \Longrightarrow \widehat{x}(k) = \widehat{x}(k-p)) \end{cases} \tag{6.1}$$

is trivially true. This is not the case, because the truth of (6.1) includes the value $k' = -1$. *The remark holds also for*

$$\begin{cases} \forall t \geq t_1', x(t) = \mu \Longrightarrow \\ \Longrightarrow (x(t) = x(t+T) \text{ and } t - T \geq t_1' \Longrightarrow x(t) = x(t-T)) \end{cases} \tag{6.2}$$

and (2.11).

Remark 75. *The requests (6.1), (6.2) derived from (2.5), (2.11) are also of eventual periodicity of* μ, *i.e. periodicity starting at* $k' \in \mathbf{N}_{\_}$ *and* $t_1' \geq t' \in I^x$. *The difference from the previous situation is given by the translations along the time axis with one period. The requests must be fulfilled for all* $k', t_1' \geq t'$, *all* p, T *and all* μ.

Remark 76. *In (2.6), (2.12) we have*

$$\begin{cases} \forall k \in \mathbf{N}_{\_}, \widehat{\sigma}^{k''}(\widehat{x})(k) = \mu \Longrightarrow (\widehat{\sigma}^{k''}(\widehat{x})(k) = \widehat{\sigma}^{k''}(\widehat{x})(k+p) \text{ and} \\ \text{and } k - p \geq -1 \Longrightarrow \widehat{\sigma}^{k''}(\widehat{x})(k) = \widehat{\sigma}^{k''}(\widehat{x})(k-p)), \end{cases}$$

$$\begin{cases} \exists t' \in I^{\sigma^{t''}(x)}, \forall t \geq t', \sigma^{t''}(x)(t) = \mu \Longrightarrow (\sigma^{t''}(x)(t) = \sigma^{t''}(x)(t+T) \text{ and} \\ \text{and } t - T \geq t' \Longrightarrow \sigma^{t''}(x)(t) = \sigma^{t''}(x)(t-T)) \end{cases}$$

i.e. the periodicity of $\widehat{x}, x$ *after having forgotten some first values.*

Remark 77. *The third group of constancy properties repeats the statements of the second group, by replacing* $\forall \mu \in \widehat{Or}(\widehat{x}), \forall \mu \in Or(x)$ *with* $\exists \mu \in \widehat{Or}(\widehat{x}), \exists \mu \in Or(x)$. *This is possible since constancy means that the orbits have exactly one point,* $\widehat{Or}(\widehat{x}) = \{\mu\}, Or(x) = \{\mu\}$. *The proofs of the implications are in general similar with those of the second group.*

Remark 78. *The properties*

$$\forall k \in \mathbf{N}_{\_}, \widehat{x}(k) = \widehat{x}(k+p),$$

$$\exists t' \in I^x, \forall t \geq t', x(t) = x(t+T)$$

from (4.1), (4.4) are of periodicity of the signals $\widehat{x}, x$ *with the period* p, T. *In order to get constancy, these properties must hold for any period* p, T.

Remark 79. *The next properties*

$$\forall k \geq k', \widehat{x}(k) = \widehat{x}(k+p),$$

$$\forall t \geq t'_1, x(t) = x(t+T)$$

that occur in (4.2), (4.5) are of eventual periodicity of the signals $\widehat{x}, x$ with the periods p, T. It is asked that periodicity starts at any limit of periodicity $k', t'_1 \geq t'$, where $t' \in I^x$ (we have periodicity so far) and that it holds for any period p, T for constancy.

Remark 80. *The properties*

$$\forall k \in \mathbf{N}_{\_}, \widehat{\sigma}^{k''}(\widehat{x})(k) = \widehat{\sigma}^{k''}(\widehat{x})(k+p),$$

$$\exists t' \in I^{\sigma^{t''}(x)}, \forall t \geq t', \sigma^{t''}(x)(t) = \sigma^{t''}(x)(t+T)$$

from (4.3), (4.6) are of eventual periodicity of $\widehat{x}, x$. They are asked to hold for any $k'' \in \mathbf{N}, t'' \in \mathbf{R}$ for periodicity and any $p \geq 1, T > 0$ for constancy.

Chapter 5

Eventually Periodic Points

We give first some statements that are equivalent with the eventual periodicity of the points and a discussion.

Section 3 shows that an eventually periodic point is accessed for time instants greater than the limit of periodicity at least once in a time interval with the length of a period. This fundamental result will be frequently used later.

The bound of the limit of periodicity and the independence of eventual periodicity on the choice of the limit of periodicity are treated in Section 4.

The property of eventual constancy that follows in Section 5 is used in Section 6 to establish the relation between the discrete time and the continuous time eventual periodicity of the points.

Section 7 highlights the relation between the support sets $\widehat{\mathbf{T}}_{\mu}^{\widehat{x}}, \mathbf{T}_{\mu}^{x}$ and the sets of the periods $\widehat{P}_{\mu}^{\widehat{x}}, P_{\mu}^{x}$.

The fact that the sum, the difference and the multiples of the periods are periods is treated in Section 8.

In Section 9 we show which is the form of $\widehat{P}_{\mu}^{\widehat{x}}, P_{\mu}^{x}$ and in particular we address the issue of the existence of the prime period.

Sections 10 and 11 give necessary and sufficient conditions of eventual periodicity and a special case of eventually periodic point is treated in Section 12, where the prime period is known.

Section 13 gives a result relating the eventually periodic points with the eventually constant signals.

1. Equivalent Properties with the Eventual Periodicity of a Point

Remark 81. *The properties that are equivalent with the eventual periodicity of the points were already used in Chapter 3 dedicated to the eventually constant signals at Theorem 17, page 27 (see also Theorem 18, page 33, and Theorem 19, page 39). To be compared (1.1),...,(1.4) to follow with $(3.1)_{page\,27}$,...,$(3.4)_{page\,27}$ and (1.5),...,(1.12) to follow with $(3.5)_{page\,28}$,...,$(3.12)_{page\,28}$. We make also the associations (1.3)-$(6.1)_{page\,40}$, (1.4)-$(6.2)_{page\,40}$ and (1.9)-$(6.3)_{page\,40}$,(1.10)-$(6.4)_{page\,40}$, (1.11)-$(6.5)_{page\,40}$, (1.12)-$(6.6)_{page\,40}$*

with the properties of eventual constancy of the signals from group four, see Theorem 20, page 40.

Theorem 27. *We consider the signals $\widehat{x} \in \widehat{S^{(n)}}, x \in S^{(n)}$.*

a) The following statements are equivalent for any $p \geq 1$ and any $\mu \in \widehat{\omega}(\widehat{x})$:

$$\begin{cases} \exists k' \in \mathbf{N}_{\_}, \forall k \in \widehat{\mathbf{T}}_{\mu}^{\widehat{x}} \cap \{k', k'+1, k'+2, ...\}, \\ \{k+zp|z \in \mathbf{Z}\} \cap \{k', k'+1, k'+2, ...\} \subset \widehat{\mathbf{T}}_{\mu}^{\widehat{x}}, \end{cases} \tag{1.1}$$

$$\exists k'' \in \mathbf{N}, \forall k \in \widehat{\mathbf{T}}_{\mu}^{\widehat{\sigma}^{k''}(\widehat{x})}, \{k+zp|z \in \mathbf{Z}\} \cap \mathbf{N}_{\_} \subset \widehat{\mathbf{T}}_{\mu}^{\widehat{\sigma}^{k''}(\widehat{x})}, \tag{1.2}$$

$$\begin{cases} \exists k' \in \mathbf{N}_{\_}, \forall k \geq k', \widehat{x}(k) = \mu \Longrightarrow \\ \Longrightarrow (\widehat{x}(k) = \widehat{x}(k+p) \text{ and } k-p \geq k' \Longrightarrow \widehat{x}(k) = \widehat{x}(k-p)), \end{cases} \tag{1.3}$$

$$\begin{cases} \exists k'' \in \mathbf{N}, \forall k \in \mathbf{N}_{\_}, \widehat{\sigma}^{k''}(\widehat{x})(k) = \mu \Longrightarrow \\ \Longrightarrow (\widehat{\sigma}^{k''}(\widehat{x})(k) = \widehat{\sigma}^{k''}(\widehat{x})(k+p) \text{ and} \\ \text{and } k-p \geq -1 \Longrightarrow \widehat{\sigma}^{k''}(\widehat{x})(k) = \widehat{\sigma}^{k''}(\widehat{x})(k-p)). \end{cases} \tag{1.4}$$

b) The following statements are also equivalent for any $T > 0$ and $\mu \in \omega(x)$:

$$\exists t' \in I^{x}, \exists t_1' \geq t', \forall t \in \mathbf{T}_{\mu}^{x} \cap [t_1', \infty), \{t+zT|z \in \mathbf{Z}\} \cap [t_1', \infty) \subset \mathbf{T}_{\mu}^{x}, \tag{1.5}$$

$$\exists t_1' \in \mathbf{R}, \forall t \in \mathbf{T}_{\mu}^{x} \cap [t_1', \infty), \{t+zT|z \in \mathbf{Z}\} \cap [t_1', \infty) \subset \mathbf{T}_{\mu}^{x}, \tag{1.6}$$

$$\begin{cases} \exists t'' \in \mathbf{R}, \exists t' \in I^{\sigma^{t''}(x)}, \\ \forall t \in \mathbf{T}_{\mu}^{\sigma^{t''}(x)} \cap [t', \infty), \{t+zT|z \in \mathbf{Z}\} \cap [t', \infty) \subset \mathbf{T}_{\mu}^{\sigma^{t''}(x)}, \end{cases} \tag{1.7}$$

$$\exists t'' \in \mathbf{R}, \exists t' \in \mathbf{R}, \forall t \in \mathbf{T}_{\mu}^{\sigma^{t''}(x)} \cap [t', \infty), \{t+zT|z \in \mathbf{Z}\} \cap [t', \infty) \subset \mathbf{T}_{\mu}^{\sigma^{t''}(x)}, \tag{1.8}$$

$$\begin{cases} \exists t' \in I^{x}, \exists t_1' \geq t', \forall t \geq t_1', x(t) = \mu \Longrightarrow \\ \Longrightarrow (x(t) = x(t+T) \text{ and } t-T \geq t_1' \Longrightarrow x(t) = x(t-T)), \end{cases} \tag{1.9}$$

$$\begin{cases} \exists t_1' \in \mathbf{R}, \forall t \geq t_1', x(t) = \mu \Longrightarrow \\ \Longrightarrow (x(t) = x(t+T) \text{ and } t-T \geq t_1' \Longrightarrow x(t) = x(t-T)), \end{cases} \tag{1.10}$$

$$\begin{cases} \exists t'' \in \mathbf{R}, \exists t' \in I^{\sigma^{t''}(x)}, \\ \forall t \geq t', \sigma^{t''}(x)(t) = \mu \Longrightarrow (\sigma^{t''}(x)(t) = \sigma^{t''}(x)(t+T) \text{ and} \\ \text{and } t-T \geq t' \Longrightarrow \sigma^{t''}(x)(t) = \sigma^{t''}(x)(t-T)), \end{cases} \tag{1.11}$$

$$\begin{cases} \exists t'' \in \mathbf{R}, \exists t' \in \mathbf{R}, \forall t \geq t', \sigma^{t''}(x)(t) = \mu \Longrightarrow \\ \Longrightarrow (\sigma^{t''}(x)(t) = \sigma^{t''}(x)(t+T) \text{ and} \\ \text{and } t-T \geq t' \Longrightarrow \sigma^{t''}(x)(t) = \sigma^{t''}(x)(t-T)). \end{cases} \tag{1.12}$$

Proof. a) The proof of the implications

$$(1.1) \Longrightarrow (1.2) \Longrightarrow (1.3) \Longrightarrow (1.4)$$

follows from the proof of Theorem 17, page 27.

(1.4)$\Longrightarrow$(1.1) From (1.4), $k'' \in \mathbf{N}$ exists making

$$\begin{cases} \forall k \in \mathbf{N}_{\_}, \widehat{\sigma}^{k''}(\widehat{x})(k) = \mu \Longrightarrow (\widehat{\sigma}^{k''}(\widehat{x})(k) = \widehat{\sigma}^{k''}(\widehat{x})(k+p) \text{ and} \\ \quad \text{and } k-p \geq -1 \Longrightarrow \widehat{\sigma}^{k''}(\widehat{x})(k) = \widehat{\sigma}^{k''}(\widehat{x})(k-p)) \end{cases} \tag{1.13}$$

true. We define $k' = k'' - 1$. The fact that $\mu \in \widehat{\omega}(\widehat{x})$ implies that $\widehat{\mathbf{T}}_{\mu}^{\widehat{x}}$ is infinite, thus $\widehat{\mathbf{T}}_{\mu}^{\widehat{x}} \cap \{k', k'+1, k'+2, ...\} \neq \varnothing$. Let $k \in \widehat{\mathbf{T}}_{\mu}^{\widehat{x}} \cap \{k', k'+1, k'+2, ...\}, z \in \mathbf{Z}$ arbitrary such that $k + zp \geq k'$. The number $k - k''$ satisfies $k - k'' = k - k' - 1 \geq -1, \widehat{x}(k - k'' + k'') = \widehat{\sigma}^{k''}(\widehat{x})(k - k'')$ and the number $k - k'' + zp$ satisfies $k - k'' + zp = k - k' - 1 + zp \geq -1$, thus we can apply (1.13). We have the following possibilities:

Case $z > 0$,

$$\mu = \widehat{x}(k) = \widehat{\sigma}^{k''}(\widehat{x})(k - k'') \overset{(1.13)}{=} \widehat{\sigma}^{k''}(\widehat{x})(k - k'' + p) \overset{(1.13)}{=}$$

$$\overset{(1.13)}{=} \widehat{\sigma}^{k''}(\widehat{x})(k - k'' + 2p) \overset{(1.13)}{=} \dots \overset{(1.13)}{=} \widehat{\sigma}^{k''}(\widehat{x})(k - k'' + zp) = \widehat{x}(k + zp);$$

Case $z = 0$,

$$\mu = \widehat{x}(k) = \widehat{x}(k + zp);$$

Case $z < 0$,

$$\mu = \widehat{x}(k) = \widehat{\sigma}^{k''}(\widehat{x})(k - k'') \overset{(1.13)}{=} \widehat{\sigma}^{k''}(\widehat{x})(k - k'' - p) \overset{(1.13)}{=}$$

$$\overset{(1.13)}{=} \widehat{\sigma}^{k''}(\widehat{x})(k - k'' - 2p) \overset{(1.13)}{=} \dots \overset{(1.13)}{=} \widehat{\sigma}^{k''}(\widehat{x})(k - k'' + zp) = \widehat{x}(k + zp).$$

It has resulted that, in all the three situations, $k + zp \in \widehat{\mathbf{T}}_{\mu}^{\widehat{x}}$, thus (1.1) holds.

b) The proof of the implications

$$(1.5) \Longrightarrow (1.6) \Longrightarrow (1.7) \Longrightarrow (1.8) \Longrightarrow (1.9) \Longrightarrow (1.10) \Longrightarrow (1.11) \Longrightarrow (1.12)$$

follows from the proof of Theorem 17, page 27.

(1.12)$\Longrightarrow$(1.5) From (1.12) we get the existence of $t'' \in \mathbf{R}$ and $t' \in \mathbf{R}$ with

$$\begin{cases} \forall t \geq t', \sigma^{t''}(x)(t) = \mu \Longrightarrow (\sigma^{t''}(x)(t) = \sigma^{t''}(x)(t+T) \text{ and} \\ \quad \text{and } t - T \geq t' \Longrightarrow \sigma^{t''}(x)(t) = \sigma^{t''}(x)(t-T)) \end{cases} \tag{1.14}$$

and on the other hand we take arbitrarily some $t''' \in I^{x}$[1]. We take an arbitrary $t_1' \geq \max\{t', t'', t'''\}$ also. We have

$$\forall t \geq t_1', \sigma^{t''}(x)(t) = x(t) \tag{1.15}$$

and, taking into account (1.14), (1.15) and Lemma 3, page 224 we infer

$$\forall t \geq t_1', x(t) = \mu \Longrightarrow (x(t) = x(t+T) \text{ and } t - T \geq t_1' \Longrightarrow x(t) = x(t-T)). \tag{1.16}$$

As $\mu \in \omega(x)$, $\mathbf{T}_{\mu}^{x}$ is superiorly unbounded thus $\mathbf{T}_{\mu}^{x} \cap [t_1', \infty) \neq \varnothing$. Let us take now $t \in \mathbf{T}_{\mu}^{x} \cap [t_1', \infty)$ and $z \in \mathbf{Z}$ arbitrarily such that $t + zT \geq t_1'$. The following possibilities exist:

[1]From this moment we prove the truth of a statement which is stronger than (1.5).

Case $z > 0$,

$$\mu = x(t) \overset{(1.16)}{=} x(t+T) \overset{(1.16)}{=} x(t+2T) \overset{(1.16)}{=} \dots \overset{(1.16)}{=} x(t+zT);$$

Case $z = 0$,

$$\mu = x(t) = x(t+zT);$$

Case $z < 0$,

$$\mu = x(t) \overset{(1.16)}{=} x(t-T) \overset{(1.16)}{=} x(t-2T) \overset{(1.16)}{=} \dots \overset{(1.16)}{=} x(t+zT).$$

It has resulted that in all these situations $x(t+zT) = \mu$, thus $t+zT \in \mathbf{T}^x_\mu$. □

Example 14. *For the signal* $x \in S^{(1)}, \forall t \in \mathbf{R}$,

$$x(t) = \chi_{(-\infty,0)}(t) \oplus \chi_{[3,4)}(t) \oplus \chi_{[5,6)}(t) \oplus \chi_{[7,8)}(t) \oplus \dots$$

neither of $0, 1 \in Or(x)$ *is periodic, but for any* $t' \in [2, \infty)$ *we get*

$$\forall t \in \mathbf{T}^x_0 \cap [t', \infty), \{t + z2 | z \in \mathbf{Z}\} \cap [t', \infty) \subset \mathbf{T}^x_0,$$

$$\forall t \in \mathbf{T}^x_1 \cap [t', \infty), \{t + z2 | z \in \mathbf{Z}\} \cap [t', \infty) \subset \mathbf{T}^x_1,$$

thus $0, 1$ *are eventually periodic with* $P^x_0 = P^x_1 = \{2, 4, 6, \dots\}$.

2. Discussion

Remark 82. *The properties (1.1), (1.3) are of eventual periodicity of* μ*, meaning that periodicity starts at a limit of periodicity* k' *which is in general bigger than the initial time* -1*. Equivalently, the properties (1.2), (1.4) are of periodicity of* μ *(starting at the initial time* -1*), however not the periodicity referring to* $\widehat{x}$*, but the periodicity referring to* $\widehat{\sigma}^{k''}(\widehat{x})$*,* $k'' \geq 0$*, meaning that* $\widehat{x}$ *might have forgotten some of its first values. The real time equivalent statements are interpreted similarly.*

Remark 83. *Note that in the statement of Theorem 27 we have asked* $\mu \in \widehat{\omega}(\widehat{x}), \mu \in \omega(x)$ *instead of* $\mu \in \widehat{Or}(\widehat{x}), \mu \in Or(x)$*, the usual demand of periodicity of* μ*. This avoids stating further requests of non-triviality* $\widehat{\mathbf{T}}^{\widehat{x}}_\mu \cap \{k', k'+1, k'+2, \dots\} \neq \emptyset$ *and* $\mathbf{T}^x_\mu \cap [t'_1, \infty) \neq \emptyset$ *that are necessary in eventual periodicity, see also Lemma 1, page 223.*

Remark 84. *The prime period of the eventually periodic point* $\mu \in \widehat{\omega}(\widehat{x})$ *always exists, but the prime period of the eventually periodic point* $\mu \in \omega(x)$ *might not exist, for example if* x *is eventually constant and equal with* μ *(i.e. if* $\lim_{t \to \infty} x(t) = \mu$*), when* $P^x_\mu = (0, \infty)$*. We shall prove in Theorem 35, page 79 that this is the only situation when the eventually periodic point* $\mu \in \omega(x)$ *has no prime period.*

3. The Accessibility of the Eventually Periodic Points

Theorem 28. *a) Let $\widehat{x}$ and $\mu \in \widehat{\omega}(\widehat{x})$ that is eventually periodic, with the period $p \geq 1$ and the limit of periodicity $k' \in \mathbf{N}_{\_}$. For any $k \geq k'$ we have $\widehat{\mathbf{T}}_{\mu}^{\widehat{x}} \cap \{k, k+1, ..., k+p-1\} \neq \varnothing$.*

b) Let x and $\mu \in \omega(x)$ that is eventually periodic with the period $T > 0$ and the limit of periodicity $t' \in \mathbf{R}$. For any $t \geq t'$, we have $\mathbf{T}_{\mu}^{x} \cap [t, t+T) \neq \varnothing$.

Proof. a) The hypothesis implies the truth of

$$\widehat{\mathbf{T}}_{\mu}^{\widehat{x}} \cap \{k', k'+1, k'+2, ...\} \neq \varnothing, \tag{3.1}$$

$$\forall k \in \widehat{\mathbf{T}}_{\mu}^{\widehat{x}} \cap \{k', k'+1, k'+2, ...\}, \{k + zp | z \in \mathbf{Z}\} \cap \{k', k'+1, k'+2, ...\} \subset \widehat{\mathbf{T}}_{\mu}^{\widehat{x}}. \tag{3.2}$$

The truth of (3.1) allows us to define $k'' = \min \widehat{\mathbf{T}}_{\mu}^{\widehat{x}} \cap \{k', k'+1, k'+2, ...\}$ and we prove that $k'' \in \widehat{\mathbf{T}}_{\mu}^{\widehat{x}} \cap \{k', k'+1, ..., k'+p-1\}$. If, against all reason, this would not be true, then we would have $k'' \geq k'+p$ and

$$k'' - p \in \{k'' + zp | z \in \mathbf{Z}\} \cap \{k', k'+1, k'+2, ...\} \overset{(3.2)}{\subset} \widehat{\mathbf{T}}_{\mu}^{\widehat{x}},$$

representing a contradiction with the definition of k''.

From (3.2) we infer that $\{k'', k''+p, k''+2p, ...\} \subset \widehat{\mathbf{T}}_{\mu}^{\widehat{x}} \cap \{k', k'+1, k'+2, ...\}$, meaning that $\forall k \geq k'$, $\widehat{\mathbf{T}}_{\mu}^{\widehat{x}} \cap \{k, k+1, ..., k+p-1\} \neq \varnothing$.

b) We have from the hypothesis that

$$\mathbf{T}_{\mu}^{x} \cap [t', \infty) \neq \varnothing, \tag{3.3}$$

$$\forall t \in \mathbf{T}_{\mu}^{x} \cap [t', \infty), \{t + zT | z \in \mathbf{Z}\} \cap [t', \infty) \subset \mathbf{T}_{\mu}^{x} \tag{3.4}$$

are fulfilled. The request (3.3) allows defining $t'' = \min \mathbf{T}_{\mu}^{x} \cap [t', \infty)$. We show that $t'' \in \mathbf{T}_{\mu}^{x} \cap [t', t'+T)$. If, against all reason, this would not be true, then we would have $t'' \geq t'+T$. This means that $t'' - T \geq t'$, thus

$$t'' - T \in \{t'' + zT | z \in \mathbf{Z}\} \cap [t', \infty) \overset{(3.4)}{\subset} \mathbf{T}_{\mu}^{x},$$

contradiction with the definition of t''.

By using (3.4) we get $\{t'', t''+T, t''+2T, ...\} \subset \mathbf{T}_{\mu}^{x} \cap [t', \infty)$. The statement of the theorem holds. □

4. The Limit of Periodicity

Theorem 29. *a) $\widehat{x} \in \widehat{S}^{(n)}, \mu \in \widehat{\omega}(\widehat{x}), p \geq 1, p' \geq 1, k' \in \mathbf{N}_{\_}, k'' \in \mathbf{N}_{\_}$ are given. If*

$$\begin{cases} \forall k \in \widehat{\mathbf{T}}_{\mu}^{\widehat{x}} \cap \{k', k'+1, k'+2, ...\}, \\ \{k + zp | z \in \mathbf{Z}\} \cap \{k', k'+1, k'+2, ...\} \subset \widehat{\mathbf{T}}_{\mu}^{\widehat{x}}, \end{cases} \tag{4.1}$$

$$\begin{cases} \forall k \in \widehat{\mathbf{T}}_{\mu}^{\widehat{x}} \cap \{k'', k''+1, k''+2, ...\}, \\ \{k + zp' | z \in \mathbf{Z}\} \cap \{k'', k''+1, k''+2, ...\} \subset \widehat{\mathbf{T}}_{\mu}^{\widehat{x}} \end{cases} \tag{4.2}$$

hold, then

$$\begin{cases} \forall k \in \widehat{\mathbf{T}}_{\mu}^{\widehat{x}} \cap \{k', k'+1, k'+2, ...\}, \\ \{k + zp' | z \in \mathbf{Z}\} \cap \{k', k'+1, k'+2, ...\} \subset \widehat{\mathbf{T}}_{\mu}^{\widehat{x}} \end{cases} \tag{4.3}$$

is true.

b) Let $x \in S^{(n)}, \mu \in \omega(x), T > 0, T' > 0, t' \in \mathbf{R}, t'' \in \mathbf{R}$. *Then*

$$\forall t \in \mathbf{T}_{\mu}^{x} \cap [t', \infty), \{t + zT | z \in \mathbf{Z}\} \cap [t', \infty) \subset \mathbf{T}_{\mu}^{x}, \tag{4.4}$$

$$\forall t \in \mathbf{T}_{\mu}^{x} \cap [t'', \infty), \{t + zT' | z \in \mathbf{Z}\} \cap [t'', \infty) \subset \mathbf{T}_{\mu}^{x} \tag{4.5}$$

imply

$$\forall t \in \mathbf{T}_{\mu}^{x} \cap [t', \infty), \{t + zT' | z \in \mathbf{Z}\} \cap [t', \infty) \subset \mathbf{T}_{\mu}^{x}. \tag{4.6}$$

Proof. b) Let $t \in \mathbf{T}_{\mu}^{x}, z \in \mathbf{Z}$ arbitrary such that $t \geq t'$ and $t + zT' \geq t'$. We have the following possibilities.

Case $t' \geq t''$

Then $t \geq t''$ and $t + zT' \geq t''$, thus $t + zT' \overset{(4.5)}{\in} \mathbf{T}_{\mu}^{x}$.

Case $t' < t''$

Then $k \in \mathbf{N}$ exists with $t + kT \geq t'', t + zT' + kT \geq t''$. Obviously $t + kT \geq t'$ and we can write

$$\mu = x(t) \overset{(4.4)}{=} x(t + kT) \overset{(4.5)}{=} x(t + zT' + kT) \overset{(4.4)}{=} x(t + zT'),$$

in other words $t + zT' \in \mathbf{T}_{\mu}^{x}$. □

Remark 85. *The previous Theorem states that the set of the limits of periodicity does not depend on the period. In particular, this justifies the notations* $\widehat{L}_{\mu}^{\widehat{x}}, L_{\mu}^{x}$ *where the period is missing.*

Example 15. *Let the signal* $x \in S^{(1)}$,

$$x(t) = \chi_{[0,1)}(t) \oplus \chi_{[4,5)}(t) \oplus \chi_{[6,7)}(t) \oplus \chi_{[8,9)}(t) \oplus \chi_{[10,11)}(t) \oplus ...$$

where $2, 4 \in P_1^x$. *We might be tempted to think that the eventual periodicity of* 1 *with the period* $T = 4$ *has the prime limit of periodicity* t' *different from its eventual periodicity with* $T' = 2$ *that has the prime limit of periodicity* $t'' = 3$. *This is not the case, the fact that* $[2, 3) \subset \mathbf{T}_1^x$ *is false shows that both prime limits of periodicity are* $t' = 3$.

Theorem 30. *a) Let* $\widehat{x} \in \widehat{S}^{(n)}$ *and* $\mu \in \widehat{\omega}(\widehat{x})$ *with the property that* $\widehat{L}_{\mu}^{\widehat{x}} \neq \emptyset$. *Then* $k' \in \mathbf{N}_{\_}$ *exists with* $\widehat{L}_{\mu}^{\widehat{x}} = \{k', k'+1, k'+2, ...\}$.

b) Let $x \in S^{(n)}$ *non constant and* $\mu \in \omega(x)$ *having the property that* $L_{\mu}^{x} \neq \emptyset$. *Then* $t' \in \mathbf{R}$ *exists such that* $L_{\mu}^{x} = [t', \infty)$.

Proof. a) The statement is a consequence of Lemma 3, page 224.

b) Because x is not constant, $t_0 \in \mathbf{R}$ exists with $I^x = (-\infty, t_0)$. Let $T \in P_{\mu}^{x}$.

b.i) We show first that $t_0 - T \notin L_{\mu}^{x}$ and we suppose against all reason that $t_0 - T \in L_{\mu}^{x}$. We have two possibilities.

Case $\mu = x(-\infty + 0)$

The hypothesis $t_0 - T \in L_\mu^x$ implies, as far as $t_0 - T \in \mathbf{T}_\mu^x$,

$$\mu = x(t_0 - T) = x(t_0),$$

representing a contradiction with the fact that $t_0 \notin I^x$.

Case $\mu \neq x(-\infty+0)$

We infer from Theorem 28, page 69 that $\mathbf{T}_\mu^x \cap [t_0 - T, t_0) \neq \varnothing$, wherefrom $\mu = x(-\infty+0)$, representing a contradiction.

b.ii) From b.i) and from Lemma 3, we draw the conclusion that L_μ^x has one of the forms $L_\mu^x = (t', \infty), L_\mu^x = [t', \infty)$, where $t' > t_0 - T$. We show that the first possibility cannot take place, thus we suppose against all reason that t' exists with $L_\mu^x = (t', \infty)$. We have the existence of $\varepsilon' > 0, \varepsilon'' > 0$ such that

$$\forall t \in (t', t' + \varepsilon'), x(t) = x(t'), \tag{4.7}$$

$$\forall t \in (t' + T, t' + T + \varepsilon''), x(t) = x(t' + T) \tag{4.8}$$

and let $\varepsilon \in (0, \min\{\varepsilon', \varepsilon''\})$. Two possibilities exist.

Case $x(t') = \mu$

We have $t' \notin L_\mu^x$, thus $x(t' + T) \neq \mu$ and $(t', t' + \varepsilon) \subset L_\mu^x$ means that

$$\forall t \in (t', t' + \varepsilon), x(t) = \mu \Longrightarrow x(t) = x(t + T). \tag{4.9}$$

Let $t \in (t', t' + \varepsilon)$ arbitrary. We can write

$$\mu = x(t') \overset{(4.7)}{=} x(t) \overset{(4.9)}{=} x(t + T) \overset{(4.8)}{=} x(t' + T),$$

contradiction.

Case $x(t') \neq \mu$

In this case two possibilities exist. The case $x(t' + T) = \mu$ when $(t', t' + \varepsilon) \subset L_\mu^x$ means the truth of (4.9). Let $t \in (t', t' + \varepsilon)$ arbitrary. We conclude that

$$\mu = x(t' + T) \overset{(4.8)}{=} x(t + T) \overset{(4.9)}{=} x(t) \overset{(4.7)}{=} x(t'),$$

representing a contradiction. And the case $x(t' + T) \neq \mu$ when $\forall k \in \mathbf{N}, x(t' + kT) \neq \mu$. As for any $t \in \mathbf{T}_\mu^x \cap (t', \infty) = \mathbf{T}_\mu^x \cap [t', \infty)$, we have $\{t + zT | z \in \mathbf{Z}\} \cap (t', \infty) = \{t + zT | z \in \mathbf{Z}\} \cap [t', \infty)$, the conclusion is $t' \in L_\mu^x$, contradiction.

It has resulted that the existence of $t' > t_0 - T$ with $L_\mu^x = [t', \infty)$ is the only possibility. □

5. A Property of Eventual Constancy

Theorem 31. *We consider the signals* $\widehat{x}, x$.

a) Let $\mu \in \widehat{\omega}(\widehat{x})$. *If* $k' \in \mathbf{N}\_$ *exists making*

$$\forall k \in \widehat{\mathbf{T}}_\mu^{\widehat{x}} \cap \{k', k'+1, k'+2, ...\}, \{k + zp | z \in \mathbf{Z}\} \cap \{k', k'+1, k'+2, ...\} \subset \widehat{\mathbf{T}}_\mu^{\widehat{x}} \tag{5.1}$$

true for $p = 1$, *then*

$$\forall k \geq k', \widehat{x}(k) = \mu \tag{5.2}$$

and (5.1) holds for any $p \geq 1$.

b) Let $\mu \in \omega(x)$ *and we suppose that* $t_0 \in \mathbf{R}, h > 0$ *exist such that x is of the form*

$$\begin{aligned} x(t) = x(-\infty+0)\cdot\chi_{(-\infty,t_0)}(t)\oplus x(t_0)\cdot\chi_{[t_0,t_0+h)}(t)\oplus ...\\ ...\oplus x(t_0+kh)\cdot\chi_{[t_0+kh,t_0+(k+1)h)}(t)\oplus ... \end{aligned} \tag{5.3}$$

If $t' \in \mathbf{R}$, $T \in (0,h)\cup(h,2h)\cup...\cup(qh,(q+1)h)\cup...$ *exist making*

$$\forall t \in \mathbf{T}_{\mu}^{x} \cap [t', \infty), \{t + zT | z \in \mathbf{Z}\} \cap [t', \infty) \subset \mathbf{T}_{\mu}^{x} \tag{5.4}$$

true, then

$$\forall t \geq t', x(t) = \mu \tag{5.5}$$

is true and in this case (5.4) holds for any $T > 0$.

c) We ask that (5.3) is fulfilled under the form

$$\begin{aligned} x(t) = \widehat{x}(-1)\cdot\chi_{(-\infty,t_0)}(t)\oplus \widehat{x}(0)\cdot\chi_{[t_0,t_0+h)}(t)\oplus ...\\ ...\oplus \widehat{x}(k)\cdot\chi_{[t_0+kh,t_0+(k+1)h)}(t)\oplus ... \end{aligned} \tag{5.6}$$

and let $\mu \in \widehat{\omega}(\widehat{x}) = \omega(x)$[2] *be arbitrary. The following statements hold:*

c.1) If $k' \in \mathbf{N}\_$ *exists making (5.1) true for* $p = 1$, *then (5.2) is true, (5.1) holds for any* $p \geq 1$ *and* $t' \in \mathbf{R}$ *exists such that (5.5) holds and (5.4) is also true for any* $T > 0$.

c.2) If $t' \in \mathbf{R}$, $T \in (0,h)\cup(h,2h)\cup...\cup(qh,(q+1)h)\cup...$ *exist making (5.4) true, then (5.5) is true, (5.4) is true for any* $T > 0$ *and* $k' \in \mathbf{N}\_$ *exists such that (5.2) is true and (5.1) is also true for any* $p \geq 1$.

Proof. a) Some $k' \in \mathbf{N}\_$ exists with (5.1) fulfilled for $p = 1$, meaning that

$$\forall k, \forall z, (k \in \widehat{\mathbf{T}}_{\mu}^{\widehat{x}} \text{ and } z \in \mathbf{Z} \text{ and } k \geq k' \text{ and } k + z \geq k') \Longrightarrow k + z \in \widehat{\mathbf{T}}_{\mu}^{\widehat{x}}$$

holds. $\mu \in \widehat{\omega}(\widehat{x})$ implies that $\widehat{\mathbf{T}}_{\mu}^{\widehat{x}}$ is infinite, thus some $k \in \widehat{\mathbf{T}}_{\mu}^{\widehat{x}}, k \geq k'$ exists indeed. We infer

$$\{k', k'+1, k'+2, ...\} \subset \widehat{\mathbf{T}}_{\mu}^{\widehat{x}}.$$

We have obtained the truth of (5.2). In these circumstances (5.1) holds for any $p \geq 1$.

b) We suppose that $t_0 \in \mathbf{R}, h > 0$ exist such that (5.3) holds and also that $t' \in \mathbf{R}$, $T \in (0,h)\cup(h,2h)\cup...\cup(qh,(q+1)h)\cup...$ exist such that (5.4) is true. Furthermore, $\mu \in \omega(x)$ implies $\mathbf{T}_{\mu}^{x} \cap [t', \infty) \neq \varnothing$, since $\mathbf{T}_{\mu}^{x}$ is unbounded from above.

We show first the existence of $\bar{t} \in \mathbf{R}$ such that

$$\forall t \geq \bar{t}, x(t) = \mu \tag{5.7}$$

is true.

[2]The fact that (5.6) implies $\widehat{\omega}(\widehat{x}) = \omega(x)$ was proved in Theorem 6, page 11.

We have the existence of $k' \in \mathbf{N}\_$ such that $t_0 + k'h \geq t'$, $x(t_0 + k'h) = \mu$ and

$$\forall t \in [t_0 + k'h, t_0 + (k'+1)h), x(t) = \mu. \tag{5.8}$$

Case $T \in (0, h)$, when

$$t_0 + k'h < t_0 + (k'+1)h - T < t_0 + (k'+1)h, \tag{5.9}$$

$$\mu \overset{(5.8)}{=} x(t_0 + k'h) \overset{(5.9)}{=} x(t_0 + (k'+1)h - T) \overset{(5.4)}{=} x(t_0 + (k'+1)h), \tag{5.10}$$

$$\forall t \in [t_0 + (k'+1)h, t_0 + (k'+2)h), x(t) \overset{(5.10)}{=} \mu; \tag{5.11}$$

$$t_0 + (k'+1)h < t_0 + (k'+2)h - T < t_0 + (k'+2)h, \tag{5.12}$$

$$\mu \overset{(5.10)}{=} x(t_0 + (k'+1)h) \overset{(5.12)}{=} x(t_0 + (k'+2)h - T) \overset{(5.4)}{=} x(t_0 + (k'+2)h), \tag{5.13}$$

$$\forall t \in [t_0 + (k'+2)h, t_0 + (k'+3)h), x(t) \overset{(5.13)}{=} \mu; \tag{5.14}$$

...

Thus the statement (5.7) holds, from (5.8), (5.11), (5.14),... for $\bar{t} = t_0 + k'h$.

Case $T \in (h, 2h)$, when

$$t_0 + (k'+1)h < t_0 + k'h + T < t_0 + (k'+2)h, \tag{5.15}$$

$$\mu \overset{(5.8)}{=} x(t_0 + k'h) \overset{(5.4)}{=} x(t_0 + k'h + T) \overset{(5.15)}{=} x(t_0 + (k'+1)h), \tag{5.16}$$

$$\forall t \in [t_0 + (k'+1)h, t_0 + (k'+2)h), x(t) \overset{(5.16)}{=} \mu; \tag{5.17}$$

$$t_0 + (k'+2)h < t_0 + (k'+1)h + T < t_0 + (k'+3)h, \tag{5.18}$$

$$\mu \overset{(5.16)}{=} x(t_0 + (k'+1)h) \overset{(5.4)}{=} x(t_0 + (k'+1)h + T) \overset{(5.18)}{=} x(t_0 + (k'+2)h), \tag{5.19}$$

$$\forall t \in [t_0 + (k'+2)h, t_0 + (k'+3)h), x(t) \overset{(5.19)}{=} \mu; \tag{5.20}$$

...

The statement (5.7) holds, from (5.8), (5.17), (5.20),... for $\bar{t} = t_0 + k'h$.

Case $T \in (2h, 3h)$. In this situation

$$t_0 + (k'+2)h < t_0 + k'h + T < t_0 + (k'+3)h, \tag{5.21}$$

$$\mu \overset{(5.8)}{=} x(t_0 + k'h) \overset{(5.4)}{=} x(t_0 + k'h + T) \overset{(5.21)}{=} x(t_0 + (k'+2)h), \tag{5.22}$$

$$\forall t \in [t_0 + (k'+2)h, t_0 + (k'+3)h), x(t) \overset{(5.22)}{=} \mu; \tag{5.23}$$

$$t_0 + (k'+4)h < t_0 + (k'+2)h + T < t_0 + (k'+5)h, \tag{5.24}$$

$$\mu \overset{(5.22)}{=} x(t_0 + (k'+2)h) \overset{(5.4)}{=} x(t_0 + (k'+2)h + T) \overset{(5.24)}{=} x(t_0 + (k'+4)h), \tag{5.25}$$

$$\forall t \in [t_0 + (k'+4)h, t_0 + (k'+5)h), x(t) \overset{(5.25)}{=} \mu; \tag{5.26}$$

$$t_0+(k'+6)h < t_0+(k'+4)h+T < t_0+(k'+7)h, \quad (5.27)$$

...

$$\forall j \in \mathbf{N}, \forall t \in [t_0+(k'+2j)h, t_0+(k'+2j+1)h), x(t)=\mu. \quad (5.28)$$

Furthermore

$$t_0+(k'+1)h < t_0+(k'+4)h-T < t_0+(k'+2)h, \quad (5.29)$$

$$\mu \overset{(5.28)}{=} x(t_0+(k'+4)h) \overset{(5.4)}{=} x(t_0+(k'+4)h-T) \overset{(5.29)}{=} x(t_0+(k'+1)h), \quad (5.30)$$

$$\forall t \in [t_0+(k'+1)h, t_0+(k'+2)h), x(t) \overset{(5.30)}{=} \mu; \quad (5.31)$$

$$t_0+(k'+3)h < t_0+(k'+1)h+T < t_0+(k'+4)h, \quad (5.32)$$

...

and we prove that

$$\forall j \in \mathbf{N}, \forall t \in [t_0+(k'+2j+1)h, t_0+(k'+2j+2)h), x(t)=\mu \quad (5.33)$$

similarly with (5.21),...,(5.28), starting from $x(t_0+(k'+1)h) \overset{(5.31)}{=} \mu$ instead of $x(t_0+k'h) \overset{(5.8)}{=} \mu$. From (5.28), (5.33) we infer that the statement (5.7) is true for $\bar{t} = t_0+k'h$.

Case $T \in (3h, 4h)$,

$$t_0+(k'+3)h < t_0+k'h+T < t_0+(k'+4)h, \quad (5.34)$$

$$\mu \overset{(5.8)}{=} x(t_0+k'h) \overset{(5.4)}{=} x(t_0+k'h+T) \overset{(5.34)}{=} x(t_0+(k'+3)h), \quad (5.35)$$

$$\forall t \in [t_0+(k'+3)h, t_0+(k'+4)h), x(t) \overset{(5.35)}{=} \mu; \quad (5.36)$$

$$t_0+(k'+6)h < t_0+(k'+3)h+T < t_0+(k'+7)h, \quad (5.37)$$

$$\mu \overset{(5.35)}{=} x(t_0+(k'+3)h) \overset{(5.4)}{=} x(t_0+(k'+3)h+T) \overset{(5.37)}{=} x(t_0+(k'+6)h), \quad (5.38)$$

$$\forall t \in [t_0+(k'+6)h, t_0+(k'+7)h), x(t) \overset{(5.38)}{=} \mu; \quad (5.39)$$

$$t_0+(k'+9)h < t_0+(k'+6)h+T < t_0+(k'+10)h, \quad (5.40)$$

...

$$\forall j \in \mathbf{N}, \forall t \in [t_0+(k'+3j)h, t_0+(k'+3j+1)h), x(t)=\mu. \quad (5.41)$$

Furthermore,

$$t_0+(k'+2)h < t_0+(k'+6)h-T < t_0+(k'+3)h, \quad (5.42)$$

$$\mu \overset{(5.41)}{=} x(t_0+(k'+6)h) \overset{(5.4)}{=} x(t_0+(k'+6)h-T) \overset{(5.42)}{=} x(t_0+(k'+2)h) \quad (5.43)$$

...

and we remake the reasoning (5.34),...,(5.41) starting from $x(t_0+(k'+2)h) \overset{(5.42)}{=} \mu$ instead of $x(t_0+k'h) \overset{(5.8)}{=} \mu$. We obtain:

$$\forall j \in \mathbf{N}, \forall t \in [t_0+(k'+3j+2)h, t_0+(k'+3j+3)h), x(t) = \mu \tag{5.44}$$

and we have also

$$t_0+(k'+1)h < t_0+(k'+5)h-T < t_0+(k'+2)h, \tag{5.45}$$

$$\mu \overset{(5.44)}{=} x(t_0+(k'+5)h) \overset{(5.4)}{=} x(t_0+(k'+5)h-T) \overset{(5.45)}{=} x(t_0+(k'+1)h), \tag{5.46}$$

$$\ldots$$

We remake the reasoning (5.34),...,(5.41) starting from $x(t_0+(k'+1)h) \overset{(5.46)}{=} \mu$ instead of $x(t_0+k'h) \overset{(5.8)}{=} \mu$. We get:

$$\forall j \in \mathbf{N}, \forall t \in [t_0+(k'+3j+1)h, t_0+(k'+3j+2)h), x(t) = \mu. \tag{5.47}$$

From (5.41), (5.44), (5.47) we have the truth of (5.7) for $\bar{t} = t_0+k'h$.

In the general case $T \in (qh,(q+1)h)$, $q \geq 2$ we prove in succession the truth of

$$\forall j \in \mathbf{N}, \forall t \in [t_0+(k'+qj)h, t_0+(k'+qj+1)h), x(t) = \mu,$$

$$\forall j \in \mathbf{N}, \forall t \in [t_0+(k'+qj+q-1)h, t_0+(k'+qj+q)h), x(t) = \mu,$$

$$\ldots$$

$$\forall j \in \mathbf{N}, \forall t \in [t_0+(k'+qj+1)h, t_0+(k'+qj+2)h), x(t) = \mu,$$

wherefrom the truth of (5.7) follows for $\bar{t} = t_0+k'h$.

We prove now that in (5.5) we can take $\bar{t} = t'$. Let us suppose, against all reason, that this is not true, i.e. $\bar{t} > t'$[3] and some $t'' \in [t',\bar{t})$ exists with $x(t'') \neq \mu$. Let $q \geq 1$ with the property that $t''+qT \geq \bar{t}$, in other words $t''+qT \in \mathbf{T}^x_\mu \cap [t',\infty)$. Then

$$t''+qT-qT \in \{t''+qT+zT \mid z \in \mathbf{Z}\} \cap [t',\infty) \subset \mathbf{T}^x_\mu$$

and we infer that $x(t'') = \mu$, contradiction. Statement (5.5) is proved and obviously (5.4) holds for any $T > 0$.

c) This is a consequence of a) and b). □

---

[3]if $\bar{t} < t'$, the other way of negating $\bar{t} = t'$, then from $\forall t \geq \bar{t}, x(t) = \mu$, we can write $\forall t \geq t', x(t) = \mu$, i.e. finally we can take $\bar{t} = t'$.

6. Discrete Time vs Real Time

Theorem 32. *We consider the signals $\widehat{x} \in \widehat{S}^{(n)}, x \in S^{(n)}$ which are not eventually constant and we suppose that*

$$x(t) = \widehat{x}(-1) \cdot \chi_{(-\infty,t_0)}(t) \oplus \widehat{x}(0) \cdot \chi_{[t_0,t_0+h)}(t) \oplus ...$$
$$... \oplus \widehat{x}(k) \cdot \chi_{[t_0+kh,t_0+(k+1)h)}(t) \oplus ... \tag{6.1}$$

is true, where $t_0 \in \mathbf{R}, h > 0$. Let $\mu \in \widehat{\omega}(\widehat{x}) = \omega(x)$.

a) If $p \geq 1$ and $k' \in \mathbf{N}_{\_}$ exist such that

$$\forall k \in \widehat{\mathbf{T}}^{\widehat{x}}_{\mu} \cap \{k', k'+1, k'+2, ...\}, \{k + zp | z \in \mathbf{Z}\} \cap \{k', k'+1, k'+2, ...\} \subset \widehat{\mathbf{T}}^{\widehat{x}}_{\mu}, \tag{6.2}$$

then $t' \in \mathbf{R}$ exists with

$$\forall t \in \mathbf{T}^{x}_{\mu} \cap [t', \infty), \{t + zT | z \in \mathbf{Z}\} \cap [t', \infty) \subset \mathbf{T}^{x}_{\mu} \tag{6.3}$$

true for $T = ph$.

b) If $T > 0$ and $t' \in \mathbf{R}$ exist for which (6.3) is true, then $\frac{T}{h} \in \{1, 2, 3, ...\}$ and $k' \in \mathbf{N}_{\_}$ exists such that (6.2) is true for $p = \frac{T}{h}$.

Proof. a) The hypothesis states the existence of $t_0 \in \mathbf{R}, h > 0$ such that (6.1) is true and also that, given $\mu \in \widehat{\omega}(\widehat{x}) = \omega(x)$, $p \geq 1$ and $k' \in \mathbf{N}_{\_}$ exist with (6.2) fulfilled.

We define $T = ph, t' = t_0 + k'h$. Let $t \in \mathbf{T}^{x}_{\mu}, z \in \mathbf{Z}$ be arbitrary with the property that $t \geq t', t + zT \geq t'$ ($\mu \in \omega(x)$ implies that $\mathbf{T}^{x}_{\mu} \cap [t', \infty) \neq \emptyset$). Some $k \geq k'$ exists then such that $t \in [t_0 + kh, t_0 + (k+1)h)$ and we can write

$$t + zT \in [t_0 + kh + zT, t_0 + (k+1)h + zT) = [t_0 + (k+zp)h, t_0 + (k+1+zp)h).$$

Obviously $t_0 + (k+zp)h \geq t_0 + k'h = t'$ implies $k + zp \geq k'$. We infer

$$\mu = x(t) = \widehat{x}(k) \overset{(6.2)}{=} \widehat{x}(k+zp) = x(t+zT),$$

in other words (6.3) holds.

b) Some $t_0 \in \mathbf{R}$ and $h > 0$ exist from the hypothesis such that (6.1) is true and, given μ, some $T > 0, t' \in \mathbf{R}$ exist also such that (6.3) holds. If $T \in (0,h) \cup (h,2h) \cup ... \cup (qh,(q+1)h) \cup ...$ then from Theorem 31 b), page 71, we have that $\lim\limits_{t\to\infty} x(t) = \lim\limits_{k\to\infty} \widehat{x}(k) = \mu$, contradiction with the hypothesis, thus $T \in \{h, 2h, 3h, ...\}$ for which we define $p = \frac{T}{h}, p \geq 1$. As $\mu \in \omega(x), \mathbf{T}^{x}_{\mu}$ is unbounded from above and $\mathbf{T}^{x}_{\mu} \cap [t', \infty) \neq \emptyset$ is true for any t'. We can suppose, by making use of Lemma 3, page 224 that in (6.3) we have $t' \geq t_0 - h$ and we denote by $k' \in \mathbf{N}_{\_}$ the number for which $t' \in [t_0 + k'h, t_0 + (k'+1)h)$.

Let now $k \in \widehat{\mathbf{T}}^{\widehat{x}}_{\mu}$ and $z \in \mathbf{Z}$ be arbitrary with $k \geq k'$ and $k + zp \geq k'$ ($\widehat{\mathbf{T}}^{\widehat{x}}_{\mu} \cap \{k', k'+1, k'+2, ...\} \neq \emptyset$ because $\mu \in \widehat{\omega}(\widehat{x}) = \omega(x)$ and $\widehat{\mathbf{T}}^{\widehat{x}}_{\mu}$ is infinite). Then

$$t' + (k-k')h \geq t', \tag{6.4}$$

$$t' + (k-k'+zp)h \geq t' \tag{6.5}$$

and on the other hand

$$t_0 + kh \leq t' + (k - k')h < t_0 + (k+1)h, \tag{6.6}$$

$$t_0 + (k + zp)h \leq t' + (k - k' + zp)h < t_0 + (k + zp + 1)h \tag{6.7}$$

are true. We conclude that

$$\mu = \widehat{x}(k) \overset{(6.6)}{=} x(t' + (k - k')h) \overset{(6.3),(6.4),(6.5)}{=}$$

$$= x(t' + (k - k')h + zT) = x(t' + (k - k' + zp)h) \overset{(6.7)}{=} \widehat{x}(k + zp),$$

thus (6.2) holds. □

Example 16. *Let the signal* $\widehat{x} \in \widehat{S}^{(1)}$ *such that*

$$\widehat{x} = \underset{-1}{0}, \underset{0}{0}, \underset{1}{0}, \underset{2}{0}, \underset{3}{1}, \underset{4}{\lambda}, \underset{5}{\lambda}, \underset{6}{1}, \underset{7}{\lambda}, \underset{8}{\lambda}, \underset{9}{1}, \underset{10}{\lambda}, \ldots$$

where $\lambda \in \mathbf{B}$. *Then* 1 *is an eventually periodic point of* $\widehat{x}$ *and* $p = 3$ *is a period. If* $\lambda = 0$ *then its prime period is* 3 *and if* $\lambda = 1$, *then its prime period is* 1 *and its prime limit of periodicity is* $k' = 1$.

7. Support Sets vs Sets of Periods

Remark 86. *Let* $x, y \in S^{(n)}$ *be two signals and* $\mu \in \omega(x) \cap \omega(y)$. *One might be tempted to think that implications of the kind*

$$\mathbf{T}^x_\mu = \mathbf{T}^y_\mu \Longrightarrow P^x_\mu = P^y_\mu, \tag{7.1}$$

$$P^x_\mu = P^y_\mu \Longrightarrow \mathbf{T}^x_\mu = \mathbf{T}^y_\mu \tag{7.2}$$

hold and the purpose of this section is that of understanding them better. We give real time examples, keeping in mind that the same statements hold in discrete time too.

Example 17. *We suppose that* $Or(x) = Or(y) = \{\mu, \mu', \mu''\}$ *and let*

$$x(t) = \mu' \cdot \chi_{(-\infty,2)}(t) \oplus \mu'' \cdot \chi_{[2,3)}(t) \oplus \mu \cdot \chi_{[3,4)}(t) \oplus \mu'' \cdot \chi_{[4,6)}(t)$$

$$\oplus \mu \cdot \chi_{[6,7)}(t) \oplus \mu'' \cdot \chi_{[7,9)}(t) \oplus \mu \cdot \chi_{[9,10)}(t) \oplus \ldots$$

$$y(t) = \mu' \cdot \chi_{(-\infty,0)}(t) \oplus \mu'' \cdot \chi_{[0,3)}(t) \oplus \mu \cdot \chi_{[3,4)}(t) \oplus \mu'' \cdot \chi_{[4,6)}(t)$$

$$\oplus \mu \cdot \chi_{[6,7)}(t) \oplus \mu'' \cdot \chi_{[7,9)}(t) \oplus \mu \cdot \chi_{[9,10)}(t) \oplus \ldots$$

We see that $I^x = (-\infty, 2)$, $I^y = (-\infty, 0)$, $\mathbf{T}^x_\mu = \mathbf{T}^y_\mu = [3,4) \cup [6,7) \cup [9,10) \cup \ldots$, $P^x_\mu = P^y_\mu = \{3, 6, 9, \ldots\}$ *and* $L^x_\mu = L^y_\mu = [1, \infty)$. *The fact that* μ *is a periodic point of* x *is expressed by the non-empty intersection* $I^x \cap L^x_\mu = [1,2)$ *and the fact that* μ *is an eventually periodic point of* y *only follows from* $I^y \cap L^y_\mu = \varnothing$. *The interpretation of (7.1) according to this example is: the implication* $\mathbf{T}^x_\mu = \mathbf{T}^y_\mu \Longrightarrow P^x_\mu = P^y_\mu$ *takes place, however* μ *may be a periodic point of* x *and an eventually periodic point of* y.

Example 18. *We take*

$$\mathbf{T}_\mu^x = (-\infty, 2) \cup [4,5) \cup [9,10) \cup [14,15) \cup ...$$

$$\mathbf{T}_\mu^y = (-\infty, 1) \cup [2,3) \cup [4,5) \cup [7,8) \cup [9,10) \cup [12,13) \cup ...$$

μ is an eventually periodic point of both x, y with $P_\mu^x = P_\mu^y = \{5, 10, 15, ...\}$ and $L_\mu^x = [2, \infty)$, $L_\mu^y = [1, \infty)$. The difference between the two signals x, y consists in the fact that in $\mathbf{T}_\mu^x$ the interval $[4,5)$ repeats periodically by translations and in $\mathbf{T}_\mu^y$ the intervals $[2,3), [4,5)$ repeat periodically by translations. The periods of μ coincide for x and y and (7.2) is false.

8. Sums, Differences and Multiples of Periods

Theorem 33. *The signals $\widehat{x}, x$ are considered.*

a) Let $p, p' \geq 1, k' \in \mathbf{N}_{\_}, \mu \in \widehat{\omega}(\widehat{x})$ and we ask that

$$\forall k \in \widehat{\mathbf{T}}_\mu^{\widehat{x}} \cap \{k', k'+1, k'+2, ...\}, \{k + zp | z \in \mathbf{Z}\} \cap \{k', k'+1, k'+2, ...\} \subset \widehat{\mathbf{T}}_\mu^{\widehat{x}}, \quad (8.1)$$

$$\forall k \in \widehat{\mathbf{T}}_\mu^{\widehat{x}} \cap \{k', k'+1, k'+2, ...\}, \{k + zp' | z \in \mathbf{Z}\} \cap \{k', k'+1, k'+2, ...\} \subset \widehat{\mathbf{T}}_\mu^{\widehat{x}} \quad (8.2)$$

hold. We have $p + p' \geq 1$,

$$\begin{cases} \forall k \in \widehat{\mathbf{T}}_\mu^{\widehat{x}} \cap \{k', k'+1, k'+2, ...\}, \\ \{k + z(p+p') | z \in \mathbf{Z}\} \cap \{k', k'+1, k'+2, ...\} \subset \widehat{\mathbf{T}}_\mu^{\widehat{x}} \end{cases} \quad (8.3)$$

and if $p > p'$, then $p - p' \geq 1$,

$$\begin{cases} \forall k \in \widehat{\mathbf{T}}_\mu^{\widehat{x}} \cap \{k', k'+1, k'+2, ...\}, \\ \{k + z(p-p') | z \in \mathbf{Z}\} \cap \{k', k'+1, k'+2, ...\} \subset \widehat{\mathbf{T}}_\mu^{\widehat{x}} \end{cases} \quad (8.4)$$

hold.

b) Let $T, T' > 0$, $t' \in \mathbf{R}$, $\mu \in \omega(x)$ be arbitrary with

$$\forall t \in \mathbf{T}_\mu^x \cap [t', \infty), \{t + zT | z \in \mathbf{Z}\} \cap [t', \infty) \subset \mathbf{T}_\mu^x, \quad (8.5)$$

$$\forall t \in \mathbf{T}_\mu^x \cap [t', \infty), \{t + zT' | z \in \mathbf{Z}\} \cap [t', \infty) \subset \mathbf{T}_\mu^x \quad (8.6)$$

fulfilled. We have on one hand that $T + T' > 0$ and

$$\forall t \in \mathbf{T}_\mu^x \cap [t', \infty), \{t + z(T+T') | z \in \mathbf{Z}\} \cap [t', \infty) \subset \mathbf{T}_\mu^x \quad (8.7)$$

are true and on the other hand that $T > T'$ implies $T - T' > 0$ and

$$\forall t \in \mathbf{T}_\mu^x \cap [t', \infty), \{t + z(T-T') | z \in \mathbf{Z}\} \cap [t', \infty) \subset \mathbf{T}_\mu^x. \quad (8.8)$$

Proof. a) We prove the second implication. We take some arbitrary, fixed $k\in\widehat{\mathbf{T}}^{\widehat{x}}_{\mu}$, $z\in\mathbf{Z}$ such that $k\geq k', k+z(p-p')\geq k'$ and we have the following possibilities:

Case $z<0$

We obtain in succession $k-zp'\geq k'$, $k-zp'\overset{(8.2)}{\in}\widehat{\mathbf{T}}^{\widehat{x}}_{\mu}$, $k-zp'+zp\overset{hyp}{\geq}k'$, $k+z(p-p')\overset{(8.1)}{\in}\widehat{\mathbf{T}}^{\widehat{x}}_{\mu}$.

Case $z=0$

$k=k+z(p-p')\in\widehat{\mathbf{T}}^{\widehat{x}}_{\mu}$ trivially.

Case $z>0$

We have $k+zp\geq k'$, $k+zp\overset{(8.1)}{\in}\widehat{\mathbf{T}}^{\widehat{x}}_{\mu}$, $k+zp-zp'\overset{hyp}{\geq}k'$, $k+z(p-p')\overset{(8.2)}{\in}\widehat{\mathbf{T}}^{\widehat{x}}_{\mu}$.

b) We prove the first implication and let $t\in\mathbf{T}^{x}_{\mu}\cap[t',\infty)$, $z\in\mathbf{Z}$ be arbitrary, fixed such that $t+z(T+T')\geq t'$.

Case $z<0$

We have in succession $t+zT\geq t+z(T+T')\overset{hyp}{\geq}t', t+zT\overset{(8.5)}{\in}\mathbf{T}^{x}_{\mu}, t+z(T+T')\overset{(8.6)}{\in}\mathbf{T}^{x}_{\mu}$.

Case $z=0$

We infer $t=t+z(T+T')\in\mathbf{T}^{x}_{\mu}$.

Case $z>0$

We have $t+zT\geq t\geq t', t+zT\overset{(8.5)}{\in}\mathbf{T}^{x}_{\mu}, t+z(T+T')\overset{hyp}{\geq}t', t+z(T+T')\overset{(8.6)}{\in}\mathbf{T}^{x}_{\mu}$. □

Theorem 34. *a) Let $p,k_1\geq 1$, $k'\in\mathbf{N}_{\_}$ and $\mu\in\widehat{\omega}(\widehat{x})$. Then $p'=k_1p$ fulfills $p'\geq 1$ and*

$$\forall k\in\widehat{\mathbf{T}}^{\widehat{x}}_{\mu}\cap\{k',k'+1,k'+2,...\},\{k+zp|z\in\mathbf{Z}\}\cap\{k',k'+1,k'+2,...\}\subset\widehat{\mathbf{T}}^{\widehat{x}}_{\mu} \tag{8.9}$$

implies

$$\forall k\in\widehat{\mathbf{T}}^{\widehat{x}}_{\mu}\cap\{k',k'+1,k'+2,...\},\{k+zp'|z\in\mathbf{Z}\}\cap\{k',k'+1,k'+2,...\}\subset\widehat{\mathbf{T}}^{\widehat{x}}_{\mu}. \tag{8.10}$$

b) Let $T>0, t'\in\mathbf{R}$, $k_1\geq 1$ and $\mu\in\omega(x)$ be arbitrary. Then $T'=k_1T$ fulfills $T'>0$ and

$$\forall t\in\mathbf{T}^{x}_{\mu}\cap[t',\infty),\{t+zT|z\in\mathbf{Z}\}\cap[t',\infty)\subset\mathbf{T}^{x}_{\mu} \tag{8.11}$$

implies

$$\forall t\in\mathbf{T}^{x}_{\mu}\cap[t',\infty),\{t+zT'|z\in\mathbf{Z}\}\cap[t',\infty)\subset\mathbf{T}^{x}_{\mu}. \tag{8.12}$$

Proof. This is a consequence of Theorem 33. □

Corollary 1. *a) For any $\widehat{x},\mu\in\widehat{\omega}(\widehat{x})$ and $p\geq 1$, if $p\in\widehat{P}^{\widehat{x}}_{\mu}$, then $\{p,2p,3p,...\}\subset\widehat{P}^{\widehat{x}}_{\mu}$;*

b) for any $x,\mu\in\omega(x)$ and $T>0$, if $T\in P^{x}_{\mu}$ then $\{T,2T,3T,...\}\subset P^{x}_{\mu}$.

Proof. The corollary is a direct consequence of Theorem 34. □

9. The Set of the Periods

Theorem 35. *a) Let $\widehat{x}\in\widehat{S}^{(n)}$ and $\mu\in\widehat{\omega}(\widehat{x})$. We ask that μ is an eventually periodic point of $\widehat{x}$. Then $\widetilde{p}\geq 1$ exists such that*

$$\widehat{P}^{\widehat{x}}_{\mu}=\{\widetilde{p},2\widetilde{p},3\widetilde{p},...\}.$$

b) We suppose that the signal $x \in S^{(n)}$ is not eventually constant and let $\mu \in \omega(x)$. We ask that μ is an eventually periodic point of x. Then $\widetilde{T} > 0$ exists such that

$$P_\mu^x = \{\widetilde{T}, 2\widetilde{T}, 3\widetilde{T}, ...\}.$$

Proof. a) We denote with $\widetilde{p}$ the least element of $\widehat{P_\mu^{\widehat{x}}}$. From Corollary 1 we have the inclusion $\{\widetilde{p}, 2\widetilde{p}, 3\widetilde{p}, ...\} \subset \widehat{P_\mu^{\widehat{x}}}$. We show that $\widehat{P_\mu^{\widehat{x}}} \subset \{\widetilde{p}, 2\widetilde{p}, 3\widetilde{p}, ...\}$. We presume against all reason that this is not true, i.e. that some $p' \in \widehat{P_\mu^{\widehat{x}}} \setminus \{\widetilde{p}, 2\widetilde{p}, 3\widetilde{p}, ...\}$ exists. In these circumstances we have the existence of $k_1 \geq 1$ with $k_1\widetilde{p} < p' < (k_1+1)\widetilde{p}$. We infer that $1 \leq p' - k_1\widetilde{p} < \widetilde{p}$ and, from Theorems 33, 34, page 78 we conclude that $p' - k_1\widetilde{p} \in \widehat{P_\mu^{\widehat{x}}}$. We have obtained a contradiction with the fact that $\widetilde{p}$ is the least element of $\widehat{P_\mu^{\widehat{x}}}$.

b) The proof is made in two steps.

b.1) We show first that $\min P_\mu^x$ exists. We suppose against all reason that this is not true, namely that a strictly decreasing sequence $T_k \in P_\mu^x, k \in \mathbf{N}$ exists that is convergent to $T = \inf P_\mu^x$. As x is not eventually constant, the following property is true:

$$\forall t \in \mathbf{R}, \exists t'' > t, x(t''-0) \neq x(t'') = \mu, \tag{9.1}$$

see Lemma 7, page 226. The hypothesis states the existence $\forall k \in \mathbf{N}$, of $t'_k \in \mathbf{R}$ with

$$\forall t \in \mathbf{T}_\mu^x \cap [t'_k, \infty), \{t + zT_k | z \in \mathbf{Z}\} \cap [t'_k, \infty) \subset \mathbf{T}_\mu^x. \tag{9.2}$$

We can suppose, as t'_k do not depend on T_k (see Theorem 29, page 69) that they have a common value t'. From (9.1) we infer that we can take some $t'' > t'$ with $x(t''-0) \neq x(t'') = \mu$ and, since $\mu \in \omega(x)$, we can apply Lemma 8, page 227 stating

$$\forall k \in \mathbf{N}, x(t''+T_k-0) \neq x(t''+T_k) = \mu. \tag{9.3}$$

We infer from Lemma 9, page 227 that $N \in \mathbf{N}$ exists with $\forall k \geq N$,

$$x(t''+T_k-0) = x(t''+T_k) = x(t''+T),$$

contradiction with (9.3). It has resulted that such a sequence $T_k, k \in \mathbf{N}$ does not exist, thus P_μ^x has a minimum that we denote by $\widetilde{T}$.

b.2) The inclusion $\{\widetilde{T}, 2\widetilde{T}, 3\widetilde{T}, ...\} \subset P_\mu^x$ results from Corollary 1, we prove the inclusion $P_\mu^x \subset \{\widetilde{T}, 2\widetilde{T}, 3\widetilde{T}, ...\}$. We suppose against all reason that some $T' \in P_\mu^x \setminus \{\widetilde{T}, 2\widetilde{T}, 3\widetilde{T}, ...\}$ exists and let $k_1 \geq 1$ with the property $T' \in (k_1\widetilde{T}, (k_1+1)\widetilde{T})$. We infer that $0 < T' - k_1\widetilde{T} < \widetilde{T}$ and, from Theorems 33, 34, we get $T' - k_1\widetilde{T} \in P_\mu^x$. We have obtained a contradiction, since $\widetilde{T}$ was defined to be the minimum of P_μ^x. $P_\mu^x = \{\widetilde{T}, 2\widetilde{T}, 3\widetilde{T}, ...\}$ holds. □

Theorem 36. *We suppose that the relation between $\widehat{x}$ and x is given by*

$$x(t) = \widehat{x}(-1) \cdot \chi_{(-\infty,t_0)}(t) \oplus \widehat{x}(0) \cdot \chi_{[t_0,t_0+h)}(t) \oplus \widehat{x}(1) \cdot \chi_{[t_0+h,t_0+2h)}(t) \oplus ...$$
$$... \oplus \widehat{x}(k) \cdot \chi_{[t_0+kh,t_0+(k+1)h)}(t) \oplus ...$$

where $t_0 \in \mathbf{R}$ and $h > 0$ and that $\mu \in \widehat{\omega}(\widehat{x}) = \omega(x)$ is an eventually periodic point of any of $\widehat{x}, x$. Then two possibilities exist:

a) $\widehat{x}$, x are both eventually constant, $\widehat{P_\mu^{\widehat{x}}} = \{1, 2, 3, ...\}$ and $P_\mu^x = (0, \infty)$;

b) none of $\widehat{x}$, x is eventually constant, $\min \widehat{P_\mu^{\widehat{x}}} = p > 1$ and $\min P_\mu^x = T = ph$.

Proof. The signals $\widehat{x},x$ are indeed simultaneously eventually constant or not, as we have noticed in Theorem 22, page 45. We suppose that they are not eventually constant and we prove b). From Theorem 32, page 76 we know that $p \in \widehat{P}_{\mu}^{\widehat{x}} \Longrightarrow T = ph \in P_{\mu}^{x}$ and conversely, $T \in P_{\mu}^{x} \Longrightarrow p = \frac{T}{h} \in \widehat{P}_{\mu}^{\widehat{x}}$. From Theorem 35 we get $\widehat{P}_{\mu}^{\widehat{x}} = \{p, 2p, 3p, ...\}$ and $P_{\mu}^{x} = \{T, 2T, 3T, ...\}$, thus $T = ph$. □

10. Necessity Conditions of Eventual Periodicity

Theorem 37. *Let $\widehat{x} \in \widehat{S}^{(n)}$ be not eventually constant. For $\mu \in \widehat{\omega}(\widehat{x})$, $p \geq 1$ and $k' \in \mathbf{N}_{\_}$ we suppose that*

$$\forall k \in \widehat{\mathbf{T}}_{\mu}^{\widehat{x}} \cap \{k', k'+1, k'+2, ...\}, \{k + zp | z \in \mathbf{Z}\} \cap \{k', k'+1, k'+2, ...\} \subset \widehat{\mathbf{T}}_{\mu}^{\widehat{x}} \tag{10.1}$$

holds. Then $n_1, n_2, ..., n_{k_1} \in \{k', k'+1, ..., k'+p-1\}, k_1 \geq 1$ exist such that

$$\widehat{\mathbf{T}}_{\mu}^{\widehat{x}} \cap \{k', k'+1, k'+2, ...\} = \bigcup_{k \in \mathbf{N}} \{n_1 + kp, n_2 + kp, ..., n_{k_1} + kp\}. \tag{10.2}$$

Proof. We apply Theorem 28, page 69 written for $k = k'$ and we obtain that $\widehat{\mathbf{T}}_{\mu}^{\widehat{x}} \cap \{k', k'+1, ..., k'+p-1\} \neq \varnothing$, wherefrom we have the existence of $n_1, n_2, ..., n_{k_1}, k_1 \geq 1$ with

$$\widehat{\mathbf{T}}_{\mu}^{\widehat{x}} \cap \{k', k'+1, ..., k'+p-1\} = \{n_1, n_2, ..., n_{k_1}\}. \tag{10.3}$$

We prove $\widehat{\mathbf{T}}_{\mu}^{\widehat{x}} \cap \{k', k'+1, k'+2, ...\} \subset \bigcup_{k \in \mathbf{N}} \{n_1 + kp, n_2 + kp, ..., n_{k_1} + kp\}$ and let $k'' \in \widehat{\mathbf{T}}_{\mu}^{\widehat{x}} \cap \{k', k'+1, k'+2, ...\}$ arbitrary. We get from (10.1) the existence of a finite sequence $k'', k''-p, ..., k''-\bar{k}p \in \widehat{\mathbf{T}}_{\mu}^{\widehat{x}}, \bar{k} \in \mathbf{N}$ with the property that $k''-\bar{k}p \in \{k', k'+1, ..., k'+p-1\}$, thus we have from (10.3) the existence of $j \in \{1, ..., k_1\}$ with $k''-\bar{k}p = n_j$. This means that $k'' = n_j + \bar{k}p \in \bigcup_{k \in \mathbf{N}} \{n_1 + kp, n_2 + kp, ..., n_{k_1} + kp\}$.

We prove that $\bigcup_{k \in \mathbf{N}} \{n_1 + kp, n_2 + kp, ..., n_{k_1} + kp\} \subset \widehat{\mathbf{T}}_{\mu}^{\widehat{x}} \cap \{k', k'+1, k'+2, ...\}$. Let $k'' \in \bigcup_{k \in \mathbf{N}} \{n_1 + kp, n_2 + kp, ..., n_{k_1} + kp\}$ arbitrary, thus $j \in \{1, ..., k_1\}$ and $k \in \mathbf{N}$ exist such that $k'' = n_j + kp$. As $n_j \in \widehat{\mathbf{T}}_{\mu}^{\widehat{x}} \cap \{k', k'+1, k'+2, ...\}$, we have $n_j + kp \geq k'$ thus we can apply (10.1) and we get $k'' \in \widehat{\mathbf{T}}_{\mu}^{\widehat{x}}$. □

Remark 87. *The hypothesis of the previous Theorem avoids the situation when $\widehat{x}$ is eventually constant. In that case $\widehat{\omega}(\widehat{x}) = \{\mu\}, p = 1, k_1 = 1, n_1 = k'$ and (10.2) takes the form $\widehat{\mathbf{T}}_{\mu}^{\widehat{x}} \cap \{k', k'+1, k'+2, ...\} = \bigcup_{k \in \mathbf{N}} \{k' + k\} = \{k', k'+1, k'+2, ...\}$.*

Remark 88. *Lemma 10, page 228 shows that we can infer from (10.2) and (10.3) that*

$$\widehat{\mathbf{T}}_{\mu}^{\widehat{x}} \cap \{k'', k''+1, ..., k''+p-1\} = \{n'_1, n'_2, ..., n'_{k_1}\},$$

$$\widehat{\mathbf{T}}_{\mu}^{\widehat{x}} \cap \{k', k'+1, k'+2, ...\} = \bigcup_{k \in \mathbf{N}} \{n_1 + kp, n_2 + kp, ..., n_{k_1} + kp\}$$
$$= \bigcup_{z \in \mathbf{Z}} \{n'_1 + zp, n'_2 + zp, ..., n'_{k_1} + zp\} \cap \{k', k'+1, k'+2, ...\}$$
$$\supset \bigcup_{k \in \mathbf{N}} \{n'_1 + kp, n'_2 + kp, ..., n'_{k_1} + kp\} = \widehat{\mathbf{T}}_{\mu}^{\widehat{x}} \cap \{k'', k''+1, k''+2, ...\},$$

where $k'' \geq k'$ is arbitrary.

Theorem 38. *We suppose that the signal $x \in S^{(n)}$ is not eventually constant and let the point $\mu \in \omega(x)$, as well as $T > 0, t' \in \mathbf{R}$ with*

$$\forall t \in \mathbf{T}_{\mu}^{x} \cap [t', \infty), \{t + zT | z \in \mathbf{Z}\} \cap [t', \infty) \subset \mathbf{T}_{\mu}^{x}. \tag{10.4}$$

Then $a_1, b_1, a_2, b_2, ..., a_{k_1}, b_{k_1} \in \mathbf{R}$, $k_1 \geq 1$ exist such that

$$t' \leq a_1 < b_1 < a_2 < b_2 < ... < a_{k_1} < b_{k_1} \leq t' + T, \tag{10.5}$$

$$\mathbf{T}_{\mu}^{x} \cap [t', \infty) = \bigcup_{k \in \mathbf{N}} ([a_1 + kT, b_1 + kT) \cup [a_2 + kT, b_2 + kT) \cup ... \cup [a_{k_1} + kT, b_{k_1} + kT)) \tag{10.6}$$

hold.

Proof. We define the intervals $[a_1, b_1), [a_2, b_2), ..., [a_{k_1}, b_{k_1})$ such that (10.5) and

$$\mathbf{T}_{\mu}^{x} \cap [t', t' + T) = [a_1, b_1) \cup [a_2, b_2) \cup ... \cup [a_{k_1}, b_{k_1}) \tag{10.7}$$

are fulfilled, by taking into account (10.4) and Theorem 28, page 69, written for $t = t'$.

We prove $\mathbf{T}_{\mu}^{x} \cap [t', \infty) \subset \bigcup_{k \in \mathbf{N}} ([a_1 + kT, b_1 + kT) \cup [a_2 + kT, b_2 + kT) \cup ... \cup [a_{k_1} + kT, b_{k_1} + kT))$ and let $t \in \mathbf{T}_{\mu}^{x} \cap [t', \infty)$ arbitrary. A finite sequence $t, t - T, t - 2T, ..., t - \bar{k}T \in \mathbf{T}_{\mu}^{x}$ exists, from (10.4), such that $t - \bar{k}T \in [t', t' + T)$, where $\bar{k} \in \mathbf{N}$. This implies the existence of $j \in \{1, ..., k_1\}$ such that $t - \bar{k}T \in [a_j, b_j)$, i.e. $t \in [a_j + \bar{k}T, b_j + \bar{k}T) \subset \bigcup_{k \in \mathbf{N}} ([a_1 + kT, b_1 + kT) \cup [a_2 + kT, b_2 + kT) \cup ... \cup [a_{k_1} + kT, b_{k_1} + kT))$.

We prove $\bigcup_{k \in \mathbf{N}} ([a_1 + kT, b_1 + kT) \cup [a_2 + kT, b_2 + kT) \cup ... \cup [a_{k_1} + kT, b_{k_1} + kT)) \subset \mathbf{T}_{\mu}^{x} \cap [t', \infty)$ and let $t \in \bigcup_{k \in \mathbf{N}} ([a_1 + kT, b_1 + kT) \cup [a_2 + kT, b_2 + kT) \cup ... \cup [a_{k_1} + kT, b_{k_1} + kT))$ arbitrary. Some $j \in \{1, ..., k_1\}$ and some $k \in \mathbf{N}$ exist such that $t \in [a_j + kT, b_j + kT)$, wherefrom $t - kT \in [a_j, b_j) \subset \mathbf{T}_{\mu}^{x} \cap [t', \infty)$. We can write

$$t \in \{t - kT + zT | z \in \mathbf{Z}\} \cap [t', \infty) \overset{(10.4)}{\subset} \mathbf{T}_{\mu}^{x}.$$

Since $t \geq t'$, (10.6) is proved. □

Remark 89. *Let us see what happens if in the hypothesis of the previous theorem x would have been eventually constant; in this case $\omega(x) = \{\mu\}, k_1 = 1, [a_1, b_1) = [t', t' + T)$ and (10.6) becomes $\mathbf{T}_{\mu}^{x} \cap [t', \infty) = [t', \infty)$.*

Remark 90. *Let $t'' \geq t'$ arbitrary. We get from Lemma 10, page 228 and equations (10.6), (10.7) that*

$$\mathbf{T}_{\mu}^{x} \cap [t'', t''+T) = [a_1', b_1') \cup [a_2', b_2') \cup \ldots \cup [a_{p_1}', b_{p_1}')$$

and

$$\mathbf{T}_{\mu}^{x} \cap [t', \infty) = \bigcup_{k \in \mathbf{N}} ([a_1+kT, b_1+kT) \cup [a_2+kT, b_2+kT) \cup \ldots \cup [a_{k_1}+kT, b_{k_1}+kT))$$

$$= \bigcup_{z \in \mathbf{Z}} ([a_1'+zT, b_1'+zT) \cup [a_2'+zT, b_2'+zT) \cup \ldots \cup [a_{p_1}'+zT, b_{p_1}'+zT)) \cap [t', \infty)$$

$$\supset \bigcup_{k \in \mathbf{N}} ([a_1'+kT, b_1'+kT) \cup [a_2'+kT, b_2'+kT) \cup \ldots \cup [a_{p_1}'+kT, b_{p_1}'+kT)) = \mathbf{T}_{\mu}^{x} \cap [t'', \infty).$$

11. Sufficiency Conditions of Eventual Periodicity

Theorem 39. *Let $\widehat{x} \in \widehat{S}^{(n)}$, $\mu \in \widehat{\omega}(\widehat{x})$, $p \geq 1, k' \in \mathbf{N}\_$ and $n_1, n_2, \ldots, n_{k_1} \in \{k', k'+1, \ldots, k'+p-1\}$, $k_1 \geq 1$ such that*

$$\widehat{\mathbf{T}}_{\mu}^{\widehat{x}} \cap \{k', k'+1, k'+2, \ldots\} = \bigcup_{k \in \mathbf{N}} \{n_1+kp, n_2+kp, \ldots, n_{k_1}+kp\}. \tag{11.1}$$

In such circumstances

$$\forall k \in \widehat{\mathbf{T}}_{\mu}^{\widehat{x}} \cap \{k', k'+1, k'+2, \ldots\}, \{k+zp | z \in \mathbf{Z}\} \cap \{k', k'+1, k'+2, \ldots\} \subset \widehat{\mathbf{T}}_{\mu}^{\widehat{x}}. \tag{11.2}$$

Proof. Let $k'' \in \widehat{\mathbf{T}}_{\mu}^{\widehat{x}} \cap \{k', k'+1, k'+2, \ldots\}$ and $z_1 \in \mathbf{Z}$ arbitrary such that $k''+z_1 p \geq k'$. Then $j \in \{1, \ldots, k_1\}$ and $\overline{k} \in \mathbf{N}$ exist with $k'' = n_j + \overline{k}p$ and we have $k''+z_1 p = n_j + (z_1+\overline{k})p \geq k'$. This means the existence of $z' \in Z$ with $n_j + z'p \geq k'$ and, as $n_j - p \leq k'-1$, we get $z' \geq 0$. In this situation $n_j + z'p \overset{(11.1)}{\in} \widehat{\mathbf{T}}_{\mu}^{\widehat{x}} \cap \{k', k'+1, k'+2, \ldots\}$, thus (11.2) holds. □

Theorem 40. *Let x, $\mu \in \omega(x)$, $T > 0, t' \in \mathbf{R}$ and the numbers $a_1, b_1, a_2, b_2, \ldots, a_{k_1}, b_{k_1} \in \mathbf{R}$, $k_1 \geq 1$ such that*

$$t' \leq a_1 < b_1 < a_2 < b_2 < \ldots < a_{k_1} < b_{k_1} \leq t'+T, \tag{11.3}$$

$$\mathbf{T}_{\mu}^{x} \cap [t', \infty) = \bigcup_{k \in \mathbf{N}} ([a_1+kT, b_1+kT) \cup [a_2+kT, b_2+kT) \cup \ldots \cup [a_{k_1}+kT, b_{k_1}+kT)) \tag{11.4}$$

hold. We infer

$$\forall t \in \mathbf{T}_{\mu}^{x} \cap [t', \infty), \{t+zT | z \in \mathbf{Z}\} \cap [t', \infty) \subset \mathbf{T}_{\mu}^{x}. \tag{11.5}$$

Proof. Let $t'' \in \mathbf{T}_{\mu}^{x} \cap [t', \infty)$ and $z_1 \in \mathbf{Z}$ arbitrary with $t''+z_1 T \geq t'$. From (11.4) we have the existence of $j \in \{1, \ldots, k_1\}$ and $\overline{k} \in \mathbf{N}$ with $t'' \in [a_j+\overline{k}T, b_j+\overline{k}T)$. We obtain that $t''+z_1 T \in [a_j+(z_1+\overline{k})T, b_j+(z_1+\overline{k})T) \subset [t', \infty)$. We get the existence of $z' \in \mathbf{Z}$ with $t''+z_1 T \in [a_j+z'T, b_j+z'T) \subset [t', \infty)$ and, since $b_j - T < t'$, we infer $z' \geq 0$. We have obtained that $t''+z_1 T \overset{(11.4)}{\in} \mathbf{T}_{\mu}^{x} \cap [t', \infty)$, thus (11.5) holds. □

12. A Special Case

Theorem 41. *Let* $\widehat{x} \in \widehat{S}^{(n)}$, $\mu \in \widehat{\omega}(\widehat{x})$, $p \geq 1, k' \in \mathbf{N}_{\_}$ *and* $n_1 \in \{k', k'+1, ..., k'+p-1\}$ *such that*

$$\widehat{\mathbf{T}}_{\mu}^{\widehat{x}} \cap \{k', k'+1, k'+2, ...\} = \{n_1, n_1+p, n_1+2p, n_1+3p, ...\}. \tag{12.1}$$

Then

a) μ *is an eventually periodic point of* $\widehat{x}$ *with the period* p :

$$\forall k \in \widehat{\mathbf{T}}_{\mu}^{\widehat{x}} \cap \{k', k'+1, k'+2, ...\}, \{k+zp | z \in \mathbf{Z}\} \cap \{k', k'+1, k'+2, ...\} \subset \widehat{\mathbf{T}}_{\mu}^{\widehat{x}}; \tag{12.2}$$

b) p *is the prime period of* μ :

$$\widehat{P}_{\mu}^{\widehat{x}} = \{p, 2p, 3p, ...\}. \tag{12.3}$$

Proof. a) This is a special case of Theorem 39, page 83, written for $k_1 = 1$.

b) We suppose against all reason that $p' \in \widehat{P}_{\mu}^{\widehat{x}}$ exists with $p' < p$. As $n_1 \in \widehat{\mathbf{T}}_{\mu}^{\widehat{x}} \cap \{k', k'+1, k'+2, ...\}$, we obtain from (12.2) that $n_1 + p' \in \widehat{\mathbf{T}}_{\mu}^{\widehat{x}} \cap \{k', k'+1, k'+2, ...\}$, contradiction with (12.1). Thus any $p' \in \widehat{P}_{\mu}^{\widehat{x}}$ fulfills $p' \geq p$. We apply Theorem 35, page 79. □

Theorem 42. *Let* x, $\mu \in \omega(x)$, $T > 0$, $t' \in \mathbf{R}$ *and the interval* $[a, b) \subset [t', t'+T)$ *such that*

$$\mathbf{T}_{\mu}^{x} \cap [t', \infty) = [a, b) \cup [a+T, b+T) \cup [a+2T, b+2T) \cup ... \tag{12.4}$$

holds. We have

a) μ *is an eventually periodic point of* x *with the period* T :

$$\forall t \in \mathbf{T}_{\mu}^{x} \cap [t', \infty), \{t+zT | z \in \mathbf{Z}\} \cap [t', \infty) \subset \mathbf{T}_{\mu}^{x}; \tag{12.5}$$

b) if x *is not eventually constant, then* T *is the prime period of* x :

$$P_{\mu}^{x} = \{T, 2T, 3T, ...\}. \tag{12.6}$$

Proof. a) This is a special case of Theorem 40, page 83, written for $k_1 = 1$.

b) We notice first of all that $b < a+T$, otherwise $\mathbf{T}_{\mu}^{x} \cap [t', \infty) = [a, \infty)$ and x is eventually constant, representing a contradiction with the hypothesis.

Let us suppose now against all reason that T is not the prime period of μ, i.e. $T' \in P_{\mu}^{x}$ exists with $T' < T$. We see that

$$\max\{a, b-T'\} < \min\{b, a+T-T'\}$$

holds, because $a < b, a < a+T-T', b-T' < b, b-T' < a+T-T'$ are all true. We take $t \in [\max\{a, b-T'\}, \min\{b, a+T-T'\})$ and we have

$$a \leq \max\{a, b-T'\} \leq t < \min\{b, a+T-T'\} \leq b,$$

$$b \leq \max\{a+T', b\} \leq t+T' < \min\{b+T', a+T\} \leq a+T.$$

We have obtained

$$\mu = x(t) = x(t+T') \overset{(12.4)}{\neq} \mu,$$

contradiction. We conclude that any $T' \in P_{\mu}^{x}$ fulfills $T' \geq T$. We apply Theorem 35, page 79. □

13. Eventually Periodic Points vs Eventually Constant Signals

Theorem 43. *a) Let the signal* $\widehat{x} \in \widehat{S}^{(n)}$ *and the point* $\mu \in \widehat{\omega}(\widehat{x})$. *We suppose that* $p \geq 1$ *and* $k', k_1, k_2 \in \mathbf{N}\_$ *exist such that*

$$k' \leq k_1 < k_2, \tag{13.1}$$

$$\forall k \in \widehat{\mathbf{T}}_{\mu}^{\widehat{x}} \cap \{k', k'+1, k'+2, ...\}, \{k+zp | z \in \mathbf{Z}\} \cap \{k', k'+1, k'+2, ...\} \subset \widehat{\mathbf{T}}_{\mu}^{\widehat{x}}, \tag{13.2}$$

$$\{k_1, k_1+1, ..., k_2\} \subset \widehat{\mathbf{T}}_{\mu}^{\widehat{x}}, \tag{13.3}$$

$$k_1 + p \leq k_2 \tag{13.4}$$

are true. Then $\{k', k'+1, k'+2, ...\} \subset \widehat{\mathbf{T}}_{\mu}^{\widehat{x}}$ *holds.*

b) The signal $x \in S^{(n)}$ *and the point* $\mu \in \omega(x)$ *are given. We suppose that* $T > 0$ *and* $t', t_1, t_2 \in \mathbf{R}$ *exist such that*

$$t' \leq t_1 < t_2, \tag{13.5}$$

$$\forall t \in \mathbf{T}_{\mu}^{x} \cap [t', \infty), \{t + zT | z \in \mathbf{Z}\} \cap [t', \infty) \subset \mathbf{T}_{\mu}^{x}, \tag{13.6}$$

$$[t_1, t_2) \subset \mathbf{T}_{\mu}^{x}, \tag{13.7}$$

$$t_1 + T \leq t_2 \tag{13.8}$$

hold. Then $[t', \infty) \subset \mathbf{T}_{\mu}^{x}$.

Proof. b) Let $t \geq t'$ be arbitrary. From (13.8) we have the existence of $z' \in \mathbf{Z}$ with $t + z'T \in [t_1, t_2)$. We infer:

$$t + z'T \overset{(13.5),(13.7)}{\in} \mathbf{T}_{\mu}^{x} \cap [t', \infty), \tag{13.9}$$

$$t \in \{t + z'T + zT | z \in \mathbf{Z}\} \cap [t', \infty) \overset{(13.6),(13.9)}{\subset} \mathbf{T}_{\mu}^{x}.$$

As t was arbitrary, we get the statement of the theorem. □

Chapter 6

Eventually Periodic Signals

In the first two sections we give properties that are equivalent with the eventual periodicity of the signals.

In Section 3 we show that, for time instants greater than the limit of periodicity, each omega limit point is accessed in a time interval with the length of at most a period.

The bound of the limit of periodicity issue is addressed in Section 4.

Sections 5 and 6 refer to a property of eventual constancy that is used in Section 7 to relate the discrete time with the real time eventually periodic signals.

The fact that the sums, the differences and the multiples of periods are periods is shown in Section 8.

Section 9 draws conclusions concerning the form of the sets $\widehat{P^{x}}, P^x$ and in particular the existence of the prime period is proved.

Sections 10, 11, 12 give necessity and sufficiency properties of eventual periodicity and also a special case, when the prime period is known.

The issue of changing the order of the quantifiers in stating eventual periodicity properties is addressed in Section 13. Since the problem is not solved so far, we introduce in Section 14 the hypothesis P stating basically that if all the points of the omega limit set are eventually periodic, then the signal is eventually periodic.

1. The First Group of Eventual Periodicity Properties

Remark 91. *These properties involve the eventual periodicity request of all the omega limit points $\mu \in \widehat{\omega}(\widehat{x}), \mu \in \omega(x)$, with a common period $p \geq 1, T > 0$ and a common limit of periodicity $k' \in \mathbf{N}$, $t' \in \mathbf{R}$. This way, we notice the associations (1.1)-(3.1)$_{page\,27}$,..., (1.4)-(3.4)$_{page\,27}$, (1.5)-(3.5)$_{page\,28}$,..., (1.12)-(3.12)$_{page\,28}$ with the statements of Theorem 17, page 27, where eventual periodicity was used to characterize eventual constancy. We make also the associations (1.1)-(1.1)$_{page\,66}$,...,(1.4)-(1.4)$_{page\,66}$ and (1.5)-(1.5)$_{page\,66}$, ...,(1.12)-(1.12)$_{page\,66}$ with the statements of Theorem 27, page 66, referring to the eventual periodicity of the points.*

Remark 92. *The statements (1.1),..., (1.4), (1.5),..., (1.12) from Theorem 44 are called of eventual periodicity of $\widehat{x}, x$ due to their equivalence with Definition 15, page 20 that will be proved in the following section, in Theorem 45.*

Theorem 44. *The signals* $\widehat{x} \in \widehat{S}^{(n)}, x \in S^{(n)}$ *are given.*

a) The following statements are equivalent for any $p \geq 1$:

$$\begin{cases} \forall \mu \in \widehat{\omega}(\widehat{x}), \exists k' \in \mathbf{N}_{\_}, \forall k \in \widehat{\mathbf{T}}_{\mu}^{\widehat{x}} \cap \{k', k'+1, k'+2, ...\}, \\ \{k + zp | z \in \mathbf{Z}\} \cap \{k', k'+1, k'+2, ...\} \subset \widehat{\mathbf{T}}_{\mu}^{\widehat{x}}, \end{cases} \tag{1.1}$$

$$\begin{cases} \forall \mu \in \widehat{\omega}(\widehat{x}), \exists k'' \in \mathbf{N}, \forall k \in \widehat{\mathbf{T}}_{\mu}^{\widehat{\sigma}^{k''}(\widehat{x})}, \\ \{k + zp | z \in \mathbf{Z}\} \cap \mathbf{N}_{\_} \subset \widehat{\mathbf{T}}_{\mu}^{\widehat{\sigma}^{k''}(\widehat{x})}, \end{cases} \tag{1.2}$$

$$\begin{cases} \forall \mu \in \widehat{\omega}(\widehat{x}), \exists k' \in \mathbf{N}_{\_}, \forall k \geq k', \widehat{x}(k) = \mu \Longrightarrow \\ \Longrightarrow (\widehat{x}(k) = \widehat{x}(k+p) \text{ and } k - p \geq k' \Longrightarrow \widehat{x}(k) = \widehat{x}(k-p)), \end{cases} \tag{1.3}$$

$$\begin{cases} \forall \mu \in \widehat{\omega}(\widehat{x}), \exists k'' \in \mathbf{N}, \forall k \in \mathbf{N}_{\_}, \widehat{\sigma}^{k''}(\widehat{x})(k) = \mu \Longrightarrow \\ \Longrightarrow (\widehat{\sigma}^{k''}(\widehat{x})(k) = \widehat{\sigma}^{k''}(\widehat{x})(k+p) \text{ and} \\ \text{and } k - p \geq -1 \Longrightarrow \widehat{\sigma}^{k''}(\widehat{x})(k) = \widehat{\sigma}^{k''}(\widehat{x})(k-p)). \end{cases} \tag{1.4}$$

b) The following statements are also equivalent for any $T > 0$:

$$\begin{cases} \forall \mu \in \omega(x), \exists t' \in I^x, \\ \exists t'_1 \geq t', \forall t \in \mathbf{T}_{\mu}^{x} \cap [t'_1, \infty), \{t + zT | z \in \mathbf{Z}\} \cap [t'_1, \infty) \subset \mathbf{T}_{\mu}^{x}, \end{cases} \tag{1.5}$$

$$\begin{cases} \forall \mu \in \omega(x), \exists t'_1 \in \mathbf{R}, \\ \forall t \in \mathbf{T}_{\mu}^{x} \cap [t'_1, \infty), \{t + zT | z \in \mathbf{Z}\} \cap [t'_1, \infty) \subset \mathbf{T}_{\mu}^{x}, \end{cases} \tag{1.6}$$

$$\begin{cases} \forall \mu \in \omega(x), \exists t'' \in \mathbf{R}, \exists t' \in I^{\sigma^{t''}(x)}, \\ \forall t \in \mathbf{T}_{\mu}^{\sigma^{t''}(x)} \cap [t', \infty), \{t + zT | z \in \mathbf{Z}\} \cap [t', \infty) \subset \mathbf{T}_{\mu}^{\sigma^{t''}(x)}, \end{cases} \tag{1.7}$$

$$\begin{cases} \forall \mu \in \omega(x), \exists t'' \in \mathbf{R}, \exists t' \in \mathbf{R}, \\ \forall t \in \mathbf{T}_{\mu}^{\sigma^{t''}(x)} \cap [t', \infty), \{t + zT | z \in \mathbf{Z}\} \cap [t', \infty) \subset \mathbf{T}_{\mu}^{\sigma^{t''}(x)}, \end{cases} \tag{1.8}$$

$$\begin{cases} \forall \mu \in \omega(x), \exists t' \in I^x, \exists t'_1 \geq t', \forall t \geq t'_1, x(t) = \mu \Longrightarrow \\ \Longrightarrow (x(t) = x(t+T) \text{ and } t - T \geq t'_1 \Longrightarrow x(t) = x(t-T)), \end{cases} \tag{1.9}$$

$$\begin{cases} \forall \mu \in \omega(x), \exists t'_1 \in \mathbf{R}, \forall t \geq t'_1, x(t) = \mu \Longrightarrow \\ \Longrightarrow (x(t) = x(t+T) \text{ and } t - T \geq t'_1 \Longrightarrow x(t) = x(t-T)), \end{cases} \tag{1.10}$$

$$\begin{cases} \forall \mu \in \omega(x), \exists t'' \in \mathbf{R}, \exists t' \in I^{\sigma^{t''}(x)}, \\ \forall t \geq t', \sigma^{t''}(x)(t) = \mu \Longrightarrow (\sigma^{t''}(x)(t) = \sigma^{t''}(x)(t+T) \text{ and} \\ \text{and } t - T \geq t' \Longrightarrow \sigma^{t''}(x)(t) = \sigma^{t''}(x)(t-T)), \end{cases} \tag{1.11}$$

$$\begin{cases} \forall \mu \in \omega(x), \exists t'' \in \mathbf{R}, \exists t' \in \mathbf{R}, \forall t \geq t', \sigma^{t''}(x)(t) = \mu \Longrightarrow \\ \Longrightarrow (\sigma^{t''}(x)(t) = \sigma^{t''}(x)(t+T) \text{ and} \\ \text{and } t - T \geq t' \Longrightarrow \sigma^{t''}(x)(t) = \sigma^{t''}(x)(t-T)). \end{cases} \tag{1.12}$$

Proof. a) The proof of the implications

$$(1.1) \Longrightarrow (1.2) \Longrightarrow (1.3) \Longrightarrow (1.4)$$

follows from Theorem 17, page 27. We prove (1.4)$\Longrightarrow$(1.1).

Let $\mu \in \widehat{\omega}(\widehat{x})$ arbitrary, fixed. Hypothesis (1.4) shows the existence of $k'' \in \mathbf{N}$ such that

$$\begin{cases} \forall k \in \mathbf{N}_{\_}, \widehat{\sigma}^{k''}(\widehat{x})(k) = \mu \Longrightarrow (\widehat{\sigma}^{k''}(\widehat{x})(k) = \widehat{\sigma}^{k''}(\widehat{x})(k+p) \text{ and} \\ \quad \text{and } k-p \geq -1 \Longrightarrow \widehat{\sigma}^{k''}(\widehat{x})(k) = \widehat{\sigma}^{k''}(\widehat{x})(k-p)). \end{cases} \tag{1.13}$$

We denote $k' = k'' - 1$, where $k' \geq -1$. We also denote $k''' = k + k' + 1$, where $k''' \geq k'$. With these notations, (1.13) becomes

$$\begin{cases} \forall k''' \geq k', \widehat{x}(k''') = \mu \Longrightarrow (\widehat{x}(k''') = \widehat{x}(k'''+p) \text{ and} \\ \quad \text{and } k'''-p \geq k' \Longrightarrow \widehat{x}(k''') = \widehat{x}(k'''-p)). \end{cases} \tag{1.14}$$

Let now $k \in \widehat{\mathbf{T}}_{\mu}^{\widehat{x}}$ and $z \in \mathbf{Z}$ arbitrary such that $k \geq k'$ and $k + zp \geq k'$. Such a k exists, since $\mu \in \widehat{\omega}(\widehat{x})$ and $\widehat{\mathbf{T}}_{\mu}^{\widehat{x}}$ is infinite. We have the following possibilities.

Case $z > 0$,

$$\mu = \widehat{x}(k) \overset{(1.14)}{=} \widehat{x}(k+p) \overset{(1.14)}{=} \widehat{x}(k+2p) \overset{(1.14)}{=} \ldots \overset{(1.14)}{=} \widehat{x}(k+zp);$$

Case $z = 0$,

$$\mu = \widehat{x}(k) = \widehat{x}(k+zp);$$

Case $z < 0$,

$$\mu = \widehat{x}(k) \overset{(1.14)}{=} \widehat{x}(k-p) \overset{(1.14)}{=} \widehat{x}(k-2p) \overset{(1.14)}{=} \ldots \overset{(1.14)}{=} \widehat{x}(k+zp).$$

We have obtained in all these situations that $k + zp \in \widehat{\mathbf{T}}_{\mu}^{\widehat{x}}$ holds, i.e. (1.1) is true.

b) The proof of the implications

$$(1.5) \Longrightarrow (1.6) \Longrightarrow (1.7) \Longrightarrow (1.8) \Longrightarrow (1.9) \Longrightarrow (1.10) \Longrightarrow (1.11) \Longrightarrow (1.12)$$

follows from Theorem 17, page 27. We prove (1.12)$\Longrightarrow$(1.5).

Let $\mu \in \omega(x)$ arbitrary. From (1.12) we get the existence of $t'' \in \mathbf{R}$ and $t''' \in \mathbf{R}$ such that

$$\begin{cases} \forall t \geq t''', \sigma^{t''}(x)(t) = \mu \Longrightarrow (\sigma^{t''}(x)(t) = \sigma^{t''}(x)(t+T) \text{ and} \\ \quad \text{and } t-T \geq t''' \Longrightarrow \sigma^{t''}(x)(t) = \sigma^{t''}(x)(t-T)). \end{cases} \tag{1.15}$$

We take some $t' \in I^x$ and let $t_1' = \max\{t', t'', t'''\}$. On one hand (1.15) is still true if we replace t''' with t_1', from Lemma 3, page 224. On the other hand, $\forall t \geq t_1', \sigma^{t''}(x)(t) = x(t)$, thus (1.15) becomes

$$\forall t \geq t_1', x(t) = \mu \Longrightarrow (x(t) = x(t+T) \text{ and } t-T \geq t_1' \Longrightarrow x(t) = x(t-T)). \tag{1.16}$$

Let $t \in \mathbf{T}_{\mu}^{x}$ and $z \in \mathbf{Z}$ arbitrary with $t \geq t_1'$ and $t + zT \geq t_1'$. We prove in all the three cases $z > 0, z = 0, z < 0$ that (1.16) implies $t + zT \in \mathbf{T}_{\mu}^{x}$. □

Example 19. *Let* $x \in S^{(2)}$,

$$x(t) = (0,1) \cdot \chi_{(-\infty,-\frac{1}{2})}(t) \oplus (0,1) \cdot \chi_{[0,1)}(t) \oplus (1,0) \cdot \chi_{[1,2)}(t) \oplus (1,1) \cdot \chi_{[2,3)}(t)$$
$$\oplus (0,1) \cdot \chi_{[3,4)}(t) \oplus (1,0) \cdot \chi_{[4,5)}(t) \oplus (1,1) \cdot \chi_{[5,6)}(t) \oplus (0,1) \cdot \chi_{[6,7)}(t) \oplus \ldots$$

x is eventually periodic and it fulfills

$$\forall t \geq 0, x(t) = x(t+3), \tag{1.17}$$

since all of $(0,1),(1,0),(1,1) \in \omega(x)$ *are eventually periodic with the period* $T = 3$ *and the limit of periodicity* $t' = 0$.

Example 20. *The signal* $\widehat{x} \in \widehat{S}^{(1)}$,

$$\widehat{x} = \underbrace{0}_{1}, \underbrace{1}_{1}, \underbrace{0,0}_{2}, \underbrace{1,1}_{2}, \underbrace{0,0,0}_{3}, \underbrace{1,1,1}_{3}, \underbrace{0,0,0,0}_{4}, \ldots$$

is not eventually periodic, because none of $0,1$ *is eventually periodic.*

2. The Second Group of Eventual Periodicity Properties

Remark 93. *This group of properties refers to signals, not to their values. These properties were presented previously in Theorem 20, page 40, as useful in characterizing the eventual constancy. We notice the associations (2.1)-(6.1)*$_{page\,40}$, *(2.2)-(6.2)*$_{page\,40}$ *and (2.3)-(6.3)*$_{page\,40}$, ...,*(2.6)-(6.6)*$_{page\,40}$.

Theorem 45. *The signals* $\widehat{x}, x$ *are given.*

a) For any $p \geq 1$, *the following statements are equivalent with the eventual periodicity of* $\widehat{x}$:

$$\exists k' \in \mathbf{N}\_, \forall k \geq k', \widehat{x}(k) = \widehat{x}(k+p), \tag{2.1}$$

$$\exists k'' \in \mathbf{N}, \forall k \in \mathbf{N}\_, \widehat{\sigma}^{k''}(\widehat{x})(k) = \widehat{\sigma}^{k''}(\widehat{x})(k+p). \tag{2.2}$$

b) For any $T > 0$, *the following statements are also equivalent with the eventual periodicity of* x:

$$\exists t' \in I^x, \exists t_1' \geq t', \forall t \geq t_1', x(t) = x(t+T), \tag{2.3}$$

$$\exists t_1' \in \mathbf{R}, \forall t \geq t_1', x(t) = x(t+T), \tag{2.4}$$

$$\exists t'' \in \mathbf{R}, \exists t' \in I^{\sigma^{t''}(x)}, \forall t \geq t', \sigma^{t''}(x)(t) = \sigma^{t''}(x)(t+T), \tag{2.5}$$

$$\exists t'' \in \mathbf{R}, \exists t' \in \mathbf{R}, \forall t \geq t', \sigma^{t''}(x)(t) = \sigma^{t''}(x)(t+T). \tag{2.6}$$

Proof. a) The implication (2.1)$\Longrightarrow$(2.2) results from Theorem 20. We prove (1.1)$\Longrightarrow$(2.1).

We suppose that $\widehat{\omega}(\widehat{x}) = \{\mu^1, \ldots, \mu^s\}, s \geq 1$. For any $i \in \{1, \ldots, s\}$, some $k_i' \in \mathbf{N}\_$ exists with the property

$$\begin{cases} \forall k \in \widehat{\mathbf{T}}^{\widehat{x}}_{\mu^i} \cap \{k_i', k_i'+1, k_i'+2, \ldots\}, \\ \{k + zp | z \in \mathbf{Z}\} \cap \{k_i', k_i'+1, k_i'+2, \ldots\} \subset \widehat{\mathbf{T}}^{\widehat{x}}_{\mu^i}. \end{cases} \tag{2.7}$$

Let $\widetilde{k} \in \mathbf{N}_{\_}$ be a time instant that fulfills $\widehat{\omega}(\widehat{x}) = \{\widehat{x}(k)|k \geq \widetilde{k}\}$. With $k' = \max\{\widetilde{k}, k'_1, ..., k'_s\}$, from Lemma 3, page 224 we have

$$\begin{cases} \forall k \in \widehat{\mathbf{T}}^{\widehat{x}}_{\mu^i} \cap \{k', k'+1, k'+2, ...\}, \\ \{k+zp|z \in \mathbf{Z}\} \cap \{k', k'+1, k'+2, ...\} \subset \widehat{\mathbf{T}}^{\widehat{x}}_{\mu^i}, \end{cases} \tag{2.8}$$

and this statement is true for all $i \in \{1, ..., s\}$.

Let now $k \geq k'$ arbitrary, for which i exists with $\widehat{x}(k) = \mu^i$. We infer

$$k+p \in \{k+zp|z \in \mathbf{Z}\} \cap \{k', k'+1, k'+2, ...\} \overset{(2.8)}{\subset} \widehat{\mathbf{T}}^{\widehat{x}}_{\mu^i}, \tag{2.9}$$

thus $\widehat{x}(k+p) = \mu^i = \widehat{x}(k)$.

(2.2)$\Longrightarrow$(1.1) Let $\mu \in \widehat{\omega}(\widehat{x})$ arbitrary. Some $k'' \in \mathbf{N}$ exists with the property that

$$\forall k \in \mathbf{N}_{\_}, \widehat{x}(k+k'') = \widehat{x}(k+k''+p). \tag{2.10}$$

We denote $k' = k'' - 1$, where $k' \in \mathbf{N}_{\_}$. We also denote $k''' = k+k'+1$, where $k''' \geq k'$. With these notations (2.10) becomes

$$\forall k''' \geq k', \widehat{x}(k''') = \widehat{x}(k'''+p). \tag{2.11}$$

Let now $k \in \widehat{\mathbf{T}}^{\widehat{x}}_{\mu}$ and $z \in \mathbf{Z}$ arbitrary such that $k \geq k'$ and $k+zp \geq k'$. The following possibilities exist.

Case $z > 0$,

$$\mu = \widehat{x}(k) \overset{(2.11)}{=} \widehat{x}(k+p) \overset{(2.11)}{=} \widehat{x}(k+2p) \overset{(2.11)}{=} ... \overset{(2.11)}{=} \widehat{x}(k+zp);$$

Case $z = 0$,

$$\mu = \widehat{x}(k) = \widehat{x}(k+zp);$$

Case $z < 0$,

$$\widehat{x}(k+zp) \overset{(2.11)}{=} \widehat{x}(k+(z+1)p) \overset{(2.11)}{=} \widehat{x}(k+(z+2)p) \overset{(2.11)}{=} ...$$

$$... \overset{(2.11)}{=} \widehat{x}(k-p) \overset{(2.11)}{=} \widehat{x}(k) = \mu.$$

In all these cases $\widehat{x}(k+zp) = \mu$, thus $k+zp \in \widehat{\mathbf{T}}^{\widehat{x}}_{\mu}$. Statement (1.1) is proved.

b) The implications

$$(2.3) \Longrightarrow (2.4) \Longrightarrow (2.5) \Longrightarrow (2.6)$$

result from Theorem 20, page 40.

(1.5)$\Longrightarrow$(2.3) We suppose that $\omega(x) = \{\mu^1, ..., \mu^s\}$ and let $i \in \{1, ..., s\}$ arbitrary. From (1.5) we have the existence of $t^i \in I^x$ and $t^i_1 \geq t^i$ with

$$\forall t \in \mathbf{T}^x_{\mu^i} \cap [t^i_1, \infty), \{t+zT|z \in \mathbf{Z}\} \cap [t^i_1, \infty) \subset \mathbf{T}^x_{\mu^i} \tag{2.12}$$

fulfilled.

We denote $t' = \max\{t^1,...,t^s\}$ and we notice that $t' \in I^x$ holds, since t' coincides with one of $t^1,...,t^s$. We put $\widetilde{t} \in \mathbf{R}$ for the time instant that fulfills

$$\omega(x) = \{x(t)|t \geq \widetilde{t}\}. \tag{2.13}$$

Let $i \in \{1,...,s\}$ arbitrary and fixed. The fact that for $t_1' = \max\{\widetilde{t},t',t_1^1,...,t_1^s\}$ the statement

$$\forall t \in \mathbf{T}^x_{\mu^i} \cap [t_1',\infty), \{t+zT|z \in \mathbf{Z}\} \cap [t_1',\infty) \subset \mathbf{T}^x_{\mu^i} \tag{2.14}$$

holds is a consequence of (2.12) and Lemma 3 b), page 224.

Let now $t \geq t_1'$ arbitrary. Some $i \in \{1,...,s\}$ exists with $x(t) = \mu^i$, thus we can write $t \in \mathbf{T}^x_{\mu^i} \cap [t_1',\infty)$ and

$$t+T \in \{t+zT|z \in \mathbf{Z}\} \cap [t_1',\infty) \overset{(2.14)}{\subset} \mathbf{T}^x_{\mu^i},$$

wherefrom $x(t+T) = \mu^i = x(t)$.

(2.6)$\Longrightarrow$(1.5) We denote with $\widetilde{t} \in \mathbf{R}$ the time instant that fulfills (2.13). The hypothesis shows the existence of $t'' \in \mathbf{R}$ and $t' \in \mathbf{R}$ such that

$$\forall t \geq t', \sigma^{t''}(x)(t) = \sigma^{t''}(x)(t+T) \tag{2.15}$$

and let $t_1' = \max\{\widetilde{t},t'',t'\}$ arbitrary. We have from (2.15):

$$\forall t \geq t_1', x(t) = x(t+T). \tag{2.16}$$

Let now $\mu \in \omega(x)$ arbitrary and $t''' \in I^x$. We suppose that $t''' \leq t_1'$, as this is always possible. Let $t \in \mathbf{T}^x_\mu \cap [t_1',\infty)$ arbitrary and let us take $z \in \mathbf{Z}$ arbitrary itself with $t+zT \geq t_1'$. We have:

Case $z > 0$,

$$\mu = x(t) \overset{(2.16)}{=} x(t+T) \overset{(2.16)}{=} x(t+2T) \overset{(2.16)}{=} ... \overset{(2.16)}{=} x(t+zT);$$

Case $z = 0$,

$$\mu = x(t) = x(t+zT);$$

Case $z < 0$,

$$x(t+zT) \overset{(2.16)}{=} ... \overset{(2.16)}{=} x(t-2T) \overset{(2.16)}{=} x(t-T) \overset{(2.16)}{=} x(t) = \mu.$$

We have obtained that in all these situations $x(t+zT) = \mu$, i.e. $t+zT \in \mathbf{T}^x_\mu$. □

Remark 94. *The eventual periodicity of the signals highlights the existence of two time instants, $\widetilde{k} \in \mathbf{N}\_$ and $k' \in \mathbf{N}\_$ given by*

$$\widehat{\omega}(\widehat{x}) = \{\widehat{x}(k)|k \geq \widetilde{k}\}, \tag{2.17}$$

$$\forall k \geq k', \widehat{x}(k) = \widehat{x}(k+p). \tag{2.18}$$

None of $\widetilde{k}, k'$ is unique, in the sense that (2.17), (2.18) may be rewritten for any $\widetilde{k}_1 \geq \widetilde{k}, k_1' \geq k'$ but if $\widetilde{k}, k'$ are chosen to be the least such that (2.17), (2.18) hold, then $\widetilde{k} \leq k'$.

The situation is also true in the real time case, if the least $\widetilde{t}$ such that

$$\omega(x) = \{x(t)|t \geq \widetilde{t}\} \tag{2.19}$$

exists; we notice that if $Or(x) = \omega(x)$, then the least $\widetilde{t}$ like at (2.19) does not exist.

Remark 95. *The statements (1.1),..., (1.4) and (1.5),..., (1.12) refer to left-and-right time shifts, while the statements (2.1), (2.2) and (2.3),..., (2.6) refer to right time shifts only.*

3. The Accessibility of the Omega Limit Set

Theorem 46. *a) If $\widehat{x} \in \widehat{S}^{(n)}$, then*

$$\bigcap_{\mu \in \widehat{\omega}(\widehat{x})} \widehat{P}_{\mu}^{\widehat{x}} \neq \varnothing \Longrightarrow \forall k' \in \bigcap_{\mu \in \widehat{\omega}(\widehat{x})} \widehat{L}_{\mu}^{\widehat{x}}, \ \widehat{\omega}(\widehat{x}) = \{\widehat{x}(k) | k \geq k'\}, \tag{3.1}$$

$$\widehat{P}^{\widehat{x}} \neq \varnothing \Longrightarrow \forall k' \in \widehat{L}^{\widehat{x}}, \ \widehat{\omega}(\widehat{x}) = \{\widehat{x}(k) | k \geq k'\} \tag{3.2}$$

hold.

b) For $x \in S^{(n)}$, we have the truth of

$$\bigcap_{\mu \in \omega(x)} P_{\mu}^{x} \neq \varnothing \Longrightarrow \forall t' \in \bigcap_{\mu \in \omega(x)} L_{\mu}^{x}, \ \omega(x) = \{x(t) | t \geq t'\}, \tag{3.3}$$

$$P^{x} \neq \varnothing \Longrightarrow \forall t' \in L^{x}, \ \omega(x) = \{x(t) | t \geq t'\}. \tag{3.4}$$

Proof. a) We prove (3.1). The hypothesis states that $\bigcap_{\mu \in \widehat{\omega}(\widehat{x})} \widehat{P}_{\mu}^{\widehat{x}} \neq \varnothing$. If $\bigcap_{\mu \in \widehat{\omega}(\widehat{x})} \widehat{L}_{\mu}^{\widehat{x}} = \varnothing$[1] then the statement is trivially true, thus we suppose that $\bigcap_{\mu \in \widehat{\omega}(\widehat{x})} \widehat{L}_{\mu}^{x} \neq \varnothing$ and let $k' \in \bigcap_{\mu \in \widehat{\omega}(\widehat{x})} \widehat{L}_{\mu}^{x}$ arbitrary. We prove $\widehat{\omega}(\widehat{x}) \subset \{\widehat{x}(k) | k \geq k'\}$. Some $\widetilde{k} \in \mathbf{N}\_$ exists such that $\widehat{\omega}(\widehat{x}) = \{\widehat{x}(k) | k \geq \widetilde{k}\}$ and we have the following possibilities.

Case $k' < \widetilde{k}$

In this case $\widehat{\omega}(\widehat{x}) \subset \{\widehat{x}(k) | k \geq k'\}$.

Case $k' \geq \widetilde{k}$

If so, we have from Theorem 7, page 11 that $\widehat{\omega}(\widehat{x}) = \{\widehat{x}(k) | k \geq k'\}$.

We prove $\{\widehat{x}(k) | k \geq k'\} \subset \widehat{\omega}(\widehat{x})$. For this we take arbitrarily $k \geq k'$ and $p \in \bigcap_{\mu \in \widehat{\omega}(\widehat{x})} \widehat{P}_{\mu}^{\widehat{x}}$. We have $\widehat{x}(k) = \widehat{x}(k+p) = \widehat{x}(k+2p) = ...$, thus $\widehat{\mathbf{T}}_{\widehat{x}(k)}^{\widehat{x}}$ is infinite and $\widehat{x}(k) \in \widehat{\omega}(\widehat{x})$.

b) We prove (3.4). If $P^{x} \neq \varnothing$, then $L^{x} \neq \varnothing$ and let $t' \in L^{x}$ arbitrary. We show that $\omega(x) \subset \{x(t) | t \geq t'\}$. From Theorem 7, page 11 we know that some $\widetilde{t} \in \mathbf{R}$ exists with $\omega(x) = \{x(t) | t \geq \widetilde{t}\}$. There are two possibilities.

Case $t' < \widetilde{t}$

If so, then $\omega(x) \subset \{x(t) | t \geq t'\}$.

Case $t' \geq \widetilde{t}$

In this case, see Theorem 7, $\omega(x) = \{x(t) | t \geq t'\}$.

We show now that $\{x(t) | t \geq t'\} \subset \omega(x)$ holds and let $t \geq t', T \in P^{x}$ arbitrary. The hypothesis shows that $x(t) = x(t+T) = x(t+2T) = ...$, i.e. $\mathbf{T}_{x(t)}^{x}$ is superiorly unbounded. This means that $x(t) \in \omega(x)$. $\square$

Theorem 47. *a) If $\widehat{x}$ is eventually periodic with the period $p \geq 1$ and the limit of periodicity $k' \in \mathbf{N}\_$:*

$$\forall k \geq k', \widehat{x}(k) = \widehat{x}(k+p), \tag{3.5}$$

[1]The hypothesis, together with Theorem 49, page 95 and Theorem 58, page 102 to follow give $\varnothing \neq \bigcap_{\mu \in \widehat{\omega}(\widehat{x})} \widehat{P}_{\mu}^{\widehat{x}} = \widehat{P}^{\widehat{x}} \Longrightarrow \varnothing \neq \bigcap_{\mu \in \widehat{\omega}(\widehat{x})} \widehat{L}_{\mu}^{\widehat{x}} = \widehat{L}^{\widehat{x}}$, thus the possibility $\bigcap_{\mu \in \widehat{\omega}(\widehat{x})} \widehat{L}_{\mu}^{\widehat{x}} = \varnothing$ does not exist.

then

$$\forall k \geq k', \widehat{\omega}(\widehat{x}) = \{\widehat{x}(i) | i \in \{k, k+1, ..., k+p-1\}\}. \tag{3.6}$$

b) If x is eventually periodic with the period $T > 0$ and the limit of periodicity $t' \in \mathbf{R}$:

$$\forall t \geq t', x(t) = x(t+T), \tag{3.7}$$

then

$$\forall t \geq t', \omega(x) = \{x(\xi) | \xi \in [t, t+T)\}. \tag{3.8}$$

Proof. a) We know from Theorem 46 that $\widehat{\omega}(\widehat{x}) = \{\widehat{x}(k) | k \geq k'\}$. Let $k \geq k'$ and $\mu \in \widehat{\omega}(\widehat{x})$ arbitrary, fixed. As μ is eventually periodic with the period p, we have from Theorem 28, page 69 that $\widehat{\mathbf{T}}_{\mu}^{\widehat{x}} \cap \{k, k+1, ..., k+p-1\} \neq \varnothing$. We get the existence of $i \in \widehat{\mathbf{T}}_{\mu}^{\widehat{x}} \cap \{k, k+1, ..., k+p-1\}$ thus $\mu = \widehat{x}(i)$. We have proved that $\widehat{\omega}(\widehat{x}) \subset \{\widehat{x}(i) | i \in \{k, k+1, ..., k+p-1\}\}$. The inverse inclusion is obvious, since any eventually periodic value of $\widehat{x}$ is an omega limit point.

b) Theorem 46 shows that $\omega(x) = \{x(t) | t \geq t'\}$. Let us fix arbitrarily $t \geq t'$ and $\mu \in \omega(x)$. As μ is eventually periodic with the period T, we infer from Theorem 28 that $\mathbf{T}_{\mu}^{x} \cap [t, t+T) \neq \varnothing$ and let $\xi \in \mathbf{T}_{\mu}^{x} \cap [t, t+T)$ thus $\mu = x(\xi)$. We have shown the inclusion $\omega(x) \subset \{x(\xi) | \xi \in [t, t+T)\}$. The inclusion $\{x(\xi) | \xi \in [t, t+T)\} \subset \omega(x)$ is obvious, since any point of the left hand set is eventually periodic and omega limit. □

Remark 96. *The previous theorem states the property that, in the case of the eventually periodic signals, all the omega limit points are accessible in a time interval with the length of a period.*

4. The Limit of Periodicity

Theorem 48. *a) $\widehat{x} \in \widehat{S}^{(n)}, p \geq 1, p' \geq 1, k' \in \mathbf{N}_{\_}, k'' \in \mathbf{N}_{\_}$ are given such that*

$$\forall k \geq k', \widehat{x}(k) = \widehat{x}(k+p), \tag{4.1}$$

$$\forall k \geq k'', \widehat{x}(k) = \widehat{x}(k+p') \tag{4.2}$$

hold. We have

$$\forall k \geq k', \widehat{x}(k) = \widehat{x}(k+p'). \tag{4.3}$$

b) We consider the signal $x \in S^{(n)}$, together with $T > 0, T' > 0, t' \in \mathbf{R}, t'' \in \mathbf{R}$ and we ask that

$$\forall t \geq t', x(t) = x(t+T), \tag{4.4}$$

$$\forall t \geq t'', x(t) = x(t+T') \tag{4.5}$$

are fulfilled. Then

$$\forall t \geq t', x(t) = x(t+T') \tag{4.6}$$

is true.

Proof. a) Let $k \geq k'$ arbitrary, fixed. We have two possibilities.

Case $k' \geq k''$

In this situation $k \geq k''$, thus we can write

$$\widehat{x}(k) \overset{(4.2)}{=} \widehat{x}(k+p').$$

Case $k' < k''$

Let us take $k_1 \in \mathbf{N}$ with the property that $k + k_1 p \geq k''$. We can write:

$$\widehat{x}(k) \overset{(4.1)}{=} \widehat{x}(k+k_1 p) \overset{(4.2)}{=} \widehat{x}(k+k_1 p+p') \overset{(4.1)}{=} \widehat{x}(k+p').$$

□

Remark 97. *The previous theorem states the fact that, if $\widehat{x}, x$ are eventually periodic, then $\widehat{L^{\widehat{x}}}, L^x$ do not depend on the choice of $p \in \widehat{P^{\widehat{x}}}, T \in P^x$.*

Theorem 49. *a) If $\widehat{x}$ is eventually periodic, then*

$$\widehat{L^{\widehat{x}}} = \bigcap_{\mu \in \widehat{\omega}(\widehat{x})} \widehat{L^{\widehat{x}}_{\mu}};$$

b) if x is eventually periodic, we have

$$L^x = \bigcap_{\mu \in \omega(x)} L^x_{\mu}.$$

Proof. a) The hypothesis states $\widehat{P^{\widehat{x}}} \neq \varnothing$ thus $\widehat{L^{\widehat{x}}} \neq \varnothing$ and let $p \in \widehat{P^{\widehat{x}}}$. We show that $\widehat{L^{\widehat{x}}} \subset \bigcap_{\mu \in \widehat{\omega}(\widehat{x})} \widehat{L^{\widehat{x}}_{\mu}}$ and we take for this $k' \in \widehat{L^{\widehat{x}}}$ arbitrary, thus

$$\forall k \geq k', \widehat{x}(k) = \widehat{x}(k+p). \tag{4.7}$$

Starting from Theorem 45, page 90, the proof of $(2.2)_{page\ 90} \Longrightarrow (1.1)_{page\ 88}$ it is shown that (4.7) implies

$$\begin{cases} \forall \mu \in \widehat{\omega}(\widehat{x}), \forall k \in \widehat{\mathbf{T}^{\widehat{x}}_{\mu}} \cap \{k', k'+1, k'+2, ...\}, \\ \{k+zp | z \in \mathbf{Z}\} \cap \{k', k'+1, k'+2, ...\} \subset \widehat{\mathbf{T}^{\widehat{x}}_{\mu}}, \end{cases} \tag{4.8}$$

wherefrom we have that $k' \in \bigcap_{\mu \in \widehat{\omega}(\widehat{x})} \widehat{L^{\widehat{x}}_{\mu}}$.

We show that $\bigcap_{\mu \in \widehat{\omega}(\widehat{x})} \widehat{L^{\widehat{x}}_{\mu}} \subset \widehat{L^{\widehat{x}}}$ and let for this $k' \in \bigcap_{\mu \in \widehat{\omega}(\widehat{x})} \widehat{L^{\widehat{x}}_{\mu}}$, i.e. (4.8) holds. Starting from the implication $(1.1)_{page\ 88} \Longrightarrow (2.1)_{page\ 90}$ of Theorem 45, page 90, it is shown the truth of (4.7), in other words $k' \in \widehat{L^{\widehat{x}}}$. □

Theorem 50. *a) Let $\widehat{x} \in \widehat{S}^{(n)}$ eventually periodic. Then $k' \in \mathbf{N}_{\_}$ exists with $\widehat{L^{\widehat{x}}} = \{k', k'+1, k'+2, ...\}$.*

b) Let $x \in S^{(n)}$ be eventually periodic and not constant. Then $t' \in \mathbf{R}$ exists such that $L^x = [t', \infty)$.

Proof. a) We put $\widehat{\omega}(\widehat{x})$ under the form $\widehat{\omega}(\widehat{x}) = \{\mu^1,...,\mu^s\}, s \geq 1$. Theorem 30, page 70 shows the existence of $k_i' \in \mathbf{N}\_$ that fulfill

$$\widehat{L_{\mu^i}^{\widehat{x}}} = \{k_i', k_i'+1, k_i'+2, ...\}, i = \overline{1,s}.$$

We apply Theorem 49 and we get

$$\widehat{L^{\widehat{x}}} = \widehat{L_{\mu^1}^{\widehat{x}}} \cap ... \cap \widehat{L_{\mu^s}^{\widehat{x}}} = \{k', k'+1, k'+2, ...\},$$

where $k' = \max\{k_1', ..., k_s'\}$. □

5. A Property of Eventual Constancy

Theorem 51. *We consider the signals $\widehat{x}, x$.*

a) If $k' \in \mathbf{N}\_$ exists making

$$\forall k \geq k', \widehat{x}(k) = \widehat{x}(k+p) \tag{5.1}$$

true for $p = 1$, then $\mu \in \widehat{\omega}(\widehat{x})$ exists with

$$\forall k \geq k', \widehat{x}(k) = \mu \tag{5.2}$$

fulfilled and in this case (5.1) holds for any $p \geq 1$.

b) We suppose that

$$\begin{aligned} x(t) = x(-\infty+0) \cdot \chi_{(-\infty,t_0)}(t) \oplus x(t_0) \cdot \chi_{[t_0,t_0+h)}(t) \oplus ... \\ ... \oplus x(t_0+kh) \cdot \chi_{[t_0+kh,t_0+(k+1)h)}(t) \oplus ... \end{aligned} \tag{5.3}$$

is true for $t_0 \in \mathbf{R}$ and $h > 0$. If

$$\forall t \geq t', x(t) = x(t+T) \tag{5.4}$$

holds for $t' \in \mathbf{R}$, $T \in (0,h) \cup (h,2h) \cup ... \cup (qh,(q+1)h) \cup ...$ then some $\mu \in \omega(x)$ exists such that

$$\forall t \geq t', x(t) = \mu \tag{5.5}$$

and in this case (5.4) is true for any $T > 0$.

c) We presume that (5.3) takes the form

$$\begin{aligned} x(t) = \widehat{x}(-1) \cdot \chi_{(-\infty,t_0)}(t) \oplus \widehat{x}(0) \cdot \chi_{[t_0,t_0+h)}(t) \oplus ... \\ ... \oplus \widehat{x}(k) \cdot \chi_{[t_0+kh,t_0+(k+1)h)}(t) \oplus ... \end{aligned} \tag{5.6}$$

and let $\mu \in \widehat{\omega}(\widehat{x}) = \omega(x)$ be an arbitrary point.

c.1) If $k' \in \mathbf{N}\_$ exists such that (5.1) is true for $p = 1$, then $\mu \in \widehat{\omega}(\widehat{x})$ exists such that (5.2) is fulfilled and $t' \in \mathbf{R}$ exists also such that (5.5) is true. In this case (5.1) holds for any $p \geq 1$ and (5.4) holds for any $T > 0$.

c.2) If $t' \in \mathbf{R}$, $T \in (0,h) \cup (h,2h) \cup ... \cup (qh,(q+1)h) \cup ...$ exist making (5.4) true, then $k' \in \mathbf{N}\_$ and $\mu \in \widehat{\omega}(\widehat{x})$ exist such that (5.2) holds and (5.5) holds too. Moreover, in this situation (5.1) is true for any $p \geq 1$ and (5.4) is true for any $T > 0$.

Proof. a) Let $k' \in \mathbf{N}\_$ be with the property that (5.1) holds for $p = 1$, i.e.

$$\widehat{x}(k') = \widehat{x}(k'+1) = \widehat{x}(k'+2) = ... \tag{5.7}$$

We denote $\widehat{x}(k')$ with μ and this obviously implies that $\mu \in \widehat{\omega}(\widehat{x})$. Equation (5.7) may be rewritten under the form (5.2) and

$$\forall k \geq k', \widehat{x}(k) = \mu = \widehat{x}(k+p)$$

holds for any $p \geq 1$.

b) The hypothesis states the existence of $t_0 \in \mathbf{R}, h > 0$ such that (5.3) holds and also the existence of $t' \in \mathbf{R}$ and $T \in (0,h) \cup (h,2h) \cup ... \cup (qh,(q+1)h) \cup ...$ such that (5.4) holds. We denote $x(t')$ with μ.

We suppose first that $T \in (0,h)$ is arbitrary. If, against all reason, x does not fulfill (5.5), the time instant $t_0' > t'$ exists such that

$$\forall t \in [t', t_0'), x(t) = \mu, \tag{5.8}$$

$$x(t_0') \neq \mu. \tag{5.9}$$

Since obviously $t_0' \geq t_0$, we have the existence of $k_0 \in \mathbf{N}$ such that $t_0' \in [t_0 + k_0 h, t_0 + (k_0 + 1)h)$. As $\forall t \in [t_0 + k_0 h, t_0 + (k_0+1)h)$, $x(t) = x(t_0 + k_0 h)$, we get $t_0' = t_0 + k_0 h$. With the notation $\widetilde{t} = \max\{t_0' - T, t'\}$, we infer $\widetilde{t} < t_0'$ and for any $t'' \in (\widetilde{t}, t_0')$, we have

$$t' < t'' < t_0' < t'' + T < t_0' + T < t_0' + h. \tag{5.10}$$

We deduce

$$x(t'' + T) = x(t_0' + T), \tag{5.11}$$

as far as both previous terms are equal with $x(t_0')$, and

$$\mu \overset{(5.8),(5.10)}{=} x(t'') \overset{(5.4),(5.10)}{=} x(t''+T) \overset{(5.11)}{=} x(t_0'+T) \overset{(5.4),(5.10)}{=} x(t_0') \overset{(5.9)}{\neq} \mu,$$

contradiction showing that a t_0' that makes true (5.8), (5.9) does not exist.

The case when for $q \geq 1$, we have that $T \in (qh, (q+1)h)$ is similar with the previous one. Property (5.4) continues to be true for some $t' \in \mathbf{R}$ and if, against all reason, x does not fulfill (5.5), we get that $t_q' > t'$ exists with

$$\forall t \in [t', t_q'), x(t) = \mu, \tag{5.12}$$

$$x(t_q') \neq \mu. \tag{5.13}$$

Thus $k_q \in \mathbf{N}$ exists such that $t_q' \in [t_0 + k_q h, t_0 + (k_q + 1)h)$ and, from the fact that $\forall t \in [t_0 + k_q h, t_0 + (k_q+1)h)$, we get $x(t) = x(t_0 + k_q h)$, the conclusion is $t_g' = t_0 + k_q h$. With the notation $\widetilde{t} = \max\{t_q' + qh - T, t'\}$, we obtain $\widetilde{t} < t_q'$ and for any $t'' \in (\widetilde{t}, t_q')$ we have

$$t' < t'' < t_q' < t_q' + qh < t'' + T < t_q' + T < t_q' + (q+1)h. \tag{5.14}$$

We infer

$$x(t'' + T) = x(t_q' + T), \tag{5.15}$$

because both previous terms are equal with $x(t'_q + qh)$ and

$$\mu \overset{(5.12),(5.14)}{=} x(t'') \overset{(5.4),(5.14)}{=} x(t'' + T) \overset{(5.15)}{=} x(t'_q + T) \overset{(5.4),(5.14)}{=} x(t'_q) \overset{(5.13)}{\neq} \mu$$

contradiction, in other words a $t'_q \in \mathbf{R}$ that makes (5.12), (5.13) true does not exist. Thus x fulfills (5.5) and in such circumstances (5.4) is true for any $T > 0$. □

6. Discussion on Eventual Constancy

Remark 98. *The point is that Theorem 31, page 71 and Theorem 51, page 96 express the same idea, meaning that in the situation when $\widehat{x}, x$ are related by*

$$x(t) = \widehat{x}(-1) \cdot \chi_{(-\infty,t_0)}(t) \oplus \widehat{x}(0) \cdot \chi_{[t_0,t_0+h)}(t) \oplus ...$$
$$... \oplus \widehat{x}(k) \cdot \chi_{[t_0+kh,t_0+(k+1)h)}(t) \oplus ...$$

any of a)

$$\begin{cases} \forall \mu \in \widehat{\omega}(\widehat{x}), \forall k \in \widehat{\mathbf{T}}^{\widehat{x}}_{\mu} \cap \{k', k'+1, k'+2, ...\}, \\ \{k + zp | z \in \mathbf{Z}\} \cap \{k', k'+1, k'+2, ...\} \subset \widehat{\mathbf{T}}^{\widehat{x}}_{\mu}, \end{cases} \tag{6.1}$$

or

$$\forall k \geq k', \widehat{x}(k) = \widehat{x}(k + p) \tag{6.2}$$

true for $p = 1$ and some $k' \in \mathbf{N}_{\_}$,

b)

$$\forall \mu \in \omega(x), \forall t \in \mathbf{T}^x_{\mu} \cap [t', \infty), \{t + zT | z \in \mathbf{Z}\} \cap [t', \infty) \subset \mathbf{T}^x_{\mu}, \tag{6.3}$$

or

$$\forall t \geq t', x(t) = x(t + T) \tag{6.4}$$

true for $T \in (0,h) \cup (h,2h) \cup ... \cup (qh,(q+1)h) \cup ...$ and some $t' \in \mathbf{R}$ implies the truth of

$$\forall k \geq k', \widehat{x}(k) = \mu, \tag{6.5}$$

$$\forall t \geq t', x(t) = \mu \tag{6.6}$$

showing in particular that $\widehat{x}, x$ are eventually equal with the same constant μ. However Theorem 45, page 90 states the equivalence, for any $p \geq 1, k' \in \mathbf{N}_{\_}$ between (6.1) and (6.2) and also the equivalence, for any $T > 0, t' \in \mathbf{R}$ between (6.3) and (6.4), thus the fact that Theorems 31 and 51 give the same conclusion is natural.

7. Discrete Time vs Real Time

Theorem 52. *We suppose that $\widehat{x}, x$ are related by*

$$\begin{aligned} x(t) = \widehat{x}(-1) \cdot \chi_{(-\infty,t_0)}(t) \oplus \widehat{x}(0) \cdot \chi_{[t_0,t_0+h)}(t) \oplus ... \\ ... \oplus \widehat{x}(k) \cdot \chi_{[t_0+kh,t_0+(k+1)h)}(t) \oplus ... \end{aligned} \tag{7.1}$$

where $t_0 \in \mathbf{R}, h > 0$. The existence of $p \geq 1$ and $k' \in \mathbf{N}_{\_}$ such that

$$\forall k \geq k', \widehat{x}(k) = \widehat{x}(k + p), \tag{7.2}$$

implies the existence of $t' \in \mathbf{R}$ such that

$$\forall t \geq t', x(t) = x(t+T) \tag{7.3}$$

is true for $T = ph$.

Proof. Equation (7.1) is true for some $t_0 \in \mathbf{R}, h > 0$ and $p \geq 1, k' \in \mathbf{N}\_$ exist having the property that (7.2) holds. We use the notations $T = ph, t' = t_0 + k'h$ and let $t \geq t'$ be arbitrary, fixed. Some $k \geq k'$ exists with the property $t \in [t_0 + kh, t_0 + (k+1)h)$, wherefrom $t + T \in [t_0 + (k+p)h, t_0 + (k+1+p)h)$ and we finally get that

$$x(t) = \widehat{x}(k) \overset{(7.2)}{=} \widehat{x}(k+p) = x(t+T).$$

Because $t \geq t'$ was arbitrarily chosen, we have inferred the truth of (7.3). □

Theorem 53. *If $\widehat{x}, x$ are not eventually constant, (7.1) holds for $t_0 \in \mathbf{R}, h > 0$ and $T > 0, t' \in \mathbf{R}$ exist such that x fulfills (7.3), then $\frac{T}{h} \in \{1,2,3,...\}$ and $k' \in \mathbf{N}\_$ exists such that (7.2) is true for $p = \frac{T}{h}$.*

Proof. Some $t_0 \in R, h > 0$ exist with (7.1) fulfilled and $T > 0, t' \in \mathbf{R}$ exist also with (7.3) true. If in (7.3) we have $T \in (0,h) \cup (h,2h) \cup ... \cup (qh,(q+1)h) \cup ...$ then, from Theorem 51, page 96, $\mu \in \widehat{\omega}(\widehat{x}) = \omega(x)$ and $k' \in \mathbf{N}\_$ exist such that $\forall k \geq k', \widehat{x}(k) = \mu$ resulting a contradiction with the hypothesis, stating that $\widehat{x}, x$ are not eventually constant. We suppose from now that $T \in \{h, 2h, 3h, ...\}$. We denote $p = \frac{T}{h}, p \geq 1$. As far as for any $t'' \geq t'$ we have

$$\forall t \geq t'', x(t) = x(t+T), \tag{7.4}$$

we can suppose without loosing the generality the existence of $k' \in \mathbf{N}\_$ with $t'' = t_0 + k'h$. In this situation for any $k \geq k'$ and any $t \geq t''$ with $t \in [t_0 + kh, t_0 + (k+1)h)$ we have

$$t + T \in [t_0 + kh + ph, t_0 + (k+1)h + ph) = [t_0 + (k+p)h, t_0 + (k+1+p)h)$$

and we can write

$$\widehat{x}(k) = x(t) \overset{(7.4)}{=} x(t+T) = \widehat{x}(k+p),$$

thus (7.2) is true. □

Example 21. *We define $\widehat{x} \in \widehat{S}^{(1)}$ by*

$$\forall k \in \mathbf{N}\_, \widehat{x}(k) = \begin{cases} 1, if\ k \in \{-1,2,4,6,8,...\} \\ 0, otherwise \end{cases}$$

and $x \in S^{(1)}$ respectively by

$$x(t) = \widehat{x}(-1) \cdot \chi_{(-\infty,-4)}(t) \oplus \widehat{x}(0) \cdot \chi_{[-4,-2)}(t) \oplus$$
$$\oplus \widehat{x}(1) \cdot \chi_{[-2,0)}(t) \oplus \widehat{x}(2) \cdot \chi_{[0,2)}(t) \oplus ...$$

We have

$$\forall t \geq -2, x(t) = x(t+4),$$
$$\forall k \geq 1, \widehat{x}(k) = \widehat{x}(k+2)$$

thus (7.3) is fulfilled with $T = 4, t' = -2$ and (7.2) is true with $p = 2, k' = 1$. Furthermore, in this example $h = 2$.

8. Sums, Differences and Multiples of Periods

Theorem 54. *Let the signals $\widehat{x}, x$.*

a) We suppose that $\widehat{x}$ has the periods $p, p' \geq 1$ and the limit of periodicity $k' \in \mathbf{N}_{\_}$:

$$\forall k \geq k', \widehat{x}(k) = \widehat{x}(k+p), \tag{8.1}$$

$$\forall k \geq k', \widehat{x}(k) = \widehat{x}(k+p'). \tag{8.2}$$

Then $p+p' \geq 1$, $\widehat{x}$ has the period $p+p'$ and the limit of periodicity k'

$$\forall k \geq k', \widehat{x}(k) = \widehat{x}(k+p+p') \tag{8.3}$$

and if $p > p'$, then $p-p' \geq 1$, $\widehat{x}$ has the period $p-p'$ and the limit of periodicity k'

$$\forall k \geq k', \widehat{x}(k) = \widehat{x}(k+p-p'). \tag{8.4}$$

b) Let $T, T' > 0$, $t' \in \mathbf{R}$ be arbitrary with

$$\forall t \geq t', x(t) = x(t+T), \tag{8.5}$$

$$\forall t \geq t', x(t) = x(t+T') \tag{8.6}$$

fulfilled. We have on one hand that $T+T' > 0$ and

$$\forall t \geq t', x(t) = x(t+T+T'), \tag{8.7}$$

and on the other hand that $T > T'$ implies $T-T' > 0$ and

$$\forall t \geq t', x(t) = x(t+T-T'). \tag{8.8}$$

Proof. a) Let $k \geq k'$ be arbitrary and fixed. Then

$$\widehat{x}(k) \overset{(8.1)}{=} \widehat{x}(k+p) \overset{(8.2)}{=} \widehat{x}(k+p+p').$$

We suppose now that $p > p'$, thus $k+p-p' \geq k'$. We can write that

$$\widehat{x}(k+p-p') \overset{(8.2)}{=} \widehat{x}(k+p) \overset{(8.1)}{=} \widehat{x}(k).$$

□

Theorem 55. *We consider the signals $\widehat{x}, x$.*

a) Let $p, k_1 \geq 1$ and $k' \in \mathbf{N}_{\_}$. Then $p' = k_1 p$ fulfills $p' \geq 1$ and

$$\forall k \geq k', \widehat{x}(k) = \widehat{x}(k+p) \tag{8.9}$$

implies

$$\forall k \geq k', \widehat{x}(k) = \widehat{x}(k+p'). \tag{8.10}$$

b) Let $T > 0, k_1 \geq 1$ and $t' \in \mathbf{R}$. Then $T' = k_1 T$ fulfills $T' > 0$ and

$$\forall t \geq t', x(t) = x(t+T) \tag{8.11}$$

implies

$$\forall t \geq t', x(t) = x(t+T'). \tag{8.12}$$

Proof. This is a consequence of Theorem 54. □

Corollary 2. *a) If $p \in \widehat{P^{x}}$, then $\{p, 2p, 3p, ...\} \subset \widehat{P^{x}}$.*

b) If $T \in P^{x}$, then $\{T, 2T, 3T, ...\} \subset P^{x}$.

Proof. This follows from Theorem 55. □

9. The Set of the Periods

Theorem 56. *a) We suppose that for $\widehat{x} \in \widehat{S}^{(n)}$, the set $\widehat{P}^{\widehat{x}}$ is not empty. Some $\widetilde{p} \geq 1$ exists then with the property*

$$\widehat{P}^{\widehat{x}} = \{\widetilde{p}, 2\widetilde{p}, 3\widetilde{p}, ...\}. \tag{9.1}$$

b) Let $x \in S^{(n)}$ be not eventually constant and we suppose that the set P^x is not empty. Then $\widetilde{T} > 0$ exists such that

$$P^x = \{\widetilde{T}, 2\widetilde{T}, 3\widetilde{T}, ...\}. \tag{9.2}$$

Proof. a) We have $\widehat{P}^{\widehat{x}} \neq \varnothing$ and we denote with $\widetilde{p} \geq 1$ its minimum. The inclusion $\{\widetilde{p}, 2\widetilde{p}, 3\widetilde{p}, ...\} \subset \widehat{P}^{\widehat{x}}$ was stated in Corollary 2. In order to prove that $\widehat{P}^{\widehat{x}} \subset \{\widetilde{p}, 2\widetilde{p}, 3\widetilde{p}, ...\}$, we suppose against all reason that $p' \in \widehat{P}^{\widehat{x}}$ exists with the property that $p' \notin \{\widetilde{p}, 2\widetilde{p}, 3\widetilde{p}, ...\}$ and consequently $k \geq 1$ exists such that $k\widetilde{p} < p' < (k+1)\widetilde{p}$. We infer that $1 \leq p' - k\widetilde{p} < \widetilde{p}$ and, from Theorem 54, page 100 and Theorem 55, page 100 that $p' - k\widetilde{p} \in \widehat{P}^{\widehat{x}}$. As $p' - k\widetilde{p} < \widetilde{p}$, this fact is in contradiction with the supposition that $\widetilde{p} = \min \widehat{P}^{\widehat{x}}$. We infer that $\widehat{P}^{\widehat{x}} \subset \{\widetilde{p}, 2\widetilde{p}, 3\widetilde{p}, ...\}$ holds.

b) We proceed in two steps. At b.i) we prove that $\min P^x$ exists and at b.ii) we prove that the only elements of P^x are the multiples of $\min P^x$.

b.i) We suppose against all reason that $\min P^x$ does not exist, namely that a strictly decreasing sequence $T_k \in P^x, k \in \mathbf{N}$ exists that is convergent to $T = \inf P^x$, where $T \notin P^x$. As x is not eventually constant, the following property

$$\forall t \in \mathbf{R}, \exists t'' > t, x(t'' - 0) \neq x(t'') \tag{9.3}$$

is true, from Lemma 7, page 226. The hypothesis states $\forall k \in \mathbf{N}$ the existence of $t'_k \in \mathbf{R}$ with

$$\forall t \geq t'_k, x(t) = x(t + T_k). \tag{9.4}$$

As t'_k do not depend on T_k, see Theorem 48, page 94, we can suppose that they are all equal with some $t' \in \mathbf{R}$. Property (9.3) implies that we can take a $t'' > t'$ such that $x(t'' - 0) \neq x(t'')$ and we can apply now Lemma 8, page 227 giving

$$\forall k \in \mathbf{N}, x(t'' + T_k - 0) \neq x(t'' + T_k). \tag{9.5}$$

We infer from Lemma 9, page 227 that $N \in \mathbf{N}$ exists with $\forall k \geq N$,

$$x(t'' + T_k - 0) = x(t'' + T_k) = x(t'' + T),$$

contradiction with (9.5). It has resulted that such a sequence $T_k, k \in \mathbf{N}$ does not exist, thus P^x, which is non-empty, has a minimum $\widetilde{T} > 0$.

b.ii) We have from Corollary 2, page 100 that $\{\widetilde{T}, 2\widetilde{T}, 3\widetilde{T}, ...\} \subset P^x$. We prove that $\forall k \geq 1, \forall T' \in (k\widetilde{T}, (k+1)\widetilde{T}), T' \notin P^x$. We suppose against all reason that $k \geq 1$ and $T' \in (k\widetilde{T}, (k+1)\widetilde{T})$ exist such that $T' \in P^x$. This means, from Theorem 54, page 100 and Theorem 55, page 100, that $T' - k\widetilde{T} \in P^x$ and since $T' - k\widetilde{T} < \widetilde{T}$, we have obtained a contradiction with the fact that $\widetilde{T} = \min P^x$. □

Theorem 57. *We suppose that the relation between* $\widehat{x}$ *and* x *is given by*

$$x(t) = \widehat{x}(-1)\cdot\chi_{(-\infty,t_0)}(t) \oplus \widehat{x}(0)\cdot\chi_{[t_0,t_0+h)}(t) \oplus \widehat{x}(1)\cdot\chi_{[t_0+h,t_0+2h)}(t) \oplus \ldots$$
$$\ldots \oplus \widehat{x}(k)\cdot\chi_{[t_0+kh,t_0+(k+1)h)}(t) \oplus \ldots$$

where $t_0 \in \mathbf{R}$ *and* $h > 0$. *If* $\widehat{x}, x$ *are eventually periodic, two possibilities exist:*

a) $\widehat{x}, x$ *are both eventually constant,* $p = 1$ *is the prime period of* $\widehat{x}$ *and* x *has no prime period*[2]*;*

b) neither of $\widehat{x}, x$ *is eventually constant,* $\widetilde{p} > 1$ *is the prime period of* $\widehat{x}$ *and* $\widetilde{T} = \widetilde{p}h$ *is the prime period of* x.

Proof. $\widehat{x}, x$ are simultaneously eventually constant or not. Let us suppose that they are not eventually constant and we prove b). Theorem 52, page 98 shows that $p \in \widehat{P^{\widehat{x}}} \Longrightarrow T = ph \in P^x$ and conversely, Theorem 53, page 99 shows that $T \in P^x \Longrightarrow p = \frac{T}{h} \in \widehat{P^{\widehat{x}}}$. From Theorem 56 we have that $\widehat{P^{\widehat{x}}} = \{\widetilde{p}, 2\widetilde{p}, 3\widetilde{p}, \ldots\}, P^x = \{\widetilde{T}, 2\widetilde{T}, 3\widetilde{T}, \ldots\}$, thus $\widetilde{T} = \widetilde{p}h$. □

Theorem 58. *We suppose that* $\widehat{x}, x$ *are not eventually constant.*

a) If $\widehat{x}$ *is eventually periodic and* $\widehat{\omega}(\widehat{x}) = \{\mu^1, \ldots, \mu^s\}$, $s \geq 2$ *then*

$$\widehat{P^{\widehat{x}}} = \widehat{P^{\widehat{x}}_{\mu^1}} \cap \ldots \cap \widehat{P^{\widehat{x}}_{\mu^s}}. \tag{9.6}$$

b) We suppose that x *is eventually periodic and* $\omega(x) = \{\mu^1, \ldots, \mu^s\}, s \geq 2$ *holds. In this case*

$$P^x = P^x_{\mu^1} \cap \ldots \cap P^x_{\mu^s}. \tag{9.7}$$

Proof. The equivalence between $(1.1)_{page\,88}$ and $(2.1)_{page\,90}$ gives the truth of (9.6), while the equivalence between $(1.5)_{page\,88}$ and $(2.3)_{page\,90}$ gives the truth of (9.7). □

10. Necessity Conditions of Eventual Periodicity

Theorem 59. *Let* $\widehat{x} \in \widehat{S}^{(n)}$ *with* $\widehat{\omega}(\widehat{x}) = \{\mu^1, \ldots, \mu^s\}$. *We suppose that* $\widehat{x}$ *is eventually periodic with the period* $p \geq 1$ *and the limit of periodicity* $k' \in \mathbf{N}_{\_}$. *Then* $n_1^i, n_2^i, \ldots, n_{k_i}^i \in \{k', k' + 1, \ldots, k' + p - 1\}$ *exist,* $k_i \geq 1$, *such that*

$$\widehat{\mathbf{T}}^{\widehat{x}}_{\mu^i} \cap \{k', k'+1, k'+2, \ldots\} = \bigcup_{k \in \mathbf{N}} \{n_1^i + kp, n_2^i + zp, \ldots, n_{k_i}^i + kp\} \tag{10.1}$$

for $i \in \{1, \ldots, s\}$.

Proof. If $\widehat{x}$ is eventually periodic with the period p and the limit of periodicity k', then every $\mu^i \in \widehat{\omega}(\widehat{x})$ is eventually periodic with the period p, from Theorem 58, and the limit of periodicity $k', i \in \{1, \ldots, s\}$, from Theorem 49, page 95 and we apply Theorem 37, page 81. □

[2]In other words $\widehat{P^{\widehat{x}}} = \{1, 2, 3, \ldots\}$ and $P^x = (0, \infty)$.

Theorem 60. *We consider the non eventually constant signal $x \in S^{(n)}$ and we put its omega limit set under the form $\omega(x) = \{\mu^1, ..., \mu^s\}$, $s \geq 2$. We suppose that x is eventually periodic with the period $T > 0$ and the limit of periodicity $t' \in \mathbf{R}$. Then $a_1^i, b_1^i, a_2^i, b_2^i, ..., a_{k_i}^i, b_{k_i}^i \in \mathbf{R}$ exist,*

$$t' \leq a_1^i < b_1^i < a_2^i < b_2^i < ... < a_{k_i}^i < b_{k_i}^i \leq t' + T, \tag{10.2}$$

with $k_i \geq 1$, $i \in \{1, ..., s\}$, such that

$$[a_1^i, b_1^i) \cup [a_2^i, b_2^i) \cup ... \cup [a_{k_i}^i, b_{k_i}^i) = \mathbf{T}_{\mu^i}^x \cap [t', t'+T), \tag{10.3}$$

$$\mathbf{T}_{\mu^i}^x \cap [t', \infty) = \bigcup_{k \in \mathbf{N}} ([a_1^i + kT, b_1^i + kT) \cup [a_2^i + kT, b_2^i + kT) \cup ... \cup [a_{k_i}^i + kT, b_{k_i}^i + kT)) \tag{10.4}$$

hold for $i \in \{1, ..., s\}$.

Proof. x is eventually periodic, with the period T and the limit of periodicity t', thus $\forall i \in \{1, ..., s\}, \mu^i$ is eventually periodic with the period T , from Theorem 58, page 102, and the limit of periodicity t', from Theorem 49, page 95. We apply Theorem 38, page 82. □

Example 22. *The eventually periodic signal $x \in S^{(1)}$,*

$$x(t) = \chi_{(-\infty,0)}(t) \oplus \chi_{[1,5)}(t) \oplus \chi_{[6,7)}(t) \oplus \chi_{[8,10)}(t) \oplus \chi_{[11,12)}(t) \oplus \chi_{[13,15)}(t) \oplus ...$$

fulfills $\mu^1 = 1, \mu^2 = 0, k_1 = k_2 = 2, T = 5, t' = 3, a_1^1 = 3, b_1^1 = 5, a_2^1 = 6, b_2^1 = 7, a_1^2 = 5, b_1^2 = 6, a_2^2 = 7, b_2^2 = 8$.

11. Sufficiency Conditions of Eventual Periodicity

Theorem 61. *Let $\widehat{x} \in \widehat{S}^{(n)}$, $\widehat{\omega}(\widehat{x}) = \{\mu^1, ..., \mu^s\}$ and $p \geq 1, k' \in \mathbf{N}_\_$. We ask that for any $i \in \{1, ..., s\}$, the numbers $n_1^i, n_2^i, ..., n_{k_i}^i \in \{k', k'+1, ..., k'+p-1\}$ exist, $k_i \geq 1$, making*

$$\widehat{\mathbf{T}}_{\mu^i}^{\widehat{x}} \cap \{k', k'+1, k'+2, ...\} = \bigcup_{k \in \mathbf{N}} \{n_1^i + kp, n_2^i + kp, ..., n_{k_i}^i + kp\} \tag{11.1}$$

true. Then $\widehat{x}$ is eventually periodic with the period $p \geq 1$ and the limit of periodicity $k' \in \mathbf{N}_\_ : \forall i \in \{1, ..., s\}$,

$$\forall k \in \widehat{\mathbf{T}}_{\mu^i}^{\widehat{x}} \cap \{k', k'+1, k'+2, ...\}, \{k + zp | z \in \mathbf{Z}\} \cap \{k', k'+1, k'+2, ...\} \subset \widehat{\mathbf{T}}_{\mu^i}^{\widehat{x}}. \tag{11.2}$$

Proof. We suppose that $\forall i \in \{1, ..., s\}, k_i \geq 1$ and $n_1^i, n_2^i, ..., n_{k_i}^i \in \{k', k'+1, ..., k'+p-1\}$ exist such that (11.1) holds. We infer from Theorem 39, page 83 that $\mu^1, ..., \mu^s$ are all eventually periodic with the period p and the limit of periodicity k', i.e. $\widehat{x}$ is eventually periodic with the period p and the limit of periodicity k': the equivalence between $(2.1)_{page\,90}$ and $(1.1)_{page\,88}$ was proved at Theorem 45, page 90. □

Theorem 62. *Let the signal $x \in S^{(n)}$, with $\omega(x) = \{\mu^1, ..., \mu^s\}$, $s \geq 2$ and $T > 0, t' \in \mathbf{R}$. For all $i \in \{1, ..., s\}$, the numbers $a_1^i, b_1^i, a_2^i, b_2^i, ..., a_{k_i}^i, b_{k_i}^i \in \mathbf{R}$, $k_i \geq 1$ are given with the property that*

$$t' \leq a_1^i < b_1^i < a_2^i < b_2^i < ... < a_{k_i}^i < b_{k_i}^i \leq t' + T, \tag{11.3}$$

$$\mathbf{T}_{\mu^i}^x \cap [t', \infty) = \bigcup_{k \in \mathbf{N}} ([a_1^i + kT, b_1^i + kT) \cup [a_2^i + kT, b_2^i + kT) \cup ... \cup [a_{k_i}^i + kT, b_{k_i}^i + kT)) \tag{11.4}$$

hold. Then x is eventually periodic with the period T and the limit of periodicity t' : $\forall i \in \{1, ..., s\}$,

$$\forall t \in \mathbf{T}_{\mu^i}^x \cap [t', \infty), \{t + zT | z \in \mathbf{Z}\} \cap [t', \infty) \subset \mathbf{T}_{\mu^i}^x. \tag{11.5}$$

Proof. Theorem 40, page 83 proves that $\mu^1, ..., \mu^s$ are eventually periodic with the period T and the limit of periodicity t'. In addition, Theorem 45, page 90 proves the equivalence between $(1.5)_{page\,88}$ and $(2.4)_{page\,90}$, showing that in these circumstances x is eventually periodic with the period T and the limit of periodicity t'. □

12. A Special Case

Theorem 63. *Let the signal $\widehat{x} \in \widehat{S}^{(n)}$ with $\widehat{\omega}(\widehat{x}) = \{\mu^1, ..., \mu^s\}$ and $p \geq 1, k' \in \mathbf{N}_\_$. We ask that $\forall i \in \{1, ..., s\}, n^i \in \{k', k'+1, ..., k'+p-1\}$ exists such that*

$$\widehat{\mathbf{T}}_{\mu^i}^{\widehat{x}} \cap \{k', k'+1, k'+2, ...\} = \{n^i, n^i + p, n^i + 2p, ...\}. \tag{12.1}$$

a) We have: $\forall i \in \{1, ..., s\}$,

$$\forall k \in \widehat{\mathbf{T}}_{\mu^i}^{\widehat{x}} \cap \{k', k'+1, k'+2, ...\}, \{k + zp | z \in \mathbf{Z}\} \cap \{k', k'+1, k'+2, ...\} \subset \widehat{\mathbf{T}}_{\mu^i}^{\widehat{x}}. \tag{12.2}$$

b) p is the prime period of $\widehat{x}$: for any p' and k'' with $\forall i \in \{1, ..., s\}$,

$$\begin{gathered}\forall k \in \widehat{\mathbf{T}}_{\mu^i}^{\widehat{x}} \cap \{k'', k''+1, k''+2, ...\}, \\ \{k + zp' | z \in \mathbf{Z}\} \cap \{k'', k''+1, k''+2, ...\} \subset \widehat{\mathbf{T}}_{\mu^i}^{\widehat{x}},\end{gathered} \tag{12.3}$$

we have $p' \in \{p, 2p, 3p, ...\}$.

Proof. This follows from Theorem 41, page 84. □

Theorem 64. *We consider the signal x with $\omega(x) = \{\mu^1, ..., \mu^s\}$ and $T > 0, t' \in \mathbf{R}$. For all $i \in \{1, ..., s\}$, the intervals $[a^i, b^i) \subset [t', t'+T)$ are given with*

$$\mathbf{T}_{\mu^i}^x \cap [t', \infty) = [a^i, b^i) \cup [a^i + T, b^i + T) \cup [a^i + 2T, b^i + 2T) \cup .... \tag{12.4}$$

true. Then

a) x is eventually periodic with the period T and the limit of periodicity t' : $\forall i \in \{1, ..., s\}$,

$$\forall t \in \mathbf{T}_{\mu^i}^x \cap [t', \infty), \{t + zT | z \in \mathbf{Z}\} \cap [t', \infty) \subset \mathbf{T}_{\mu^i}^x; \tag{12.5}$$

b) if x is not eventually constant (i.e. if $s \geq 2$), T is the prime period of x, i.e. for any T' and t'' with $\forall i \in \{1, ..., s\}$,

$$\forall t \in \mathbf{T}_{\mu^i}^x \cap [t'', \infty), \{t + zT' | z \in \mathbf{Z}\} \cap [t'', \infty) \subset \mathbf{T}_{\mu^i}^x, \tag{12.6}$$

we infer $T' \in \{T, 2T, 3T, ...\}$.

Proof. This follows from Theorem 42, page 84. □

Example 23. *The signal* $x \in S^{(1)}$,

$$x(t) = \chi_{[1,3)}(t) \oplus \chi_{[6,8)}(t) \oplus \chi_{[11,13)}(t) \oplus \chi_{[16,18)}(t) \oplus ...$$

fulfills $\omega(x) = \{0,1\}, T = 5, t' = -2, \mathbf{T}_0^x \cap [-2,\infty) = [-2,1) \cup [3,6) \cup [8,11) \cup ...$, $\mathbf{T}_1^x \cap [-2,\infty) = [1,3) \cup [6,8) \cup [11,13) \cup ...$ *and* $P^x = \{5,10,15,...\}$.

13. Changing the Order of the Quantifiers

Theorem 65. [3]*a) The statements*

$$\begin{cases} \exists p \geq 1, \exists k' \in \mathbf{N}_\_, \forall \mu \in \widehat{\omega}(\widehat{x}), \forall k \in \widehat{\mathbf{T}}_\mu^{\widehat{x}} \cap \{k', k'+1, k'+2, ...\}, \\ \{k + zp | z \in \mathbf{Z}\} \cap \{k', k'+1, k'+2, ...\} \subset \widehat{\mathbf{T}}_\mu^{\widehat{x}}, \end{cases} \tag{13.1}$$

$$\begin{cases} \exists p \geq 1, \forall \mu \in \widehat{\omega}(\widehat{x}), \exists k' \in \mathbf{N}_\_, \forall k \in \widehat{\mathbf{T}}_\mu^{\widehat{x}} \cap \{k', k'+1, k'+2, ...\}, \\ \{k + zp | z \in \mathbf{Z}\} \cap \{k', k'+1, k'+2, ...\} \subset \widehat{\mathbf{T}}_\mu^{\widehat{x}}, \end{cases} \tag{13.2}$$

$$\begin{cases} \exists k' \in \mathbf{N}_\_, \forall \mu \in \widehat{\omega}(\widehat{x}), \exists p \geq 1, \forall k \in \widehat{\mathbf{T}}_\mu^{\widehat{x}} \cap \{k', k'+1, k'+2, ...\}, \\ \{k + zp | z \in \mathbf{Z}\} \cap \{k', k'+1, k'+2, ...\} \subset \widehat{\mathbf{T}}_\mu^{\widehat{x}}, \end{cases} \tag{13.3}$$

$$\begin{cases} \forall \mu \in \widehat{\omega}(\widehat{x}), \exists p \geq 1, \exists k' \in \mathbf{N}_\_, \forall k \in \widehat{\mathbf{T}}_\mu^{\widehat{x}} \cap \{k', k'+1, k'+2, ...\}, \\ \{k + zp | z \in \mathbf{Z}\} \cap \{k', k'+1, k'+2, ...\} \subset \widehat{\mathbf{T}}_\mu^{\widehat{x}} \end{cases} \tag{13.4}$$

are equivalent.

b) The real time statements

$$\begin{cases} \exists T > 0, \exists t' \in \mathbf{R}, \forall \mu \in \omega(x), \\ \forall t \in \mathbf{T}_\mu^x \cap [t', \infty), \{t + zT | z \in \mathbf{Z}\} \cap [t', \infty) \subset \mathbf{T}_\mu^x), \end{cases} \tag{13.5}$$

$$\begin{cases} \exists T > 0, \forall \mu \in \omega(x), \exists t' \in \mathbf{R}, \\ \forall t \in \mathbf{T}_\mu^x \cap [t', \infty), \{t + zT | z \in \mathbf{Z}\} \cap [t', \infty) \subset \mathbf{T}_\mu^x, \end{cases} \tag{13.6}$$

$$\begin{cases} \exists t' \in \mathbf{R}, \forall \mu \in \omega(x), \exists T > 0, \\ \forall t \in \mathbf{T}_\mu^x \cap [t', \infty), \{t + zT | z \in \mathbf{Z}\} \cap [t', \infty) \subset \mathbf{T}_\mu^x, \end{cases} \tag{13.7}$$

$$\begin{cases} \forall \mu \in \omega(x), \exists T > 0, \exists t' \in \mathbf{R}, \\ \forall t \in \mathbf{T}_\mu^x \cap [t', \infty), \{t + zT | z \in \mathbf{Z}\} \cap [t', \infty) \subset \mathbf{T}_\mu^x \end{cases} \tag{13.8}$$

are also equivalent.

Proof. a) The implications (13.1) $\Longrightarrow$ (13.2) $\Longrightarrow$ (13.4), (13.1) $\Longrightarrow$ (13.3) $\Longrightarrow$ (13.4) are obvious, thus we give the proof of (13.4)$\Longrightarrow$(13.1). Let $\widehat{\omega}(\widehat{x}) = \{\mu^1, ..., \mu^s\}$. Property (13.4) states that for an arbitrary $i \in \{1, ..., s\}$, some $p_i \geq 1$ and $k_i' \in \mathbf{N}_\_$ exist such that

$$\forall k \in \widehat{\mathbf{T}}_{\mu^i}^{\widehat{x}} \cap \{k_i', k_i'+1, k_i'+2, ...\}, \{k + zp_i | z \in \mathbf{Z}\} \cap \{k_i', k_i'+1, k_i'+2, ...\} \subset \widehat{\mathbf{T}}_{\mu^i}^{\widehat{x}}. \tag{13.9}$$

[3]This theorem is partially without proof.

We denote $k' = \max\{k'_1, ..., k'_s\}$. From (13.9) and from Lemma 3, page 224 we infer that

$$\forall k \in \widehat{\mathbf{T}}^{\widehat{x}}_{\mu^i} \cap \{k', k'+1, k'+2, ...\}, \{k + zp_i | z \in \mathbf{Z}\} \cap \{k', k'+1, k'+2, ...\} \subset \widehat{\mathbf{T}}^{\widehat{x}}_{\mu^i} \quad (13.10)$$

is true for any $i \in \{1, ..., s\}$. We fix $i \in \{1, ..., s\}$ and $k \in \widehat{\mathbf{T}}^{\widehat{x}}_{\mu^i} \cap \{k', k'+1, k'+2, ...\}$ arbitrary. We denote $p = p_1 \cdot ... \cdot p_s$. We have $p \geq 1$ and

$$\{k + zp | z \in \mathbf{Z}\} \cap \{k', k'+1, k'+2, ...\} \subset$$

$$\subset \{k + zp_i | z \in \mathbf{Z}\} \cap \{k', k'+1, k'+2, ...\} \overset{(13.10)}{\subset} \widehat{\mathbf{T}}^{\widehat{x}}_{\mu^i},$$

thus (13.1) is true.

b) The implications (13.5)$\Longrightarrow$(13.6)$\Longrightarrow$(13.8), (13.5)$\Longrightarrow$(13.7)$\Longrightarrow$(13.8) are obvious. □

Remark 99. *Stating periodicity properties may depend in general on the order of the quantifiers. This issue is trivial when quantifiers of the same kind occur ($\exists, \exists$ or $\forall, \forall$) and it is not trivial when quantifiers of different kinds occur ($\exists, \forall$ or $\forall, \exists$). Our aim in the previous Theorem is to show that the eventual periodicity properties are independent on the order of the quantifiers. However the fact that any of (13.6), (13.7), (13.8) implies (13.5) could not be proved so far. Such a proof would be important, since we are tempted to define the eventual periodicity of the signals by (13.4), (13.8) (each point of the omega limit set is eventually periodic) and to use (13.1) or (13.5) instead (a common period exists for all the points of the omega limit set).*

Remark 100. *Let $\widehat{\omega}(\widehat{x}) = \{\mu^1, ..., \mu^s\}$. The implication (13.4)$\Longrightarrow$(13.1) of Theorem 65 showed that if $\mu^1, ..., \mu^s$ are all eventually periodic: $\widehat{P}^{\widehat{x}}_{\mu^1} \neq \emptyset$ and ... and $\widehat{P}^{\widehat{x}}_{\mu^s} \neq \emptyset$ then $\widehat{P}^{\widehat{x}}_{\mu^1} \cap ... \cap \widehat{P}^{\widehat{x}}_{\mu^s} \neq \emptyset$. Since the equality*

$$\widehat{P}^{\widehat{x}} = \widehat{P}^{\widehat{x}}_{\mu^1} \cap ... \cap \widehat{P}^{\widehat{x}}_{\mu^s}$$

is always true, even when the left hand term and the right hand term are both empty, we conclude that the eventual periodicity of $\widehat{x}$ expressed by (13.1) (i.e. $\widehat{P}^{\widehat{x}} \neq \emptyset$) and the eventual periodicity of all the points of the orbit expressed by (13.4) (i.e. $\widehat{P}^{\widehat{x}}_{\mu^1} \neq \emptyset$ and ... and $\widehat{P}^{\widehat{x}}_{\mu^s} \neq \emptyset$) are equivalent. We could not prove that this is true in the real time case, even if, for $\omega(x) = \{\mu^1, ..., \mu^s\}$,

$$P^x = P^x_{\mu^1} \cap ... \cap P^x_{\mu^s}$$

is also true, see Theorem 58, page 102.

Remark 101. *From the previous remark we infer that we have, in particular, the property*

$$\forall \mu \in \widehat{\omega}(\widehat{x}), \forall \mu' \in \widehat{\omega}(\widehat{x}), (\widehat{P}^{\widehat{x}}_{\mu} \neq \emptyset \text{ and } \widehat{P}^{\widehat{x}}_{\mu'} \neq \emptyset) \Longrightarrow \widehat{P}^{\widehat{x}}_{\mu} \cap \widehat{P}^{\widehat{x}}_{\mu'} \neq \emptyset,$$

while the truth of the implication

$$\forall \mu \in \omega(x), \forall \mu' \in \omega(x), (P^x_{\mu} \neq \emptyset \text{ and } P^x_{\mu'} \neq \emptyset) \Longrightarrow P^x_{\mu} \cap P^x_{\mu'} \neq \emptyset$$

was not proved so far.

14. The Hypothesis P

Definition 20. *We consider the signal x. If*

$$(\forall \mu \in \omega(x), P_\mu^x \neq \varnothing) \Longrightarrow \bigcap_{\mu \in \omega(x)} P_\mu^x \neq \varnothing,$$

we say that x ***fulfills the hypothesis*** *P.*

Theorem 66. *a) We suppose that the signal $\widehat{x} \in \widehat{S}^{(n)}$ is not eventually constant, we denote $\widehat{\omega}(\widehat{x}) = \{\mu^1, ..., \mu^s\}$ and we ask that $\forall i \in \{1, ..., s\}$, the set $\widehat{P}_{\mu^i}^{\widehat{x}}$ is not empty. We denote with $\widetilde{p}_i \geq 1, \widetilde{p} \geq 1$ the numbers that fulfill*

$$\widehat{P}_{\mu^i}^{\widehat{x}} = \{\widetilde{p}_i, 2\widetilde{p}_i, 3\widetilde{p}_i, ...\}, \tag{14.1}$$

$$\widehat{P}^{\widehat{x}} = \{\widetilde{p}, 2\widetilde{p}, 3\widetilde{p}, ...\}. \tag{14.2}$$

Then

$$\widetilde{p} = n_1 \widetilde{p}_1 = ... = n_s \widetilde{p}_s, \tag{14.3}$$

where $n_1 \geq 1, ..., n_s \geq 1$ are relatively prime ($\widetilde{p}$ is the least common multiple of $\widetilde{p}_1, ..., \widetilde{p}_s$).

b) We suppose that the signal $x \in S^{(n)}$ is not eventually constant and that it fulfills the hypothesis P. We denote $\omega(x) = \{\mu^1, ..., \mu^s\}$ and we ask that $\forall i \in \{1, ..., s\}$, the set $P_{\mu^i}^x$ is not empty. We denote with $\widetilde{T}_i > 0, \widetilde{T} > 0$ the numbers that satisfy

$$P_{\mu^i}^x = \{\widetilde{T}_i, 2\widetilde{T}_i, 3\widetilde{T}_i, ...\}, \tag{14.4}$$

$$P^x = \{\widetilde{T}, 2\widetilde{T}, 3\widetilde{T}, ...\}. \tag{14.5}$$

We have

$$\widetilde{T} = n_1 \widetilde{T}_1 = ... = n_s \widetilde{T}_s, \tag{14.6}$$

where $n_1 \geq 1, ..., n_s \geq 1$ are relatively prime.

Proof. a) Any $p \in \widehat{P}^{\widehat{x}}$ belongs to $\widehat{P}_{\mu^1}^{\widehat{x}} \cap ... \cap \widehat{P}_{\mu^s}^{\widehat{x}}$, thus $n_1' \geq 1, p_1 \in \widehat{P}_{\mu^1}^{\widehat{x}}, ..., n_s' \geq 1, p_s \in \widehat{P}_{\mu^s}^{\widehat{x}}$ exist such that

$$p = n_1' p_1 = ... = n_s' p_s. \tag{14.7}$$

But each p_i is a multiple of $\widetilde{p}_i$, thus $n_1'' \geq 1, ..., n_s'' \geq 1$ exist with

$$p_1 = n_1'' \widetilde{p}_1, ..., p_s = n_s'' \widetilde{p}_s. \tag{14.8}$$

We replace the equations (14.8) in (14.7) and we get

$$p = n_1 \widetilde{p}_1 = ... = n_s \widetilde{p}_s, \tag{14.9}$$

where $n_1 = n_1' n_1'', ..., n_s = n_s' n_s''$. When $n_1, ..., n_s$ are relatively prime, $p = \min \widehat{P}^{\widehat{x}}$.

b) As x fulfills the hypothesis P and $P_{\mu^1}^x \neq \varnothing, ..., P_{\mu^s}^x \neq \varnothing$, we have that in the equation

$$P^x = P_{\mu^1}^x \cap ... \cap P_{\mu^s}^x$$

both terms are non-empty. From this moment the reasoning is the same like at a). □

Chapter 7

Periodic Points

First we give in Section 1 several properties that are equivalent to the periodicity of a point. These properties were previously used to characterize the constancy of the signals. A discussion of these properties is made in Section 2.

Section 3 shows that the periodic points are accessed at least once in a time interval with the length of a period.

The independence of the real time periodicity of μ on the initial time t' of $x =$ limit of periodicity of μ and also the bounds of t' are the topics of Section 4.

The property of constancy from Section 5 is interesting by itself, but it is also useful in treating the discrete time vs the real time periodic points, representing the topic of Section 6.

One might be tempted to think that the relation between $\widehat{\mathbf{T}}_{\mu}^{\widehat{x}}, \mathbf{T}_{\mu}^{x}$ and $\widehat{P}_{\mu}^{\widehat{x}}, P_{\mu}^{x}$ is closer than it really is. Some examples and comments on this relation are given in Section 7.

The fact that the sums, the differences and the multiples of the periods are periods is formalized in Section 8.

The important topic of existence of the prime period is treated in Section 9, together with the form of $\widehat{P}_{\mu}^{\widehat{x}}, P_{\mu}^{x}$.

Necessary conditions, respectively sufficient conditions of periodicity of μ, related with the form of $\widehat{\mathbf{T}}_{\mu}^{\widehat{x}}, \mathbf{T}_{\mu}^{x}$ are given in Sections 10, respectively 11.

Section 12 deals with a special case of periodicity, applying results from Section 10 and Section 11. The point is that in this special case we know the precise value of the prime period.

In Section 13 we show that by forgetting some first values of $\widehat{x}, x$ we get the same sets of periods $\widehat{P}_{\mu}^{\widehat{x}}, P_{\mu}^{x}$. This natural observation connects the periodicity of μ with its eventual periodicity.

Some ideas concerning further research on the periodic points are presented in Section 14.

1. Properties which Are Equivalent to the Periodicity of a Point

Remark 102. *The properties of periodicity of the points were present in the second group of constancy properties of the signals from Theorem 23, page 47 (and the third group, The-*

orem 24, page 52), thus (1.1),...,(1.6) will be compared with $(2.1)_{page\,47},\dots,(2.6)_{page\,47}$ *and (1.7),...,(1.12) will be compared with* $(2.7)_{page\,47},\dots,(2.12)_{page\,47}$. *We make also the associations* (1.4)-$(4.1)_{page\,59},\dots,(1.6)$-$(4.3)_{page\,59}$ *and* (1.10)-$(4.4)_{page\,59},\dots,(1.12)$-$(4.6)_{page\,59}$ *with the fourth group of constancy properties from Theorem 25, page 59.*

Theorem 67. *We consider the signals* $\widehat{x}\in\widehat{S}^{(n)},x\in S^{(n)}$.

a) The following statements are equivalent for any $p\geq 1$ *and* $\mu\in\widehat{Or}(\widehat{x})$:

$$\forall k\in\widehat{\mathbf{T}}_{\mu}^{\widehat{x}},\{k+zp|z\in\mathbf{Z}\}\cap\mathbf{N}_{\_}\subset\widehat{\mathbf{T}}_{\mu}^{\widehat{x}},\tag{1.1}$$

$$\begin{cases}\forall k'\in\mathbf{N}_{\_},\forall k\in\widehat{\mathbf{T}}_{\mu}^{\widehat{x}}\cap\{k',k'+1,k'+2,...\},\\ \{k+zp|z\in\mathbf{Z}\}\cap\{k',k'+1,k'+2,...\}\subset\widehat{\mathbf{T}}_{\mu}^{\widehat{x}},\end{cases}\tag{1.2}$$

$$\forall k''\in\mathbf{N},\forall k\in\widehat{\mathbf{T}}_{\mu}^{\widehat{\sigma}^{k''}(\widehat{x})},\{k+zp|z\in\mathbf{Z}\}\cap\mathbf{N}_{\_}\subset\widehat{\mathbf{T}}_{\mu}^{\widehat{\sigma}^{k''}(\widehat{x})},\tag{1.3}$$

$$\begin{cases}\forall k\in\mathbf{N}_{\_},\widehat{x}(k)=\mu\Longrightarrow\\ \Longrightarrow(\widehat{x}(k)=\widehat{x}(k+p)\ and\ k-p\geq-1\Longrightarrow\widehat{x}(k)=\widehat{x}(k-p)),\end{cases}\tag{1.4}$$

$$\begin{cases}\forall k'\in\mathbf{N}_{\_},\forall k\geq k',\widehat{x}(k)=\mu\Longrightarrow\\ \Longrightarrow(\widehat{x}(k)=\widehat{x}(k+p)\ and\ k-p\geq k'\Longrightarrow\widehat{x}(k)=\widehat{x}(k-p)),\end{cases}\tag{1.5}$$

$$\begin{cases}\forall k''\in\mathbf{N},\forall k\in\mathbf{N}_{\_},\widehat{\sigma}^{k''}(\widehat{x})(k)=\mu\Longrightarrow\\ \Longrightarrow(\widehat{\sigma}^{k''}(\widehat{x})(k)=\widehat{\sigma}^{k''}(\widehat{x})(k+p)\ and\\ and\ k-p\geq-1\Longrightarrow\widehat{\sigma}^{k''}(\widehat{x})(k)=\widehat{\sigma}^{k''}(\widehat{x})(k-p)).\end{cases}\tag{1.6}$$

b) The following statements are also equivalent for any $T>0$ *and* $\mu\in Or(x)$:

$$\exists t'\in I^{x},\forall t\in\mathbf{T}_{\mu}^{x}\cap[t',\infty),\{t+zT|z\in\mathbf{Z}\}\cap[t',\infty)\subset\mathbf{T}_{\mu}^{x},\tag{1.7}$$

$$\exists t'\in I^{x},\forall t_1'\geq t',\forall t\in\mathbf{T}_{\mu}^{x}\cap[t_1',\infty),\{t+zT|z\in\mathbf{Z}\}\cap[t_1',\infty)\subset\mathbf{T}_{\mu}^{x},\tag{1.8}$$

$$\begin{cases}\forall t''\in\mathbf{R},\exists t'\in I^{\sigma^{t''}(x)},\\ \forall t\in\mathbf{T}_{\mu}^{\sigma^{t''}(x)}\cap[t',\infty),\{t+zT|z\in\mathbf{Z}\}\cap[t',\infty)\subset\mathbf{T}_{\mu}^{\sigma^{t''}(x)},\end{cases}\tag{1.9}$$

$$\begin{cases}\exists t'\in I^{x},\forall t\geq t',\\ x(t)=\mu\Longrightarrow(x(t)=x(t+T)\ and\ t-T\geq t'\Longrightarrow x(t)=x(t-T)),\end{cases}\tag{1.10}$$

$$\begin{cases}\exists t'\in I^{x},\forall t_1'\geq t',\forall t\geq t_1',x(t)=\mu\Longrightarrow\\ \Longrightarrow(x(t)=x(t+T)\ and\ t-T\geq t_1'\Longrightarrow x(t)=x(t-T)),\end{cases}\tag{1.11}$$

$$\begin{cases}\forall t''\in\mathbf{R},\exists t'\in I^{\sigma^{t''}(x)},\\ \forall t\geq t',\sigma^{t''}(x)(t)=\mu\Longrightarrow(\sigma^{t''}(x)(t)=\sigma^{t''}(x)(t+T)\ and\\ and\ t-T\geq t'\Longrightarrow\sigma^{t''}(x)(t)=\sigma^{t''}(x)(t-T)).\end{cases}\tag{1.12}$$

Proof. The proof of the implications

$$(1.1)\Longrightarrow(1.2)\Longrightarrow(1.3)\Longrightarrow(1.4)\Longrightarrow(1.5)\Longrightarrow(1.6)$$

follows from the proof of Theorem 23, page 47.

(1.6)$\Longrightarrow$(1.1) We can use (1.4) that is a special case of (1.6) when $k''=0$. Let $k\in\widehat{\mathbf{T}}_{\mu}^{\widehat{x}}$ and $z\in\mathbf{Z}$ with $k+zp\geq -1$ and we have the following possibilities.

Case $z>0$,

$$\mu=\widehat{x}(k)\overset{(1.4)}{=}\widehat{x}(k+p)\overset{(1.4)}{=}\widehat{x}(k+2p)\overset{(1.4)}{=}\ldots\overset{(1.4)}{=}\widehat{x}(k+zp);$$

Case $z=0$,

$$\mu=\widehat{x}(k)=\widehat{x}(k+zp);$$

Case $z<0$,

$$\mu=\widehat{x}(k)\overset{(1.4)}{=}\widehat{x}(k-p)\overset{(1.4)}{=}\widehat{x}(k-2p)\overset{(1.4)}{=}\ldots\overset{(1.4)}{=}\widehat{x}(k+zp).$$

In all these cases $k+zp\in\widehat{\mathbf{T}}_{\mu}^{\widehat{x}}$.

b) The proof of the implications

$$(1.7)\Longrightarrow(1.8)\Longrightarrow(1.9)\Longrightarrow(1.10)\Longrightarrow(1.11)\Longrightarrow(1.12)$$

follows from the proof of Theorem 23, page 47.

(1.12)$\Longrightarrow$(1.7) We write (1.12) in the special case when t'' fulfills $\forall t\leq t'', x(t)=x(-\infty+0)$ and consequently $\sigma^{t''}(x)=x$, $t'\in I^x$ and

$$\forall t\geq t', x(t)=\mu\Longrightarrow(x(t)=x(t+T)\text{ and }t-T\geq t'\Longrightarrow x(t)=x(t-T))\qquad(1.13)$$

hold. We have $\mathbf{T}_{\mu}^{x}\cap[t',\infty)\neq\emptyset$, so let $t\in\mathbf{T}_{\mu}^{x}\cap[t',\infty)$ and $z\in\mathbf{Z}$ arbitrary with $t+zT\geq t'$. We get the following possibilities.

Case $z>0$,

$$\mu=x(t)\overset{(1.13)}{=}x(t+T)\overset{(1.13)}{=}x(t+2T)\overset{(1.13)}{=}\ldots\overset{(1.13)}{=}x(t+zT);$$

Case $z=0$,

$$\mu=x(t)=x(t+zT);$$

Case $z<0$,

$$\mu=x(t)\overset{(1.13)}{=}x(t-T)\overset{(1.13)}{=}x(t-2T)\overset{(1.13)}{=}\ldots\overset{(1.13)}{=}x(t+zT)$$

and consequently in all these situations $t+zT\in\mathbf{T}_{\mu}^{x}$. Property (1.7) is true. □

Example 24. *We give the following example of signal $\widehat{x}\in\widehat{S}^{(1)}$ where none of $0,1\in\widehat{Or}(\widehat{x})$ is periodic or eventually periodic:*

$$\widehat{x}=0,\underbrace{1}_{1},0,\underbrace{1,1}_{2},0,\underbrace{1,1,1}_{3},0,\underbrace{1,1,1,1}_{4},0,\ldots$$

This signal is similar with that of Example 20, page 90.

Example 25. *Let the signal* $\widehat{x} \in \widehat{S}^{(1)}$ *and we presume that*

$$\widehat{x} = 0, \widehat{x}(0), \widehat{x}(1), 0, \widehat{x}(3), \widehat{x}(4), 0, \widehat{x}(6), ...$$

Then 0 *is a periodic point of* $\widehat{Or}(\widehat{x})$ *and it has the period* 3. *In particular if* $\widehat{x}(0), \widehat{x}(1), \widehat{x}(3), \widehat{x}(4), \widehat{x}(6), ...$ *are all equal with* 1 *then* 3 *is the prime period of* 0 *and if they are all equal with* 0 *then* 1 *is the prime period of* 0.

Example 26. *The signal* $x \in S^{(1)}$,

$$x(t) = \chi_{(-\infty,0)}(t) \oplus \chi_{[1,3)}(t) \oplus \chi_{[7,9)}(t) \oplus \chi_{[10,12)}(t) \oplus \chi_{[16,18)}(t) \oplus \chi_{[19,21)}(t) \oplus ...$$

fulfills $-1 \in I^x$ *and*

$$\forall t \in \mathbf{T}_1^x \cap [-1, \infty), \{t + z9 | z \in \mathbf{Z}\} \cap [-1, \infty) \subset \mathbf{T}_1^x,$$

i.e. the point 1 *has the period* 9. *To be noticed how the couple* $[1,3), [7,9)$ *of intervals 'generates' periodicity.*

2. Discussion

Remark 103. *From the periodicity properties (1.1), (1.7), we have that* $\widehat{\mathbf{T}}_\mu^{\widehat{x}}$ *is infinite and* $\mathbf{T}_\mu^x$ *is superiorly unbounded, thus the periodic points* $\mu \in \widehat{Or}(\widehat{x}), \mu \in Or(x)$ *satisfy in fact* $\mu \in \widehat{\omega}(\widehat{x}), \mu \in \omega(x)$. *This was noticed since the first introduction of the periodic points, see Remark 31 from page 22, and several times afterwards.*

Remark 104. *The prime period of the periodic point* $\mu \in \widehat{Or}(\widehat{x})$ *always exists, but the prime period of the periodic point* $\mu \in Or(x)$ *does not exist if* x *is constant and equal with* μ, *see Theorem 23, page 47 where* $P_\mu^x = (0, \infty)$. *We shall prove later (Theorem 73, page 121) that this is the only situation when the periodic point* $\mu \in Or(x)$ *has no prime period.*

Remark 105. *Two more compact forms of writing (1.4) and (1.10) are*

$$\exists p \in \mathbf{Z}^*, \forall k \in \mathbf{N}_\_, (\widehat{x}(k) = \mu \text{ and } k + p \geq -1) \Longrightarrow \widehat{x}(k) = \widehat{x}(k+p), \tag{2.1}$$

$$\exists T \in \mathbf{R}^*, \exists t' \in I^x, \forall t \geq t', (x(t) = \mu \text{ and } t + T \geq t') \Longrightarrow x(t) = x(t+T), \tag{2.2}$$

where we have denoted $\mathbf{Z}^* = \mathbf{Z} \setminus \{0\}$ *and* $\mathbf{R}^* = \mathbf{R} \setminus \{0\}$. *Such statements accept the existence of negative periods. We shall suppose in the rest of our presentation that* $p \geq 1, T > 0$.

Remark 106. *A temptation exists to write (1.4) and (1.10) in a wrong way, recalling the periodicity* $(4.1)_{page\ 22}$, $(4.2)_{page\ 23}$ *of the signals, under the form*

$$\exists p \geq 1, \forall k \in \mathbf{N}_\_, \widehat{x}(k) = \mu \Longrightarrow \widehat{x}(k) = \widehat{x}(k+p), \tag{2.3}$$

$$\exists T > 0, \exists t' \in I^x, \forall t \geq t', x(t) = \mu \Longrightarrow x(t) = x(t+T), \tag{2.4}$$

that accepts only right time shifts in the definition of periodicity. We give the discrete time example of

$$\widehat{x} = 0, 0, 1, 0, 1, 0, 1, 0, 1, ...$$

that fulfills (2.3) with $\mu = 1, p = 2$. *For this signal* μ *is not periodic and the left time shift requirement* $\widehat{x}(1) = 1 \Longrightarrow \widehat{x}(1-2) = 1$ *shows where the problem is. In fact, if* $\mu \in \widehat{Or}(\widehat{x}), \mu \in Or(x)$ *then (2.3), (2.4) are requirements of eventual periodicity, not of periodicity.*

3. The Accessibility of the Periodic Points

Remark 107. *Let $\widehat{x}$, $\mu \in \widehat{Or}(\widehat{x})$ and $p \geq 1$. We have $\widehat{\mathbf{T}}_{\mu}^{\widehat{x}} \neq \varnothing$ and if*

$$\forall k \in \widehat{\mathbf{T}}_{\mu}^{\widehat{x}}, \{k+zp|z \in \mathbf{Z}\} \cap \mathbf{N}\_ \subset \widehat{\mathbf{T}}_{\mu}^{\widehat{x}},$$

then we deduce from Theorem 28, page 69 that for any $k \in \mathbf{N}\_$ we have $\widehat{\mathbf{T}}_{\mu}^{\widehat{x}} \cap \{k, k+1, ..., k+p-1\} \neq \varnothing$. Similarly, x, $\mu \in Or(x)$, $T > 0$, $t' \in I^x$ are given. If

$$\forall t \in \mathbf{T}_{\mu}^{x} \cap [t', \infty), \{t+zT|z \in \mathbf{Z}\} \cap [t', \infty) \subset \mathbf{T}_{\mu}^{x},$$

as far as $\mathbf{T}_{\mu}^{x} \cap [t', \infty) \neq \varnothing$ (from Lemma 2, page 223), we can use Theorem 28 again and get $\forall t \geq t'$, $\mathbf{T}_{\mu}^{x} \cap [t, t+T) \neq \varnothing$.

4. The Limit of Periodicity

Example 27. *We consider x, $\mu = x(-\infty+0)$, $T > 0$, $t_0, t_1, t_2, t_3 \in \mathbf{R}$ and we suppose that*

$$t_0 < t_1 < t_2 < t_3 < t_0 + T,$$

$$\mathbf{T}_{\mu}^{x} = (-\infty, t_0) \cup [t_1, t_2) \cup [t_3, t_0+T) \cup [t_1+T, t_2+T) \cup [t_3+T, t_0+2T) \cup$$
$$\cup [t_1+2T, t_2+2T) \cup [t_3+2T, t_0+3T) \cup ...$$

hold. For $t' \in [t_3 - T, t_0)$ we have $t' \in I^x$ and

$$\forall t \in \mathbf{T}_{\mu}^{x} \cap [t', \infty), \{t+zT|z \in \mathbf{Z}\} \cap [t', \infty) \subset \mathbf{T}_{\mu}^{x} \tag{4.1}$$

fulfilled, thus the property of periodicity of μ with the period T is true. For $t' < t_3 - T$, let us take an arbitrary $t \in [\max\{t', t_2 - T\}, t_3 - T)$. On one hand $t \in \mathbf{T}_{\mu}^{x} \cap [t', \infty)$ and on the other hand the truth of

$$t+T \in \{t+zT|z \in \mathbf{Z}\} \cap [t', \infty) \overset{(4.1)}{\subset} \mathbf{T}_{\mu}^{x}$$

should indicate that $x(t+T) = \mu$. But this is false, since $t+T \in [t_2, t_3)$. We have shown that $L_{\mu}^{x} = [t_3 - T, \infty)$. We notice also that choosing $t' \geq t_0$ is not possible, since $I^x = (-\infty, t_0)$. We conclude that the exact bounds of the initial time=limit of periodicity t' are given by $t' \in I^x \cap L_{\mu}^{x} = [t_3 - T, t_0)$.

Theorem 68. *Let the non constant signal x be given, together with $\mu \in Or(x)$, $T > 0$ and $t' \in I^x$ having the property that*

$$\forall t \in \mathbf{T}_{\mu}^{x} \cap [t', \infty), \{t+zT|z \in \mathbf{Z}\} \cap [t', \infty) \subset \mathbf{T}_{\mu}^{x}. \tag{4.2}$$

Then $t'_0, t_0 \in \mathbf{R}$ exist, $t'_0 < t_0$ such that $\forall t'' \in [t'_0, t_0)$, we have that $t'' \in I^x$,

$$\forall t \in \mathbf{T}_{\mu}^{x} \cap [t'', \infty), \{t+zT|z \in \mathbf{Z}\} \cap [t'', \infty) \subset \mathbf{T}_{\mu}^{x} \tag{4.3}$$

hold and for any $t'' \notin [t'_0, t_0)$, at least one of $t'' \in I^x$, (4.3) is false. In other words $[t'_0, t_0) = I^x \cap L_{\mu}^{x}$.

Remark 108. *We give two proofs of the previous Theorem: the first is short and easy to understand, the second is longer and generates a Corollary.*

Proof. The first proof of Theorem 68.

From the fact that x is not constant we get the existence of $t_0 \in \mathbf{R}$ with $I^x = (-\infty, t_0)$. From $\mu \in Or(x)$ and $t' \in I^x$ we have that $\mathbf{T}_\mu^x \cap [t', \infty) \neq \varnothing$ and this, taking into account (4.2) also, implies $\mu \in \omega(x)$. The existence of $t' \in \mathbf{R}$ such that (4.2) holds shows the fact that $L_\mu^x \neq \varnothing$, thus we can apply Theorem 30, page 70. The existence of $t_0' \in \mathbf{R}$ has resulted with $L_\mu^x = [t_0', \infty)$.

The existence of $t' \in I^x$ making (4.2) true shows furthermore that $t_0' < t_0$ and $[t_0', t_0) = I^x \cap L_\mu^x$. □

Proof. The second proof of Theorem 68.

The fact that x is not constant shows the existence of t_0 that is defined by

$$\forall t < t_0, x(t) = x(-\infty + 0), \tag{4.4}$$

$$x(t_0) \neq x(-\infty + 0). \tag{4.5}$$

From (4.4), (4.5), $t' \in I^x$ we infer that $I^x = (-\infty, t_0)$, $t' < t_0$ hold. We have the following possibilities.

a) Case $\mu = x(-\infty + 0)$

We show first that $[t' + T, t_0 + T) \subset \mathbf{T}_\mu^x$ and let for this an arbitrary $t \in [t' + T, t_0 + T)$. We have $t - T \geq t', t \geq t'$ and $t - T \in [t', t_0) \subset \mathbf{T}_\mu^x$, so that we can apply (4.2):

$$t \in \{t - T + zT | z \in \mathbf{Z}\} \cap [t', \infty) \subset \mathbf{T}_\mu^x.$$

The inclusion $[t' + T, t_0 + T) \subset \mathbf{T}_\mu^x$ is proved.

We get the existence of $t_1 \leq t' + T$ with

$$[t_1, t_0 + T) \subset \mathbf{T}_\mu^x, \tag{4.6}$$

$$x(t_1 - 0) \neq \mu. \tag{4.7}$$

We have $t_1 > t_0$, because the other possibility $t_0 \geq t_1$ is in contradiction with (4.4). The conclusion is that

$$t_1 - T \leq t' < t_0 < t_1 < t_0 + T. \tag{4.8}$$

From Lemma 4, page 224 and (4.4) we infer

$$(-\infty, t_0) \cup [t_1, t_0 + T) \cup [t_1 + T, t_0 + 2T) \cup [t_1 + 2T, t_0 + 3T) \cup \ldots \subset \mathbf{T}_\mu^x. \tag{4.9}$$

We claim that $t_0' = t_1 - T$ fulfills the statement of the Theorem, in particular that

$$(-\infty, t_1 - T] \subset \mathbf{T}_{x(-\infty+0)}^x, \tag{4.10}$$

$$\forall t \in \mathbf{T}_\mu^x \cap [t_1 - T, \infty), \{t + zT | z \in \mathbf{Z}\} \cap [t_1 - T, \infty) \subset \mathbf{T}_\mu^x \tag{4.11}$$

hold. We notice that the truth of (4.10) is trivial (from (4.4) and (4.8)) and, in order to prove the satisfaction of (4.11), let $t \in \mathbf{T}_\mu^x \cap [t_1 - T, \infty)$ arbitrary. We have the following sub-cases.

a.1) Case $t \in [t_1 - T, t_0) \cup [t_1, t_0 + T) \cup [t_1 + T, t_0 + 2T) \cup \ldots$

Some $k \in \mathbf{N}\_$ exists with $t \in [t_1 + kT, t_0 + (k+1)T)$. Then

$$\{t + zT | z \in \mathbf{Z}\} \cap [t_1 - T, \infty) = \{t + (-k-1)T, t + (-k)T, t + (-k+1)T, ...\}$$

$$\subset [t_1 - T, t_0) \cup [t_1, t_0 + T) \cup [t_1 + T, t_0 + 2T) \cup [t_1 + 2T, t_0 + 3T) \cup ... \overset{(4.9)}{\subset} \mathbf{T}^x_\mu.$$

a.2) Case $t \in \mathbf{T}^x_\mu \cap ([t_0, t_1) \cup [t_0 + T, t_1 + T) \cup [t_0 + 2T, t_1 + 2T) \cup ...)$

Then $t \in \mathbf{T}^x_\mu \cap [t', \infty)$ and $k \in \mathbf{N}$ exists such that $t \in [t_0 + kT, t_1 + kT)$. We have, since $t + (-k-1)T < t_1 - T$, that

$$\{t + zT | z \in \mathbf{Z}\} \cap [t_1 - T, \infty) = \{t + (-k)T, t + (-k+1)T, t + (-k+2)T, ...\}$$

$$= \{t + zT | z \in \mathbf{Z}\} \cap [t', \infty) \overset{(4.2)}{\subset} \mathbf{T}^x_\mu.$$

This ends proving the truth of (4.11). For any $t'' \in [t'_0, t_0)$, we have that $t'' \in I^x$, (4.3) are fulfilled, see also Lemma 3, page 224 (the statement $\mu \in \omega(x)$ from the hypothesis of the lemma results from $t' \in I^x$, giving $\mathbf{T}^x_\mu \cap [t', \infty) \neq \varnothing$, and from (4.2)).

In order to prove the last statement of the theorem, let $t'' \in I^x$, (4.3) be true with arbitrary, fixed t''. We suppose against all reason that $t'' < t_1 - T$ and let $\varepsilon > 0$ with the property that

$$\forall \xi \in (t_1 - \varepsilon, t_1), x(\xi) = x(t_1 - 0). \tag{4.12}$$

We take an arbitrary $t \in (\max\{t'', t_1 - T - \varepsilon\}, t_1 - T)$ for which we can write that $x(t) = \mu$ and, on the other hand,

$$t + T \in \{t + zT | z \in \mathbf{Z}\} \cap [t'', \infty) \overset{(4.3)}{\subset} \mathbf{T}^x_\mu,$$

thus $x(t+T) = \mu$. But $t + T \in (t_1 - \varepsilon, t_1)$, wherefrom $x(t+T) \overset{(4.12)}{=} x(t_1 - 0)$ and finally $x(t_1 - 0) = \mu$, contradiction with (4.7). We have obtained that $L^x_\mu = [t_1 - T, \infty)$. The supposition that $t'' \geq t_0$ is in contradiction with the hypothesis $t'' \in I^x$, since $I^x = (-\infty, t_0)$.

b) Case $\mu \neq x(-\infty + 0)$

We show that $[t' + T, t_0 + T) \cap \mathbf{T}^x_\mu = \varnothing$ and we suppose against all reason that $t \in [t' + T, t_0 + T) \cap \mathbf{T}^x_\mu$ exists, thus $t - T \in [t', t_0)$ and $t \geq t'$ hold. We infer

$$t - T \in \{t + zT | z \in \mathbf{Z}\} \cap [t', \infty) \overset{(4.2)}{\subset} \mathbf{T}^x_\mu \tag{4.13}$$

wherefrom the contradiction

$$x(-\infty + 0) \overset{(4.4)}{=} x(t - T) \overset{(4.13)}{=} \mu \neq x(-\infty + 0).$$

We infer from here, taking into account Theorem 28, page 69 also (the statement $\mu \in \omega(x)$ from the hypothesis of the theorem follows, like previously, from $t' \in I^x$, implying that $\mathbf{T}^x_\mu \cap [t', \infty) \neq \varnothing$, and from (4.2)), written for $t = t_0$, stating that $\mathbf{T}^x_\mu \cap [t_0, t_0 + T) \neq \varnothing$, the existence of $t_1, t_2 \in \mathbf{R}$ with

$$t_2 - T \leq t' < t_0 \leq t_1 < t_2 \leq t' + T < t_0 + T, \tag{4.14}$$

$$x(t_1-0)\neq\mu, \tag{4.15}$$

$$[t_1,t_2)\subset\mathbf{T}^x_\mu, \tag{4.16}$$

$$[t_2,t_0+T)\cap\mathbf{T}^x_\mu=\varnothing. \tag{4.17}$$

We claim that the statement of the theorem is fulfilled by $t'_0=t_2-T$ and in particular that

$$(-\infty,t_2-T]\subset\mathbf{T}^x_{x(-\infty+0)}, \tag{4.18}$$

$$\forall t\in\mathbf{T}^x_\mu\cap[t_2-T,\infty),\{t+zT|z\in\mathbf{Z}\}\cap[t_2-T,\infty)\subset\mathbf{T}^x_\mu \tag{4.19}$$

hold. We notice that (4.18) results from (4.4) and (4.14). Let $t\in\mathbf{T}^x_\mu\cap[t_2-T,\infty)$ arbitrary. We easily see that $((-\infty,t_0)\cup[t_2,t_0+T)\cup[t_2+T,t_0+2T)\cup[t_2+2T,t_0+3T)\cup...)\cap\mathbf{T}^x_\mu=\varnothing$ since by supposing against all reason that this is not true we get a contradiction, thus $\mathbf{T}^x_\mu\subset[t_0,t_2)\cup[t_0+T,t_2+T)\cup[t_0+2T,t_2+2T)\cup...$ and let $k\in\mathbf{N}$ with $t\in[t_0+kT,t_2+kT)$. This means that, on one hand $t\geq t_0>t'$ and on the other hand

$$t+(-k-1)T<t_2-T\leq t'<t_0\leq t-kT<t_2,$$

thus

$$\{t+zT|z\in\mathbf{Z}\}\cap[t_2-T,\infty)=\{t+(-k)T,t+(-k+1)T,t+(-k+2)T,...\}$$

$$=\{t+zT|z\in\mathbf{Z}\}\cap[t',\infty)\overset{(4.2)}{\subset}\mathbf{T}^x_\mu.$$

This ends proving (4.19). For any $t''\in[t'_0,t_0)$, we have that $t''\in I^x$, (4.3) are true, see also Lemma 3, page 224.

The supposition that $t''<t_2-T$ makes, from (4.16), that (4.3) is false, thus $L^x_\mu=[t_2-T,\infty)$. The supposition that $t''\geq t_0$ makes $t''\in I^x$ be false. □

Remark 109. *We use to think that the property of periodicity of $\mu\in Or(x)$ is independent on the choice of the initial time=limit of periodicity in the terms given by Theorem 68.*

Corollary 3. *We suppose that μ is a periodic point of the non constant signal x, that $T>0$ is its period and that t' is the initial time of x and the limit of periodicity of μ at the same time.*

a) If $\mu=x(-\infty+0)$ and $t_0<t_1$ are defined by

$$\forall t<t_0,x(t)=x(-\infty+0), \tag{4.20}$$

$$x(t_0)\neq x(-\infty+0), \tag{4.21}$$

$$[t_1,t_0+T)\subset\mathbf{T}^x_\mu, \tag{4.22}$$

$$x(t_1-0)\neq\mu, \tag{4.23}$$

then $t'\in[t_1-T,t_0)$;

b) if $\mu\neq x(-\infty+0)$ and $t_0<t_2$ are defined by (4.20), (4.21),

$$x(t_2-0)=\mu, \tag{4.24}$$

$$[t_2,t_0+T)\cap\mathbf{T}^x_\mu=\varnothing, \tag{4.25}$$

then $t'\in[t_2-T,t_0)$.

Proof. These are consequences of the second proof of Theorem 68, page 113. □

5. A Property of Constancy

Theorem 69. *The signals* $\widehat{x} \in \widehat{S}^{(n)}, x \in S^{(n)}$ *are considered.*

a) If $\mu \in \widehat{Or}(\widehat{x})$ *and the statement*

$$\forall k \in \widehat{\mathbf{T}}_{\mu}^{\widehat{x}}, \{k + zp | z \in \mathbf{Z}\} \cap \mathbf{N}_{\_} \subset \widehat{\mathbf{T}}_{\mu}^{\widehat{x}} \tag{5.1}$$

is true for $p = 1$, *then we have*

$$\forall k \in \mathbf{N}_{\_}, \widehat{x}(k) = \mu \tag{5.2}$$

and (5.1) is true for any $p \geq 1$.

b) Let $\mu \in Or(x)$ *be some point and we suppose that* $t_0 \in \mathbf{R}, h > 0$ *exist such that* x *has the form*

$$\begin{aligned} x(t) = x(-\infty + 0) \cdot \chi_{(-\infty, t_0)}(t) \oplus x(t_0) \cdot \chi_{[t_0, t_0+h)}(t) \oplus \ldots \\ \ldots \oplus x(t_0 + kh) \cdot \chi_{[t_0+kh, t_0+(k+1)h)}(t) \oplus \ldots \end{aligned} \tag{5.3}$$

If the statement

$$\forall t \in \mathbf{T}_{\mu}^{x} \cap [t', \infty), \{t + zT | z \in \mathbf{Z}\} \cap [t', \infty) \subset \mathbf{T}_{\mu}^{x} \tag{5.4}$$

is true for some $t' \in I^x$, $T \in (0, h) \cup (h, 2h) \cup \ldots \cup (qh, (q+1)h) \cup \ldots$, *then*

$$\forall t \in \mathbf{R}, x(t) = \mu \tag{5.5}$$

holds and (5.4) is true for any $t' \in \mathbf{R}$ *and any* $T > 0$.

c) If (5.3) is true under the form

$$\begin{aligned} x(t) = \widehat{x}(-1) \cdot \chi_{(-\infty, t_0)}(t) \oplus \widehat{x}(0) \cdot \chi_{[t_0, t_0+h)}(t) \oplus \ldots \\ \ldots \oplus \widehat{x}(k) \cdot \chi_{[t_0+kh, t_0+(k+1)h)}(t) \oplus \ldots \end{aligned} \tag{5.6}$$

and $\mu \in \widehat{Or}(\widehat{x}) = Or(x)$[1] *is arbitrary, then*

c.1) the satisfaction of (5.1) for $p = 1$ *implies that (5.2), (5.5) are true, (5.1) holds for any* $p \geq 1$ *and (5.4) holds for any* $t' \in \mathbf{R}$ *and any* $T > 0$;

c.2) the satisfaction of (5.4) for some $t' \in I^x$, $T \in (0, h) \cup (h, 2h) \cup \ldots \cup (qh, (q+1)h) \cup \ldots$ *implies also that (5.2), (5.5) are true, (5.1) holds for any* $p \geq 1$ *and (5.4) holds for any* $t' \in \mathbf{R}$ *and any* $T > 0$.

Proof. a) Statement (5.1) written for $p = 1$,

$$\forall k, \forall z, (k \in \widehat{\mathbf{T}}_{\mu}^{\widehat{x}} \text{ and } z \in \mathbf{Z} \text{ and } k + z \geq -1) \Longrightarrow k + z \in \widehat{\mathbf{T}}_{\mu}^{\widehat{x}} \tag{5.7}$$

together with $\widehat{\mathbf{T}}_{\mu}^{\widehat{x}} \neq \varnothing$ (since $\mu \in \widehat{Or}(\widehat{x})$) implies that $\widehat{\mathbf{T}}_{\mu}^{\widehat{x}} = \mathbf{N}_{\_}$, meaning the truth of (5.2). In these circumstances (5.1) is true for any $p \geq 1$.

b) If $\mu \in Or(x)$ and $t' \in I^x$, then $\mathbf{T}_{\mu}^{x} \cap [t', \infty) \neq \varnothing$, and from (5.4) we have $\mu \in \omega(x)$. The hypothesis asks furthermore that $T \in (0, h) \cup (h, 2h) \cup \ldots \cup (qh, (q+1)h) \cup \ldots$ and $t_0 \in \mathbf{R}, h > 0$ exist making (5.3) true. In this situation, Theorem 31, page 71 states that

$$\forall t \geq t', x(t) = \mu, \tag{5.8}$$

[1]The fact that (5.6) implies $\widehat{Or}(\widehat{x}) = Or(x)$ is shown in Theorem 6 a), page 11.

and on the other hand we have

$$\forall t \leq t', x(t) = \mu. \tag{5.9}$$

Statement (5.5) is true. In these conditions $I^x = \mathbf{R}$, $P^x_\mu = (0,\infty)$, thus (5.4) holds for any $t' \in \mathbf{R}$ and any $T > 0$.

c) This follows from a) and b). □

6. Discrete Time vs Real Time

Theorem 70. *Let the non constant signals $\widehat{x} \in \widehat{S}^{(n)}, x \in S^{(n)}$ be related by*

$$\begin{aligned} x(t) = \widehat{x}(-1) \cdot \chi_{(-\infty,t_0)}(t) \oplus \widehat{x}(0) \cdot \chi_{[t_0,t_0+h)}(t) \oplus \ldots \\ \ldots \oplus \widehat{x}(k) \cdot \chi_{[t_0+kh,t_0+(k+1)h)}(t) \oplus \ldots \end{aligned} \tag{6.1}$$

where $t_0 \in \mathbf{R}, h > 0$ and let $\mu \in \widehat{Or}(\widehat{x}) = Or(x)$.

a) If $p \geq 1$ exists such that

$$\forall k \in \widehat{\mathbf{T}}^{\widehat{x}}_\mu, \{k + zp | z \in \mathbf{Z}\} \cap \mathbf{N}\_ \subset \widehat{\mathbf{T}}^{\widehat{x}}_\mu \tag{6.2}$$

is true, then

$$\exists t' \in I^x, \forall t \in \mathbf{T}^x_\mu \cap [t',\infty), \{t + zT | z \in \mathbf{Z}\} \cap [t',\infty) \subset \mathbf{T}^x_\mu \tag{6.3}$$

holds for $T = ph$.

b) We presume that (6.3) is true for some $T > 0$. Then $\frac{T}{h} \in \{1,2,3,\ldots\}$ and $k' \in \mathbf{N}\_$ exists such that

$$\widehat{\mathbf{T}}^{\widehat{x}}_\mu \cap \{k', k'+1, k'+2, \ldots\} \neq \varnothing, \tag{6.4}$$

$$\left\{ \begin{array}{c} \forall k \in \widehat{\mathbf{T}}^{\widehat{x}}_\mu \cap \{k', k'+1, k'+2, \ldots\}, \\ \{k + zp | z \in \mathbf{Z}\} \cap \{k', k'+1, k'+2, \ldots\} \subset \widehat{\mathbf{T}}^{\widehat{x}}_\mu \end{array} \right. \tag{6.5}$$

hold for $p = \frac{T}{h}$.

Proof. a) The existence of $p \geq 1$ such that (6.2) is true shows that $\mu \in \widehat{\omega}(\widehat{x})$, thus, as far as $\widehat{\omega}(\widehat{x}) = \omega(x)$, we infer $\mu \in \omega(x)$. The fact that $\mu \in \widehat{\omega}(\widehat{x})$ is eventually periodic with the period p implies, from Theorem 32, page 76, that μ is eventually periodic with the period $T = ph$. As x is not constant and $L^x_\mu \neq \varnothing$, we have, see Theorem 30, page 70 the existence of $t'_0 \in \mathbf{R}$ with $L^x_\mu = [t'_0, \infty)$. We claim that $t_0 - h \geq t'_0$.

Let us suppose against all reason that this is not the case, i.e. that

$$\forall t \in \mathbf{T}^x_\mu \cap [t_0 - h, \infty), \{t + zT | z \in \mathbf{Z}\} \cap [t_0 - h, \infty) \subset \mathbf{T}^x_\mu \tag{6.6}$$

is false. This means the existence of $t_1 \in \mathbf{T}^x_\mu, z_1 \in \mathbf{Z}$ with $t_1 \geq t_0 - h$, $t_1 + z_1 ph \geq t_0 - h$ and $t_1 + z_1 ph \notin \mathbf{T}^x_\mu$. Then $k_1 \in \mathbf{N}\_$ exists such that $t_1 \in [t_0 + k_1 h, t_0 + (k_1+1)h)$, $t_1 + z_1 ph \in [t_0 + (k_1 + z_1 p)h, t_0 + (k_1 + z_1 p + 1)h)$. We have, as far as $k_1 + z_1 p \geq -1$:

$$\mu = x(t_1) = \widehat{x}(k_1) \overset{(6.2)}{=} \widehat{x}(k_1 + z_1 p) = x(t_1 + z_1 ph) = x(t_1 + z_1 T),$$

contradiction with the way that t_1 was defined.

As $t_0 - h \geq t_0'$ we get the truth of a), since $(-\infty, t_0) \subset I^x, [t_0 - h, \infty) \subset L_\mu^x$ and $\varnothing \neq [t_0 - h, t_0) \subset I^x \cap L_\mu^x$.

b) If $\mu \in Or(x)$ satisfies (6.3), then $\mu \in \omega(x) = \widehat{\omega}(\widehat{x})$. The fact that $\frac{T}{h} \in \{1, 2, 3, ...\}$ and the existence of $k' \in \mathbf{N}\_$ such that (6.4), (6.5) are fulfilled for $p = \frac{T}{h}$ result from Theorem 32, page 76. □

Remark 110. *Theorem 70 states, in a manner that updates Theorem 32 to periodic points, that the discrete time and the real time periodicity of the points are not equivalent when (6.1) is true: if $\mu \in \widehat{Or}(\widehat{x})$ is periodic with the period p, then $\mu \in Or(x)$ is periodic with the period ph, while the converse implication takes place under the form: if $\mu \in Or(x)$ is periodic with the period T, then $\frac{T}{h} \in \{1, 2, 3, ...\}$ and $\mu \in \widehat{Or}(\widehat{x})$ is eventually periodic with the period $\frac{T}{h}$.*

7. Support Sets vs Sets of Periods

Remark 111. *Let the signals $\widehat{x}, \widehat{y} \in \widehat{S}^{(n)}, x, y \in S^{(n)}$ with $\mu \in \widehat{Or}(\widehat{x}) \cap \widehat{Or}(\widehat{y})$. The implications*

$$\widehat{\mathbf{T}}_\mu^{\widehat{x}} = \widehat{\mathbf{T}}_\mu^{\widehat{y}} \Longrightarrow \widehat{P}_\mu^{\widehat{x}} = \widehat{P}_\mu^{\widehat{y}}, \tag{7.1}$$

$$\mathbf{T}_\mu^x = \mathbf{T}_\mu^y \Longrightarrow P_\mu^x = P_\mu^y \tag{7.2}$$

are not true, in the sense given by Example 17, page 77 and its discrete time counterpart: $\widehat{\mathbf{T}}_\mu^{\widehat{x}} = \widehat{\mathbf{T}}_\mu^{\widehat{y}}$ may take place and μ may be a periodic point of $\widehat{x}$ and an eventually periodic point of $\widehat{y}$. If so, the equality $\widehat{P}_\mu^{\widehat{x}} = \widehat{P}_\mu^{\widehat{y}}$ refers to eventual periodicity, not to periodicity.

Remark 112. *Note that the implications*

$$\widehat{P}_\mu^{\widehat{x}} = \widehat{P}_\mu^{\widehat{y}} \Longrightarrow \widehat{\mathbf{T}}_\mu^{\widehat{x}} = \widehat{\mathbf{T}}_\mu^{\widehat{y}}, \tag{7.3}$$

$$P_\mu^x = P_\mu^y \Longrightarrow \mathbf{T}_\mu^x = \mathbf{T}_\mu^y \tag{7.4}$$

are not true. For this, we take $x, y \in S^{(1)}$,

$$x(t) = \chi_{[4,5)}(t) \oplus \chi_{[9,10)}(t) \oplus \chi_{[14,15)}(t) \oplus ...$$

$$y(t) = \chi_{[2,3)}(t) \oplus \chi_{[4,5)}(t) \oplus \chi_{[7,8)}(t) \oplus \chi_{[9,10)}(t) \oplus \chi_{[12,13)}(t) \oplus ...$$

$\mu = 1$ is a periodic point of both x, y with $I^x = (-\infty, 4), I^y = (-\infty, 2), P_\mu^x = P_\mu^y = \{5, 10, 15, ...\}$ and $L_\mu^x = L_\mu^y = [0, \infty)$. In $\mathbf{T}_\mu^x$ the interval $[4, 5)$ repeats periodically by translations and in $\mathbf{T}_\mu^y$ the intervals $[2, 3), [4, 5)$ repeat periodically by translations. The periods T coincide for x and y and (7.3), (7.4) are false in general.

8. Sums, Differences and Multiples of Periods

Theorem 71. *The signals $\widehat{x}, x$ are considered.*

a) Let $p, p' \geq 1, \mu \in \widehat{Or}(\widehat{x})$ such that

$$\forall k \in \widehat{\mathbf{T}}_\mu^{\widehat{x}}, \{k + zp | z \in \mathbf{Z}\} \cap \mathbf{N}\_ \subset \widehat{\mathbf{T}}_\mu^{\widehat{x}}, \tag{8.1}$$

$$\forall k \in \widehat{\mathbf{T}}_{\mu}^{\widehat{x}}, \{k+zp'|z \in \mathbf{Z}\} \cap \mathbf{N}_{\_} \subset \widehat{\mathbf{T}}_{\mu}^{\widehat{x}} \tag{8.2}$$

are true. We have $p+p' \geq 1$,

$$\forall k \in \widehat{\mathbf{T}}_{\mu}^{\widehat{x}}, \{k+z(p+p')|z \in \mathbf{Z}\} \cap \mathbf{N}_{\_} \subset \widehat{\mathbf{T}}_{\mu}^{\widehat{x}} \tag{8.3}$$

and if $p > p'$, *then* $p-p' \geq 1$ *and*

$$\forall k \in \widehat{\mathbf{T}}_{\mu}^{\widehat{x}}, \{k+z(p-p')|z \in \mathbf{Z}\} \cap \mathbf{N}_{\_} \subset \widehat{\mathbf{T}}_{\mu}^{\widehat{x}}. \tag{8.4}$$

b) Let $T, T' > 0$, $t' \in I^x$ *and* $\mu \in Or(x)$ *be arbitrary with*

$$\forall t \in \mathbf{T}_{\mu}^{x} \cap [t', \infty), \{t+zT|z \in \mathbf{Z}\} \cap [t', \infty) \subset \mathbf{T}_{\mu}^{x}, \tag{8.5}$$

$$\forall t \in \mathbf{T}_{\mu}^{x} \cap [t', \infty), \{t+zT'|z \in \mathbf{Z}\} \cap [t', \infty) \subset \mathbf{T}_{\mu}^{x} \tag{8.6}$$

satisfied. Then $T+T' > 0$ *and*

$$\forall t \in \mathbf{T}_{\mu}^{x} \cap [t', \infty), \{t+z(T+T')|z \in \mathbf{Z}\} \cap [t', \infty) \subset \mathbf{T}_{\mu}^{x} \tag{8.7}$$

are true; in addition $T > T'$ *implies* $T-T' > 0$ *and*

$$\forall t \in \mathbf{T}_{\mu}^{x} \cap [t', \infty), \{t+z(T-T')|z \in \mathbf{Z}\} \cap [t', \infty) \subset \mathbf{T}_{\mu}^{x}. \tag{8.8}$$

Proof. This is the special case of Theorem 33, page 78 when the eventually periodic points are periodic. □

Theorem 72. *We consider the signals* $\widehat{x}, x$.

a) Let $p, k' \geq 1$ *and* $\mu \in \widehat{Or}(\widehat{x})$. *Then* $p' = k'p$ *fulfills* $p' \geq 1$ *and*

$$\forall k \in \widehat{\mathbf{T}}_{\mu}^{\widehat{x}}, \{k+zp|z \in \mathbf{Z}\} \cap \mathbf{N}_{\_} \subset \widehat{\mathbf{T}}_{\mu}^{\widehat{x}} \tag{8.9}$$

implies

$$\forall k \in \widehat{\mathbf{T}}_{\mu}^{\widehat{x}}, \{k+zp'|z \in \mathbf{Z}\} \cap \mathbf{N}_{\_} \subset \widehat{\mathbf{T}}_{\mu}^{\widehat{x}}. \tag{8.10}$$

b) Let $T > 0, t' \in I^x$, $k' \geq 1$ *and* $\mu \in Or(x)$ *be arbitrary. Then* $T' = k'T$ *fulfills* $T' > 0$ *and*

$$\forall t \in \mathbf{T}_{\mu}^{x} \cap [t', \infty), \{t+zT|z \in \mathbf{Z}\} \cap [t', \infty) \subset \mathbf{T}_{\mu}^{x}$$

implies

$$\forall t \in \mathbf{T}_{\mu}^{x} \cap [t', \infty), \{t+zT'|z \in \mathbf{Z}\} \cap [t', \infty) \subset \mathbf{T}_{\mu}^{x}.$$

Proof. This is a direct consequence of Theorem 71. □

Remark 113. *Another way of expressing the statements of Theorem 72 is: if* $p \in \widehat{P}_{\mu}^{\widehat{x}}$, *then* $\{p, 2p, 3p, ...\} \subset \widehat{P}_{\mu}^{\widehat{x}}$ *and if* $T \in P_{\mu}^{x}$, *then* $\{T, 2T, 3T, ...\} \subset P_{\mu}^{x}$.

9. The Set of the Periods

Theorem 73. *a) Let the signal $\widehat{x} \in \widehat{S}^{(n)}$ and $\mu \in \widehat{Or}(\widehat{x})$. We ask that μ is a periodic point of $\widehat{x}$. Then some $\widetilde{p} \geq 1$ exists such that*

$$\widehat{P_{\mu}^{\widehat{x}}} = \{\widetilde{p}, 2\widetilde{p}, 3\widetilde{p}, ...\}.$$

b) We suppose that the signal $x \in S^{(n)}$ is not constant and we take some $\mu \in Or(x)$. We ask that μ is a periodic point of x. Then there is $\widetilde{T} > 0$ such that

$$P_{\mu}^{x} = \{\widetilde{T}, 2\widetilde{T}, 3\widetilde{T}, ...\}.$$

Proof. This is a special case of Theorem 35, page 79. □

Remark 114. *An asymmetry occurs here, we have not asked in the hypothesis of Theorem 73, item a) that $\widehat{x}$ is not constant; when $\widehat{x}$ is constant and equal with μ we have $\widetilde{p} = 1$ and $\widehat{P_{\mu}^{\widehat{x}}} = \{1, 2, 3, ...\}$, thus the theorem is still true. Like in the case of the eventually periodic points, item b) of the theorem does not hold if x is constant and equal with μ, since in that case $P_{\mu}^{x} = (0, \infty)$.*

Theorem 74. *The signals $\widehat{x}$, x are given such that*

$$x(t) = \widehat{x}(-1) \cdot \chi_{(-\infty, t_0)}(t) \oplus \widehat{x}(0) \cdot \chi_{[t_0, t_0+h)}(t) \oplus \widehat{x}(1) \cdot \chi_{[t_0+h, t_0+2h)}(t) \oplus ...$$
$$... \oplus \widehat{x}(k) \cdot \chi_{[t_0+kh, t_0+(k+1)h)}(t) \oplus ...$$

is true, where $t_0 \in \mathbf{R}$ and $h > 0$ and we ask that $\mu \in \widehat{\omega}(\widehat{x}) = \omega(x)$ is a periodic point of any of $\widehat{x}, x$. Then two possibilities exist:

a) $\widehat{x}$, x are both constant, $\widehat{P_{\mu}^{\widehat{x}}} = \{1, 2, 3, ...\}$ and $P_{\mu}^{x} = (0, \infty)$;

b) none of $\widehat{x}$, x is constant, $\min \widehat{P_{\mu}^{\widehat{x}}} = p > 1$ and $\min P_{\mu}^{x} = T = ph$.

Proof. The fact that $\widehat{x}, x$ are simultaneously constant or non constant is obvious. We suppose that they are both non constant and we prove b). From Theorem 70, page 118 we have that $p \in \widehat{P_{\mu}^{\widehat{x}}} \Longrightarrow T = ph \in P_{\mu}^{x}$ and conversely, $T \in P_{\mu}^{x} \Longrightarrow p = \frac{T}{h} \in \widehat{P_{\mu}^{\widehat{x}}}$. In particular, if $\widehat{P_{\mu}^{\widehat{x}}} = \{p, 2p, 3p, ...\}$ and $P_{\mu}^{x} = \{T, 2T, 3T, ...\}$ (from Theorem 73, page 121), then $T = ph$. □

10. Necessity Conditions of Periodicity

Theorem 75. *Let $\widehat{x} \in \widehat{S}^{(n)}$ non constant. For $\mu \in \widehat{Or}(\widehat{x}), p \geq 1$ we suppose that*

$$\forall k \in \widehat{\mathbf{T}}_{\mu}^{\widehat{x}}, \{k + zp | z \in \mathbf{Z}\} \cap \mathbf{N}_{\_} \subset \widehat{\mathbf{T}}_{\mu}^{\widehat{x}} \tag{10.1}$$

takes place. Then $n_1, n_2, ..., n_{k_1} \in \{-1, 0, ..., p-2\}$, $k_1 \geq 1$, exist such that

$$\widehat{\mathbf{T}}_{\mu}^{\widehat{x}} = \bigcup_{k \in \mathbf{N}} \{n_1 + kp, n_2 + kp, ..., n_{k_1} + kp\} \tag{10.2}$$

holds.

Proof. $\mu \in \widehat{Or}(\widehat{x})$ and (10.1) imply $\mu \in \widehat{\omega}(\widehat{x})$. We apply Theorem 37, page 81 written for $k' = -1$. □

Remark 115. *If $\widehat{x}$ is constant, then the previous theorem takes the form $\widehat{Or}(\widehat{x}) = \{\mu\}, p = 1, k_1 = 1, n_1 = -1$ and (10.2) becomes $\widehat{\mathbf{T}}_{\mu}^{\widehat{x}} = \bigcup_{k \in \mathbf{N}} \{-1+k\} = \mathbf{N}_{\_}$.*

Theorem 76. *The non constant signal $x \in S^{(n)}$ is considered and let the point $\mu = x(-\infty+0)$ be given, together with $T > 0, t' \in I^x$ such that*

$$\forall t \in \mathbf{T}_{\mu}^{x} \cap [t', \infty), \{t + zT | z \in \mathbf{Z}\} \cap [t', \infty) \subset \mathbf{T}_{\mu}^{x} \tag{10.3}$$

holds. Then $t_0, a_1, b_1, a_2, b_2, ..., a_{k_1}, b_{k_1} \in \mathbf{R}$, $k_1 \geq 1$ exist such that

$$\forall t < t_0, x(t) = \mu, \tag{10.4}$$

$$x(t_0) \neq \mu, \tag{10.5}$$

$$t_0 < a_1 < b_1 < a_2 < b_2 < ... < a_{k_1} < b_{k_1} = t_0 + T, \tag{10.6}$$

$$[a_1, b_1) \cup [a_2, b_2) \cup ... \cup [a_{k_1}, b_{k_1}) = \mathbf{T}_{\mu}^{x} \cap [t_0, t_0 + T), \tag{10.7}$$

$$\mathbf{T}_{\mu}^{x} = \begin{array}{l} (-\infty, t_0) \cup \bigcup\limits_{k \in \mathbf{N}} ([a_1 + kT, b_1 + kT) \cup [a_2 + kT, b_2 + kT) \cup ... \\ \quad ... \cup [a_{k_1} + kT, b_{k_1} + kT)) \end{array} \tag{10.8}$$

hold.

Proof. A t_0 like at (10.4), (10.5) exists because x is not constant and we infer $I^x = (-\infty, t_0)$, $t' < t_0$. We have from Lemma 2, page 223 that $\mathbf{T}_{\mu}^{x} \cap [t', \infty) \neq \varnothing$, thus $\mu \in \omega(x)$ from (10.3) and the fact that $\mathbf{T}_{\mu}^{x} \cap [t_0, t_0 + T) \neq \varnothing$ follows from Theorem 28, page 69.

We have on one hand the existence of $\varepsilon > 0$ with

$$\forall t \in [t_0, t_0 + \varepsilon), x(t) = x(t_0) \overset{(10.5)}{\neq} \mu, \tag{10.9}$$

showing that $a_1 = \min \mathbf{T}_{\mu}^{x} \cap [t_0, t_0 + T) > t_0$. On the other hand we must show the existence of b_{k_1} like at (10.6), (10.7). Indeed, we suppose against all reason that such a b_{k_1} does not exist and consequently that $a_{k_1} < b_{k_1} < t_0 + T$, $[a_{k_1}, b_{k_1}) \subset \mathbf{T}_{\mu}^{x}$ and $[b_{k_1}, t_0 + T) \cap \mathbf{T}_{\mu}^{x} = \varnothing$. Let then $t \in [\max\{b_{k_1}, t' + T\}, t_0 + T)$ arbitrary. We get

$$b_{k_1} \leq \max\{b_{k_1}, t' + T\} \leq t < t_0 + T$$

i.e. $t \notin \mathbf{T}_{\mu}^{x}$. We have also $t > t - T \geq t'$ and $t - T \in [t', t_0) \subset \mathbf{T}_{\mu}^{x}$, thus

$$t \in \{t - T + zT | z \in \mathbf{Z}\} \cap [t', \infty) \overset{(10.3)}{\subset} \mathbf{T}_{\mu}^{x},$$

contradiction. The existence of $t_0, a_1, b_1, a_2, b_2, ..., a_{k_1}, b_{k_1}$ like at (10.4),...,(10.7) is proved.

We prove $\mathbf{T}_{\mu}^{x} \subset (-\infty, t_0) \cup \bigcup_{k \in \mathbf{N}} ([a_1 + kT, b_1 + kT) \cup [a_2 + kT, b_2 + kT) \cup ... \cup [a_{k_1} + kT, b_{k_1} + kT))$ and let $t \in \mathbf{T}_{\mu}^{x}$ arbitrary. If $t < t_0$ the inclusion is obvious (from (10.4)), so we can suppose now that $t \geq t_0$. We get from (10.3) the existence of a finite sequence

$t, t-T, ..., t-\bar{k}T \in \mathbf{T}_\mu^x, \bar{k} \geq 0$ with the property that $t-\bar{k}T \in [t_0, t_0+T)$. We infer from (10.7) the existence of $j \in \{1, ..., k_1\}$ with $t-\bar{k}T \in [a_j, b_j)$ and we conclude that $t \in [a_j+\bar{k}T, b_j+\bar{k}T) \subset (-\infty, t_0) \cup \bigcup_{k \in \mathbf{N}} ([a_1+kT, b_1+kT) \cup [a_2+kT, b_2+kT) \cup ... \cup [a_{k_1}+kT, b_{k_1}+kT))$.

We prove $(-\infty, t_0) \cup \bigcup_{k \in \mathbf{N}} ([a_1+kT, b_1+kT) \cup [a_2+kT, b_2+kT) \cup ... \cup [a_{k_1}+kT, b_{k_1}+kT)) \subset \mathbf{T}_\mu^x$. The fact that $(-\infty, t_0) \subset \mathbf{T}_\mu^x$ coincides with (10.4) and we take an arbitrary $t \in \bigcup_{k \in \mathbf{N}} ([a_1+kT, b_1+kT) \cup [a_2+kT, b_2+kT) \cup ... \cup [a_{k_1}+kT, b_{k_1}+kT))$. Some $k \in \mathbf{N}$ and some $j \in \{1, ..., k_1\}$ exist with $t \in [a_j+kT, b_j+kT)$, thus $t-kT \in [a_j, b_j) \subset \mathbf{T}_\mu^x \cap [t_0, t_0+T) \subset \mathbf{T}_\mu^x \cap [t', \infty)$. In particular we can see that $t \geq t-kT \geq t'$. We have

$$t \in \{t-kT+zT | z \in \mathbf{Z}\} \cap [t', \infty) \overset{(10.3)}{\subset} \mathbf{T}_\mu^x,$$

wherefrom we get $t \in \mathbf{T}_\mu^x$. Property (10.8) is proved. □

Theorem 77. *The signal $x \in S^{(n)}$ is not constant and let the point $\mu \in Or(x), \mu \neq x(-\infty+0)$, as well as $T > 0, t' \in I^x$ with*

$$\forall t \in \mathbf{T}_\mu^x \cap [t', \infty), \{t+zT | z \in \mathbf{Z}\} \cap [t', \infty) \subset \mathbf{T}_\mu^x \tag{10.10}$$

fulfilled. Then $t_0, a_1, b_1, a_2, b_2, ..., a_{k_1}, b_{k_1} \in \mathbf{R}$, $k_1 \geq 1$ exist such that

$$\forall t < t_0, x(t) = x(-\infty+0), \tag{10.11}$$

$$x(t_0) \neq x(-\infty+0), \tag{10.12}$$

$$t_0 \leq a_1 < b_1 < a_2 < b_2 < ... < a_{k_1} < b_{k_1} < t_0+T, \tag{10.13}$$

$$[a_1, b_1) \cup [a_2, b_2) \cup ... \cup [a_{k_1}, b_{k_1}) = \mathbf{T}_\mu^x \cap [t_0, t_0+T), \tag{10.14}$$

$$\mathbf{T}_\mu^x = \bigcup_{k \in \mathbf{N}} ([a_1+kT, b_1+kT) \cup [a_2+kT, b_2+kT) \cup ... \cup [a_{k_1}+kT, b_{k_1}+kT)) \tag{10.15}$$

are fulfilled.

Proof. As x is not constant we get the existence of t_0 like in (10.11), (10.12) and if we take in consideration that $I^x = (-\infty, t_0)$, we get $t' < t_0$.

We have from Lemma 2, page 223 that $\mathbf{T}_\mu^x \cap [t', \infty) \neq \varnothing$ and, as $\mu \in \omega(x)$ (from (10.10)), the fact that $\mathbf{T}_\mu^x \cap [t_0, t_0+T) \neq \varnothing$ results from Theorem 28, page 69. We show that $b_{k_1} < t_0+T$ and for this we suppose against all reason that $b_{k_1} = t_0+T$. Let $t \in [\max\{a_{k_1}, t'+T\}, t_0+T)$ arbitrary, fixed. We have $t > t-T \geq t'$ and $t \in [a_{k_1}, t_0+T) \subset \mathbf{T}_\mu^x$, thus we can apply (10.10):

$$t-T \in \{t+zT | z \in \mathbf{Z}\} \cap [t', \infty) \subset \mathbf{T}_\mu^x.$$

Since $t-T \in [t', t_0)$, we have reached the contradiction

$$\mu = x(t) = x(t-T) = x(-\infty+0).$$

The fact that $a_1, b_1, a_2, b_2, ..., a_{k_1}, b_{k_1}$ exist making (10.13), (10.14) true is proved. The proof of the equation (10.15) is made like in the proof of Theorem 76. □

Remark 116. *The proofs of Theorem 76 and Theorem 77 are similar with the proof of Theorem 38, page 82 stating necessary conditions of eventual periodicity of the points* $\mu \in \omega(x), \mu \in \widehat{\omega}(\widehat{x})$.

Remark 117. *Theorem 76 and Theorem 77 are not special cases, written for periodicity, of Theorem 38, but rather versions of that theorem. To be compared* $(10.5)_{page\,82}$ *with* $(10.6)_{page\,122}$ *and* $(10.13)_{page\,123}$.

Example 28. *We take* $x \in S^{(1)}$,

$$x(t) = \chi_{(-\infty,0)}(t) \oplus \chi_{[1,2)}(t) \oplus \chi_{[3,5)}(t) \oplus \chi_{[6,7)}(t) \oplus \chi_{[8,10)}(t) \oplus \chi_{[11,12)}(t) \oplus \ldots$$

In this example, see Theorem 76, $\mu = 1, t_0 = 0, k_1 = 2, T = 5$ *and* $t' \in [-2,0)$.

11. Sufficiency Conditions of Periodicity

Theorem 78. *Let* $\widehat{x} \in \widehat{S^{(n)}}$, $\mu \in \widehat{Or}(\widehat{x})$, $p \geq 1$ *and* $n_1, n_2, \ldots, n_{k_1} \in \{-1, 0, \ldots, p-2\}$, $k_1 \geq 1$, *such that*

$$\widehat{\mathbf{T}}^{\widehat{x}}_{\mu} = \bigcup_{k \in \mathbf{N}} \{n_1 + kp, n_2 + kp, \ldots, n_{k_1} + kp\}. \tag{11.1}$$

We have

$$\forall k \in \widehat{\mathbf{T}}^{\widehat{x}}_{\mu}, \{k + zp | z \in \mathbf{Z}\} \cap \mathbf{N}_{\_} \subset \widehat{\mathbf{T}}^{\widehat{x}}_{\mu}. \tag{11.2}$$

Proof. This is a special case of Theorem 39, page 83 written for $k' = -1$. □

Theorem 79. *The signal* $x \in S^{(n)}$ *is given with* $\mu = x(-\infty + 0)$, $T > 0$ *and the numbers* $t_0, a_1, b_1, a_2, b_2, \ldots, a_{k_1}, b_{k_1} \in \mathbf{R}$, $k_1 \geq 1$ *that fulfill*

$$t_0 < a_1 < b_1 < a_2 < b_2 < \ldots < a_{k_1} < b_{k_1} = t_0 + T, \tag{11.3}$$

$$\mathbf{T}^x_{\mu} = \begin{array}{l} (-\infty, t_0) \cup \bigcup_{k \in \mathbf{N}} ([a_1 + kT, b_1 + kT) \cup [a_2 + kT, b_2 + kT) \cup \ldots \\ \ldots \cup [a_{k_1} + kT, b_{k_1} + kT)). \end{array} \tag{11.4}$$

For any $t' \in [a_{k_1} - T, t_0)$, *the properties* $t' \in I^x$ *and*

$$\forall t \in \mathbf{T}^x_{\mu} \cap [t', \infty), \{t + zT | z \in \mathbf{Z}\} \cap [t', \infty) \subset \mathbf{T}^x_{\mu} \tag{11.5}$$

hold.

Proof. Let $t' \in [a_{k_1} - T, t_0)$ arbitrary. From (11.3), (11.4) we get $I^x = (-\infty, t_0)$, thus $t' \in I^x$. We infer

$$\mathbf{T}^x_{\mu} \cap [t', \infty) = [t', t_0) \cup [a_1, b_1) \cup \ldots \cup [a_{k_1}, b_{k_1}) \cup [a_1 + T, b_1 + T) \cup \ldots$$

and we take an arbitrary $t \in \mathbf{T}^x_{\mu} \cap [t', \infty)$. We have several possibilities.

a) Case $t \in [t', t_0)$, when

$$\{t + zT | z \in \mathbf{Z}\} \cap [t', \infty) = \{t, t + T, t + 2T, \ldots\} \subset$$

$$\subset [t',t_0)\cup[t'+T,b_{k_1})\cup[t'+2T,b_{k_1}+T)\cup...\subset$$
$$\subset(-\infty,t_0)\cup[a_{k_1},b_{k_1})\cup[a_{k_1}+T,b_{k_1}+T)\cup...\subset \mathbf{T}^x_\mu.$$

We have used the fact that

$$t-T<t_0-T<a_{k_1}-T\leq t'\leq t<t_0.$$

b) Case $t\in[a_j+kT,b_j+kT),k\geq 0,j\in\{1,2,...,k_1-1\}$,

$$\{t+zT|z\in\mathbf{Z}\}\cap[t',\infty)=\{t+(-k)T,t+(-k+1)T,t+(-k+2)T,...\}\subset$$
$$\subset[a_j,b_j)\cup[a_j+T,b_j+T)\cup[a_j+2T,b_j+2T)\cup...\subset\mathbf{T}^x_\mu$$

and we have used

$$t+(-k-1)T<t'<t_0<a_j\leq t+(-k)T<b_j<t'+T.$$

c) Case $t\in[a_{k_1}+kT,b_{k_1}+kT)$, $k\geq 0$ when there are two sub-cases,
c.1) Case $t\in[t'+(k+1)T,b_{k_1}+kT)$,

$$\{t+zT|z\in\mathbf{Z}\}\cap[t',\infty)=\{t+(-k-1)T,t+(-k)T,t+(-k+1)T,...\}\subset$$
$$\subset[t',t_0)\cup[t'+T,b_{k_1})\cup[t'+2T,b_{k_1}+T)\cup...\subset$$
$$\subset(-\infty,t_0)\cup[a_{k_1},b_{k_1})\cup[a_{k_1}+T,b_{k_1}+T)\cup...\subset\mathbf{T}^x_\mu$$

and we have used the fact that

$$t+(-k-2)T<t_0-T<a_{k_1}-T\leq t'\leq t+(-k-1)T<t_0.$$

c.2) Case $t\in[a_{k_1}+kT,t'+(k+1)T)$,

$$\{t+zT|z\in\mathbf{Z}\}\cap[t',\infty)=\{t+(-k)T,t+(-k+1)T,t+(-k+2)T,...\}\subset$$
$$\subset[a_{k_1},t'+T)\cup[a_{k_1}+T,t'+2T)\cup[a_{k_1}+2T,t'+3T)\cup...\subset$$
$$\subset[a_{k_1},b_{k_1})\cup[a_{k_1}+T,b_{k_1}+T)\cup[a_{k_1}+2T,b_{k_1}+2T)\cup...\subset\mathbf{T}^x_\mu$$

and we have used

$$t+(-k-1)T<t'<t_0<a_{k_1}\leq t+(-k)T<t'+T.$$

Statement (11.5) holds. □

Theorem 80. *Let* x, $\mu\in Or(x)$, $\mu\neq x(-\infty+0)$, $T>0$ *and the numbers* $t_0,a_1,b_1,a_2,b_2,...,a_{k_1},b_{k_1}\in\mathbf{R}$, $k_1\geq 1$, *with the property that*

$$\forall t<t_0,x(t)=x(-\infty+0), \tag{11.6}$$

$$x(t_0)\neq x(-\infty+0), \tag{11.7}$$

$$b_{k_1}-T<t_0\leq a_1<b_1<a_2<b_2<...<a_{k_1}<b_{k_1}, \tag{11.8}$$

$$\mathbf{T}^x_\mu=\bigcup_{k\in\mathbf{N}}([a_1+kT,b_1+kT)\cup[a_2+kT,b_2+kT)\cup...\cup[a_{k_1}+kT,b_{k_1}+kT)). \tag{11.9}$$

For any $t'\in[b_{k_1}-T,t_0)$, *we have* $t'\in I^x$,

$$\forall t\in\mathbf{T}^x_\mu\cap[t',\infty),\{t+zT|z\in\mathbf{Z}\}\cap[t',\infty)\subset\mathbf{T}^x_\mu. \tag{11.10}$$

Proof. Let $t' \in [b_{k_1} - T, t_0)$ be arbitrary. From (11.6), (11.7) we infer $I^x = (-\infty, t_0)$, thus $t' \in I^x$.

We get $\mathbf{T}^x_\mu \cap [t', \infty) = \mathbf{T}^x_\mu$ and we take an arbitrary $t \in \mathbf{T}^x_\mu \cap [t', \infty)$. Then $k \geq 0$ and $j \in \{1, 2, ..., k_1\}$ exist such that $t \in [a_j + kT, b_j + kT)$. We have:

$$\{t + zT | z \in \mathbf{Z}\} \cap [t', \infty) = \{t + (-k)T, t + (-k+1)T, t + (-k+2)T, ...\} \subset$$

$$\subset [a_j, b_j) \cup [a_j + T, b_j + T) \cup [a_j + 2T, b_j + 2T) \cup ... \subset \mathbf{T}^x_\mu,$$

where

$$t + (-k-1)T < t' < t_0 \leq a_j \leq t + (-k)T < b_j \leq t' + T.$$

Statement (11.10) holds. □

Remark 118. *The proofs of Theorem 78, page 124, Theorem 79, page 124 and Theorem 80, page 125 are similar with the proofs of Theorem 37, page 81 and Theorem 38, page 82 that state sufficient conditions of eventual periodicity of the points $\mu \in \widehat{\omega}(\widehat{x}), \mu \in \omega(x)$.*

12. A Special Case

Theorem 81. *Let $\widehat{x} \in \widehat{S^{(n)}}$, $\mu \in \widehat{Or}(\widehat{x})$, $p \geq 1$ and $n_1 \in \{-1, 0, ..., p-2\}$ such that*

$$\widehat{\mathbf{T}}^{\widehat{x}}_\mu = \{n_1, n_1 + p, n_1 + 2p, n_1 + 3p, ...\}. \tag{12.1}$$

Then

a) μ is a periodic point of $\widehat{x}$ with the period p :

$$\forall k \in \widehat{\mathbf{T}}^{\widehat{x}}_\mu, \{k + zp | z \in \mathbf{Z}\} \cap \mathbf{N}\_ \subset \widehat{\mathbf{T}}^{\widehat{x}}_\mu; \tag{12.2}$$

b) p is the prime period of μ : for any p' with

$$\forall k \in \widehat{\mathbf{T}}^{\widehat{x}}_\mu, \{k + zp' | z \in \mathbf{Z}\} \cap \mathbf{N}\_ \subset \widehat{\mathbf{T}}^{\widehat{x}}_\mu, \tag{12.3}$$

we infer $p' \in \{p, 2p, 3p, ...\}$.

Proof. a) This is a special case of Theorem 78, page 124, written for $k_1 = 1$.

b) We suppose against all reason that $p' \in \widehat{P}^{\widehat{x}}_\mu$ exists with $p' < p$. As $n_1 \in \widehat{\mathbf{T}}^{\widehat{x}}_\mu$, we obtain that $n_1 + p' \in \widehat{\mathbf{T}}^{\widehat{x}}_\mu$, contradiction with (12.1). Thus any $p' \in \widehat{P}^{\widehat{x}}_\mu$ fulfills $p' \geq p$. We apply Theorem 73, page 121. □

Theorem 82. *Let $x \in S^{(n)}$, $\mu = x(-\infty + 0)$, $T > 0$ and the points $t_0, a_1, b_1 \in \mathbf{R}$ having the property that*

$$t_0 < a_1 < b_1 = t_0 + T, \tag{12.4}$$

$$\mathbf{T}^x_\mu = (-\infty, t_0) \cup [a_1, b_1) \cup [a_1 + T, b_1 + T) \cup [a_1 + 2T, b_1 + 2T) \cup ... \tag{12.5}$$

hold.

a) For any $t' \in [a_1 - T, t_0)$, the properties $t' \in I^x$ and

$$\forall t \in \mathbf{T}^x_\mu \cap [t', \infty), \{t + zT | z \in \mathbf{Z}\} \cap [t', \infty) \subset \mathbf{T}^x_\mu \tag{12.6}$$

are fulfilled.

b) Let $t'' \in [a_1 - T, t_0)$ arbitrary. For any $T' > 0$ such that

$$\forall t \in \mathbf{T}^x_\mu \cap [t'', \infty), \{t + zT' | z \in \mathbf{Z}\} \cap [t'', \infty) \subset \mathbf{T}^x_\mu \tag{12.7}$$

holds, we have $T' \in \{T, 2T, 3T, ...\}$.

Proof. a) This is a special case of Theorem 79, page 124, written for $k_1 = 1$.

b) We suppose against all reason that $T' < T$. Let us note in the beginning that

$$\max\{a_1, b_1 - T'\} < \min\{b_1, a_1 + T - T'\}$$

is true, since all of $a_1 < b_1, a_1 < a_1 + T - T', b_1 - T' < b_1, b_1 - T' < a_1 + T - T'$ hold. We infer that any $t \in [\max\{a_1, b_1 - T'\}, \min\{b_1, a_1 + T - T'\})$ fulfills $t \in [a_1, b_1) \subset \mathbf{T}^x_\mu \cap [t'', \infty)$ and

$$t + T' \in \{t + zT' | z \in \mathbf{Z}\} \cap [t'', \infty) \overset{(12.7)}{\subset} \mathbf{T}^x_\mu,$$

and on the other hand we have

$$b_1 \leq \max\{a_1 + T', b_1\} \leq t + T' < \min\{b_1 + T', a_1 + T\} \leq a_1 + T,$$

meaning that $t + T' \notin \mathbf{T}^x_\mu$, contradiction. We conclude that $T' \geq T$.

We get $T = \min P^x_\mu$ and, as $P^x_\mu = \{T, 2T, 3T, ...\}$ from Theorem 73, page 121, we have that $T' \in \{T, 2T, 3T, ...\}$. □

Theorem 83. *Let $x \in S^{(n)}$, $\mu \in Or(x)$, $\mu \neq x(-\infty + 0)$, $T > 0$ and the points $t_0, a_1, b_1 \in \mathbf{R}$ with the property that*

$$\forall t < t_0, x(t) = x(-\infty + 0), \tag{12.8}$$

$$x(t_0) \neq x(-\infty + 0), \tag{12.9}$$

$$b_1 - T < t_0 \leq a_1 < b_1, \tag{12.10}$$

$$\mathbf{T}^x_\mu = [a_1, b_1) \cup [a_1 + T, b_1 + T) \cup [a_1 + 2T, b_1 + 2T) \cup ... \tag{12.11}$$

hold.

a) For any $t' \in [b_1 - T, t_0)$, the following properties: $t' \in I^x$ and

$$\forall t \in \mathbf{T}^x_\mu \cap [t', \infty), \{t + zT | z \in \mathbf{Z}\} \cap [t', \infty) \subset \mathbf{T}^x_\mu \tag{12.12}$$

are fulfilled.

b) Let $t'' \in [b_1 - T, t_0)$ arbitrary. For any $T' > 0$ such that

$$\forall t \in \mathbf{T}^x_\mu \cap [t'', \infty), \{t + zT' | z \in \mathbf{Z}\} \cap [t'', \infty) \subset \mathbf{T}^x_\mu \tag{12.13}$$

is true, we have $T' \in \{T, 2T, 3T, ...\}$.

Proof. a) This is a special case of Theorem 80, page 125, written for $k_1 = 1$.

b) We suppose against all reason now that $T' < T$. Let us notice the truth of

$$\max\{a_1, b_1 - T'\} < \min\{b_1, a_1 + T - T'\}.$$

We infer that $t \in [\max\{a_1, b_1 - T'\}, \min\{b_1, a_1 + T - T'\})$ satisfies $t \in [a_1, b_1) \subset \mathbf{T}^x_\mu \cap [t'', \infty)$ and

$$t + T' \in \{t + zT' | z \in \mathbf{Z}\} \cap [t'', \infty) \overset{(12.13)}{\subset} \mathbf{T}^x_\mu,$$

thus $t + T' \in \mathbf{T}^x_\mu$; on the other hand

$$b_1 \leq \max\{a_1 + T', b_1\} \leq t + T' < \min\{b_1 + T', a_1 + T\} \leq a_1 + T,$$

wherefrom $t + T' \notin \mathbf{T}^x_\mu$, contradiction. We have proved that $T' \geq T$.

We get that $T = \min P^x_\mu$. Theorem 73, page 121 shows that $P^x_\mu = \{T, 2T, 3T, ...\}$, wherefrom $T' \in \{T, 2T, 3T, ...\}$. □

Remark 119. *Theorems 82, 83 represent the same phenomenon and their proof is formally the same: when $\mathbf{T}^x_\mu$ has one of the forms (12.5), (12.11), the prime period of μ is T. The difference between the theorems is given by the fact that $\mu = x(-\infty + 0)$ in the first case and $\mu \neq x(-\infty + 0)$ in the second case.*

13. Periodic Points vs Eventually Periodic Points

Theorem 84. *a) Let $\widehat{x}$ and the periodic point $\mu \in \widehat{Or}(\widehat{x})$; for any $\widetilde{k} \in \mathbf{N}$, we have $\widehat{P}^{\widehat{x}}_\mu = \widehat{P}^{\widehat{\sigma}^{\widetilde{k}}(\widehat{x})}_\mu$.*

b) We consider x and the periodic point $\mu \in Or(x)$; for any $\widetilde{t} \in \mathbf{R}$, we have $P^x_\mu = P^{\sigma^{\widetilde{t}}(x)}_\mu$.

Proof. a) The hypothesis states that $\widehat{P}^{\widehat{x}}_\mu \neq \varnothing$ and let $\widetilde{k} \in \mathbf{N}$ arbitrary.

We prove $\widehat{P}^{\widehat{x}}_\mu \subset \widehat{P}^{\widehat{\sigma}^{\widetilde{k}}(\widehat{x})}_\mu$. Let $p \in \widehat{P}^{\widehat{x}}_\mu$ arbitrary, thus

$$\forall k \in \widehat{\mathbf{T}}^{\widehat{x}}_\mu, \{k + zp | z \in \mathbf{Z}\} \cap \mathbf{N}_\_ \subset \widehat{\mathbf{T}}^{\widehat{x}}_\mu \tag{13.1}$$

holds and we must show that $\mu \in \widehat{Or}(\widehat{\sigma}^{\widetilde{k}}(\widehat{x}))$ and

$$\forall k \in \widehat{\mathbf{T}}^{\widehat{\sigma}^{\widetilde{k}}(\widehat{x})}_\mu, \{k + zp | z \in \mathbf{Z}\} \cap \mathbf{N}_\_ \subset \widehat{\mathbf{T}}^{\widehat{\sigma}^{\widetilde{k}}(\widehat{x})}_\mu. \tag{13.2}$$

If $\mu \in \widehat{Or}(\widehat{x})$, then $\widehat{\mathbf{T}}^{\widehat{x}}_\mu \neq \varnothing$ and from (13.1) we infer that $\mu \in \widehat{\omega}(\widehat{x})$. Theorem 5, page 9 shows that $\widehat{\omega}(\widehat{\sigma}^{\widetilde{k}}(\widehat{x})) = \widehat{\omega}(\widehat{x})$, hence $\mu \in \widehat{\omega}(\widehat{\sigma}^{\widetilde{k}}(\widehat{x})) \subset \widehat{Or}(\widehat{\sigma}^{\widetilde{k}}(\widehat{x}))$ and $\widehat{\mathbf{T}}^{\widehat{\sigma}^{\widetilde{k}}(\widehat{x})}_\mu \neq \varnothing$.

Let now $k \in \widehat{\mathbf{T}}^{\widehat{\sigma}^{\widetilde{k}}(\widehat{x})}_\mu$ and $z \in \mathbf{Z}$ with $k + zp \geq -1$, meaning that $\widehat{x}(k + \widetilde{k}) = \mu$. We have $k + \widetilde{k} \in \widehat{\mathbf{T}}^{\widehat{x}}_\mu$ and $k + \widetilde{k} + zp \geq -1$, thus we can apply (13.1). We infer that $k + \widetilde{k} + zp \in \widehat{\mathbf{T}}^{\widehat{x}}_\mu$, wherefrom $\mu = \widehat{x}(k + \widetilde{k} + zp) = \widehat{\sigma}^{\widetilde{k}}(\widehat{x})(k + zp)$ and, finally, $k + zp \in \widehat{\mathbf{T}}^{\widehat{\sigma}^{\widetilde{k}}(\widehat{x})}_\mu$.

We prove $\widehat{P}^{\widehat{\sigma}^{\widetilde{k}}(\widehat{x})}_\mu \subset \widehat{P}^{\widehat{x}}_\mu$. Let $p \in \widehat{P}^{\widehat{x}}_\mu$, thus (13.1) is true. We suppose against all reason that $\widehat{P}^{\widehat{\sigma}^{\widetilde{k}}(\widehat{x})}_\mu \subset \widehat{P}^{\widehat{x}}_\mu$ is false, i.e. some $p' \in \widehat{P}^{\widehat{\sigma}^{\widetilde{k}}(\widehat{x})}_\mu \setminus \widehat{P}^{\widehat{x}}_\mu$ exists. This means the truth of

$$\forall k \in \widehat{\mathbf{T}}^{\widehat{x}}_\mu \cap \{\widetilde{k} - 1, \widetilde{k}, \widetilde{k} + 1, ...\}, \{k + zp' | z \in \mathbf{Z}\} \cap \{\widetilde{k} - 1, \widetilde{k}, \widetilde{k} + 1, ...\} \subset \widehat{\mathbf{T}}^{\widehat{x}}_\mu, \tag{13.3}$$

$$\exists k_1 \in \widehat{\mathbf{T}}_{\mu}^{\widehat{x}}, \exists z_1 \in \mathbf{Z}, k_1 + z_1 p' \geq -1 \text{ and } k_1 + z_1 p' \notin \widehat{\mathbf{T}}_{\mu}^{\widehat{x}}. \tag{13.4}$$

Let $\bar{k} \in \mathbf{N}$ having the property that $k_1 + \bar{k}p \geq \widetilde{k} - 1$, $k_1 + \bar{k}p + z_1 p' \geq \widetilde{k} - 1$. We have

$$\mu = \widehat{x}(k_1) \overset{(13.1)}{=} \widehat{x}(k_1 + \bar{k}p) \overset{(13.3)}{=} \widehat{x}(k_1 + \bar{k}p + z_1 p') \overset{(13.1)}{=} \widehat{x}(k_1 + z_1 p'). \tag{13.5}$$

The statements (13.4), (13.5) are contradictory.

b) We suppose that $P_{\mu}^{x} \neq \varnothing$ and let $\widetilde{t} \in \mathbf{R}$ arbitrary, fixed.

We prove $P_{\mu}^{x} \subset P_{\mu}^{\sigma^{\widetilde{t}}(x)}$. Let $T \in P_{\mu}^{x}$ arbitrary, thus $t' \in I^{x}$ exists such that

$$\forall t \in \mathbf{T}_{\mu}^{x} \cap [t', \infty), \{t + zT | z \in \mathbf{Z}\} \cap [t', \infty) \subset \mathbf{T}_{\mu}^{x}. \tag{13.6}$$

We must show that $\mu \in Or(\sigma^{\widetilde{t}}(x))$ and $t'' \in I^{\sigma^{\widetilde{t}}(x)}$ exists such that

$$\forall t \in \mathbf{T}_{\mu}^{\sigma^{\widetilde{t}}(x)} \cap [t'', \infty), \{t + zT | z \in \mathbf{Z}\} \cap [t'', \infty) \subset \mathbf{T}_{\mu}^{\sigma^{\widetilde{t}}(x)}. \tag{13.7}$$

From $\mu \in Or(x)$, $t' \in I^{x}$ and Lemma 2, page 223 we have that $\mathbf{T}_{\mu}^{x} \cap [t', \infty) \neq \varnothing$; from (13.6) we infer that $\mathbf{T}_{\mu}^{x}$ is superiorly unbounded, wherefrom we have that $\mu \in \omega(x)$.

Theorem 5, page 9 shows that $\omega(x) = \omega(\sigma^{\widetilde{t}}(x))$ and we infer $\mu \in \omega(\sigma^{\widetilde{t}}(x)) \subset Or(\sigma^{\widetilde{t}}(x))$. In particular $\mathbf{T}_{\mu}^{\sigma^{\widetilde{t}}(x)}$ is superiorly unbounded and $\mathbf{T}_{\mu}^{\sigma^{\widetilde{t}}(x)} \cap [t'', \infty) \neq \varnothing$ is true for any $t'' \in \mathbf{R}$.

If x is constant, then $\sigma^{\widetilde{t}}(x) = x$ and $t'' \in I^{\sigma^{\widetilde{t}}(x)}$, (13.7) take place trivially for any t'', thus we shall suppose from now that x is not constant and consequently some $t_0 \in \mathbf{R}$ exists with

$$\forall t < t_0, x(t) = x(-\infty + 0), \tag{13.8}$$

$$x(t_0) \neq x(-\infty + 0). \tag{13.9}$$

From (13.8), (13.9) we have $I^{x} = (-\infty, t_0)$ and since $t' \in I^{x}$, we get $t' < t_0$. Two possibilities exist.

Case $\widetilde{t} \leq t_0$

In this situation $\sigma^{\widetilde{t}}(x) = x$, thus $t'' \in I^{\sigma^{\widetilde{t}}(x)}$, (13.7) take place with $t'' = t'$.

Case $\widetilde{t} > t_0$

Some $\varepsilon > 0$ exists with $\forall t \in (\widetilde{t} - \varepsilon, \widetilde{t}), x(t) = x(\widetilde{t} - 0)$ and we infer from here that $\widetilde{t} - \varepsilon \geq t_0 > t'$. We take $t'' \in (\widetilde{t} - \varepsilon, \widetilde{t})$ arbitrary, fixed and we have

$$\sigma^{\widetilde{t}}(x)(t) = \begin{cases} x(\widetilde{t} - 0), t < \widetilde{t}, \\ x(t), t \geq t''. \end{cases}$$

The statement $t'' \in I^{\sigma^{\widetilde{t}}(x)}$ is also true. In order to prove the fulfillment of (13.7), let $t \in \mathbf{T}_{\mu}^{\sigma^{\widetilde{t}}(x)} \cap [t'', \infty)$ and $z \in \mathbf{Z}$ arbitrary, with $t + zT \geq t''$. We have $t \in \mathbf{T}_{\mu}^{x}$, $t \geq t'' > t_0 > t'$ and $t + zT \geq t'' > t_0 > t'$ thus (13.6) can be applied. We get $t + zT \in \mathbf{T}_{\mu}^{x}$. As far as $\mu = x(t + zT) = \sigma^{\widetilde{t}}(x)(t + zT)$, we conclude that $t + zT \in \mathbf{T}_{\mu}^{\sigma^{\widetilde{t}}(x)}$.

We prove $P_{\mu}^{\sigma^{\widetilde{t}}(x)} \subset P_{\mu}^{x}$. Let $T \in P_{\mu}^{x}$ arbitrary, thus $t' \in I^{x}$ exists with (13.6) fulfilled. We suppose against all reason that $P_{\mu}^{\sigma^{\widetilde{t}}(x)} \subset P_{\mu}^{x}$ is false, i.e. $T' \in P_{\mu}^{\sigma^{\widetilde{t}}(x)} \setminus P_{\mu}^{x}$ exists. This means the existence of $t'' \in I^{\sigma^{\widetilde{t}}(x)}$ with

$$\forall t \in \mathbf{T}_{\mu}^{\sigma^{\widetilde{t}}(x)} \cap [t'', \infty), \{t + zT' | z \in \mathbf{Z}\} \cap [t'', \infty) \subset \mathbf{T}_{\mu}^{\sigma^{\widetilde{t}}(x)}, \tag{13.10}$$

$$\forall t''' \in I^x, \exists t_1 \in \mathbf{T}_\mu^x \cap [t''', \infty), \exists z_1 \in \mathbf{Z}, t_1 + z_1 T' \geq t''' \text{ and } t_1 + z_1 T' \notin \mathbf{T}_\mu^x. \tag{13.11}$$

Let $t''' \geq t'$ arbitrary such that $(-\infty, t'''] \subset \mathbf{T}^x_{x(-\infty+0)}$ and $\bar{k} \in N$ with the property that $t_1 + \bar{k}T \geq \max\{t'', \widetilde{t}\}, t_1 + \bar{k}T + z_1 T' \geq \max\{t'', \widetilde{t}\}$. We have

$$\begin{aligned} \mu = x(t_1) &\overset{(13.6)}{=} x(t_1 + \bar{k}T) = \sigma^{\widetilde{t}}(x)(t_1 + \bar{k}T) \\ &\overset{(13.10)}{=} \sigma^{\widetilde{t}}(x)(t_1 + \bar{k}T + z_1 T') = x(t_1 + \bar{k}T + z_1 T') \overset{(13.6)}{=} x(t_1 + z_1 T'). \end{aligned} \tag{13.12}$$

Statements (13.11), (13.12) are contradictory. □

Remark 120. *In Theorem 84, the statements about the eventual periodicity of* $\mu \in \widehat{Or}(\widehat{x}), \mu \in Or(x)$ *are in fact statements about the periodicity of* $\mu \in \widehat{Or}(\widehat{\sigma}^{\bar{k}}(\widehat{x})), \mu \in Or(\sigma^{\widetilde{t}}(x))$.

Corollary 4. *a) We suppose that* $\mu \in \widehat{\omega}(\widehat{x})$ *fulfills* $\widehat{P}_\mu^{\widehat{x}} \neq \emptyset$. *Then* $\exists p \geq 1, \exists p' \geq 1, \exists k' \in \mathbf{N}, \exists k'' \in \mathbf{N}$ *with*

$$\forall k \in \widehat{\mathbf{T}}_\mu^{\widehat{\sigma}^{k'}(\widehat{x})}, \{k + zp | z \in \mathbf{Z}\} \cap \mathbf{N}\_ \subset \widehat{\mathbf{T}}_\mu^{\widehat{\sigma}^{k'}(\widehat{x})}, \tag{13.13}$$

$$\forall k \in \widehat{\mathbf{T}}_\mu^{\widehat{\sigma}^{k''}(\widehat{x})}, \{k + zp' | z \in \mathbf{Z}\} \cap \mathbf{N}\_ \subset \widehat{\mathbf{T}}_\mu^{\widehat{\sigma}^{k''}(\widehat{x})} \tag{13.14}$$

implies $\widehat{P}_\mu^{\widehat{\sigma}^{k'}(\widehat{x})} = \widehat{P}_\mu^{\widehat{\sigma}^{k''}(\widehat{x})}$.

b) If $\mu \in \omega(x)$ *fulfills* $P_\mu^x \neq \emptyset$, *then* $\exists T > 0, \exists T' > 0, \exists t' \in \mathbf{R}, \exists t'' \in \mathbf{R}$ *such that*

$$\forall t \in \mathbf{T}_\mu^x \cap [t', \infty), \{t + zT | z \in \mathbf{Z}\} \cap [t', \infty) \subset \mathbf{T}_\mu^x, \tag{13.15}$$

$$\forall t \in \mathbf{T}_\mu^x \cap [t'', \infty), \{t + zT' | z \in \mathbf{Z}\} \cap [t'', \infty) \subset \mathbf{T}_\mu^x \tag{13.16}$$

implies $P_\mu^{\sigma^{t'}(x)} = P_\mu^{\sigma^{t''}(x)}$.

Proof. a) We have $\widehat{P}_\mu^{\widehat{\sigma}^{k'}(\widehat{x})} = \widehat{P}_\mu^{\widehat{x}} = \widehat{P}_\mu^{\widehat{\sigma}^{k''}(\widehat{x})}$. □

14. Further Research

Remark 121. *Item b) in Theorem 82, page 126 does not work in the general case, when* $\mathbf{T}_\mu^x$ *is given by (11.4)*$_{page\ 124}$, *instead of (12.5)*$_{page\ 126}$. *In order to understand the phenomenon, one may consider the case of x from Example 28, page 124:* $x \in S^{(1)}$,

$$x(t) = \chi_{(-\infty,0)}(t) \oplus \chi_{[1,2)}(t) \oplus \chi_{[3,5)}(t) \oplus \chi_{[6,7)}(t) \oplus \chi_{[8,10)}(t) \oplus \chi_{[11,12)}(t) \oplus \ldots$$

with the periodic point $\mu = 1$, $T \in \{5, 10, 15, \ldots\}$ *and* $k_1 \in \{2, 4, 6, \ldots\}$. *The same is true for item b) in Theorem 83, page 127 as we can see by observing the behavior of the periodic point* $\mu = 0$ *of the previous function for* $T \in \{5, 10, 15, \ldots\}$ *and* $k_1 \in \{2, 4, 6, \ldots\}$. *A generalization of Theorem 82 b) and Theorem 83 b) is required.*

Remark 122. *Theorem 70, page 118 referring to the periodic point* $\mu \in Or(x)$ *is continued by Theorem 93, page 144 and Theorem 94 to follow. The statement (7.7)*$_{page\ 144}$ *of Theorem 94 suggests that Theorem 70 b) can be strengthened.*

Remark 123. *Let the non constant signals $\widehat{x}, x$ and we think if the compatibility properties*

$$\forall \mu \in \widehat{Or}(\widehat{x}), \forall \mu' \in \widehat{Or}(\widehat{x}), (\widehat{P_{\mu}^{x}} \neq \varnothing \text{ and } \widehat{P_{\mu'}^{x}} \neq \varnothing) \Longrightarrow \widehat{P_{\mu}^{x}} \cap \widehat{P_{\mu'}^{x}} \neq \varnothing,$$

$$\forall \mu \in Or(x), \forall \mu' \in Or(x), (P_{\mu}^{x} \neq \varnothing \text{ and } P_{\mu'}^{x} \neq \varnothing) \Longrightarrow P_{\mu}^{x} \cap P_{\mu'}^{x} \neq \varnothing$$

hold, see Remark 101, page 106. Proving the first one is trivial, while the second one has no proof so far. Taking into account the form of the sets of periods, the above statements give the suggestions that, see Theorem 66, page 107:

a) $\widehat{P_{\mu}^{x}} = \{p, 2p, 3p, ...\}, \widehat{P_{\mu'}^{x}} = \{p', 2p', 3p', ...\}$ *imply the existence of* $n_1, n_2 \geq 1$ *relatively prime such that* $n_1 p = n_2 p'$ *and if we denote this value with* p'', *then* $\widehat{P_{\mu}^{x}} \cap \widehat{P_{\mu'}^{x}} = \{p'', 2p'', 3p'', ...\}$;

b) $P_{\mu}^{x} = \{T, 2T, 3T, ...\}, P_{\mu'}^{x} = \{T', 2T', 3T', ...\} \Longrightarrow \exists n_1 \geq 1, \exists n_2 \geq 1$ *relatively prime with* $n_1 T = n_2 T'$ *and for* T'' *equal with the previous value we get* $P_{\mu}^{x} \cap P_{\mu'}^{x} = \{T'', 2T'', 3T'', ...\}$.

The limit case consists in signals that have all their points periodic, the periodic signals.

Chapter 8

Periodic Signals

We give in Section 1 and Section 2 properties that are equivalent with the periodicity of the signals, structured in two groups.

The purpose of Section 3 is that of showing that all the values of the orbit of a periodic signal are accessible in an interval with the length of a period.

Section 4 proves the independence of periodicity on the choice of t' =initial time of x and limit of periodicity of x and gives the bounds of t'.

The property of constancy from Section 5 is interesting by itself and it is also a useful result in the exposure. The discussion from Section 6 shows the relation between stating the constancy of a signal and the corresponding statement that refers to the periodicity of its points.

When the relation between $\widehat{x}$ and x is

$$x(t) = \widehat{x}(-1) \cdot \chi_{(-\infty,t_0)}(t) \oplus \widehat{x}(0) \cdot \chi_{[t_0,t_0+h)}(t) \oplus \ldots$$
$$\ldots \oplus \widehat{x}(k) \cdot \chi_{[t_0+kh,t_0+(k+1)h)}(t) \oplus \ldots$$

we are interested to see how the periodicity of $\widehat{x}$ determines the periodicity of x and vice versa. This is made in Section 7.

The fact that the sums, the differences and the multiples of the periods are periods is proved in Section 8.

Section 9 characterizes the form of $\widehat{P^{\widehat{x}}}, P^x$, in particular the existence of the prime period is proved.

Sections 10, 11 give necessary and sufficient conditions of periodicity, stated in terms of support sets. These conditions are inspired by those of the periodic points and use the fact that if all the values of a signal are periodic with the same period, then the signal is periodic.

A special case of periodicity is presented in Section 12. In this case the exact value of the prime period is known.

By forgetting some first values of the periodic signals $\widehat{x}, x$ we get signals with the same period. This is the topic of Section 13.

In Section 14 we put the problem of changing the order of some quantifiers in stating the periodicity of the signals.

1. The First Group of Periodicity Properties

Remark 124. *These properties involve the periodicity and the eventual periodicity of all the points $\mu \in \widehat{Or}(\widehat{x}), \mu \in Or(x)$. The properties (1.1),...,(1.6) are associated with the periodicity properties $(1.1)_{page\ 110}$,...,$(1.6)_{page\ 110}$ and the properties (1.7),...,(1.12) are associated with $(1.7)_{page\ 110}$,...,$(1.12)_{page\ 110}$ from Theorem 67, page 110. One should compare also these properties with $(2.1)_{page\ 47}$,...,$(2.6)_{page\ 47}$ and $(2.7)_{page\ 47}$,...,$(2.12)_{page\ 47}$ from Theorem 23, page 47.*

Theorem 85. *The signals $\widehat{x} \in \widehat{S}^{(n)}, x \in S^{(n)}$ are given.*

a) The following statements are equivalent for any $p \geq 1$:

$$\forall \mu \in \widehat{Or}(\widehat{x}), \forall k \in \widehat{\mathbf{T}}_{\mu}^{\widehat{x}}, \{k + zp | z \in \mathbf{Z}\} \cap \mathbf{N}_{\_} \subset \widehat{\mathbf{T}}_{\mu}^{\widehat{x}}, \tag{1.1}$$

$$\begin{cases} \forall \mu \in \widehat{Or}(\widehat{x}), \forall k' \in \mathbf{N}_{\_}, \forall k \in \widehat{\mathbf{T}}_{\mu}^{\widehat{x}} \cap \{k', k'+1, k'+2, ...\}, \\ \{k + zp | z \in \mathbf{Z}\} \cap \{k', k'+1, k'+2, ...\} \subset \widehat{\mathbf{T}}_{\mu}^{\widehat{x}}, \end{cases} \tag{1.2}$$

$$\forall \mu \in \widehat{Or}(\widehat{x}), \forall k'' \in \mathbf{N}, \forall k \in \widehat{\mathbf{T}}_{\mu}^{\widehat{\sigma}^{k''}(\widehat{x})}, \{k + zp | z \in \mathbf{Z}\} \cap \mathbf{N}_{\_} \subset \widehat{\mathbf{T}}_{\mu}^{\widehat{\sigma}^{k''}(\widehat{x})}, \tag{1.3}$$

$$\begin{cases} \forall \mu \in \widehat{Or}(\widehat{x}), \forall k \in \mathbf{N}_{\_}, \widehat{x}(k) = \mu \Longrightarrow \\ \Longrightarrow (\widehat{x}(k) = \widehat{x}(k+p) \text{ and } k - p \geq -1 \Longrightarrow \widehat{x}(k) = \widehat{x}(k-p)), \end{cases} \tag{1.4}$$

$$\begin{cases} \forall \mu \in \widehat{Or}(\widehat{x}), \forall k' \in \mathbf{N}_{\_}, \forall k \geq k', \widehat{x}(k) = \mu \Longrightarrow \\ \Longrightarrow (\widehat{x}(k) = \widehat{x}(k+p) \text{ and } k - p \geq k' \Longrightarrow \widehat{x}(k) = \widehat{x}(k-p)), \end{cases} \tag{1.5}$$

$$\begin{cases} \forall \mu \in \widehat{Or}(\widehat{x}), \forall k'' \in \mathbf{N}, \forall k \in \mathbf{N}_{\_}, \widehat{\sigma}^{k''}(\widehat{x})(k) = \mu \Longrightarrow \\ \Longrightarrow (\widehat{\sigma}^{k''}(\widehat{x})(k) = \widehat{\sigma}^{k''}(\widehat{x})(k+p) \text{ and} \\ \text{and } k - p \geq -1 \Longrightarrow \widehat{\sigma}^{k''}(\widehat{x})(k) = \widehat{\sigma}^{k''}(\widehat{x})(k-p)). \end{cases} \tag{1.6}$$

b) The following statements are also equivalent for any $T > 0$:

$$\forall \mu \in Or(x), \exists t' \in I^x, \forall t \in \mathbf{T}_{\mu}^{x} \cap [t', \infty), \{t + zT | z \in \mathbf{Z}\} \cap [t', \infty) \subset \mathbf{T}_{\mu}^{x}, \tag{1.7}$$

$$\begin{cases} \forall \mu \in Or(x), \exists t' \in I^x, \\ \forall t_1' \geq t', \forall t \in \mathbf{T}_{\mu}^{x} \cap [t_1', \infty), \{t + zT | z \in \mathbf{Z}\} \cap [t_1', \infty) \subset \mathbf{T}_{\mu}^{x}, \end{cases} \tag{1.8}$$

$$\begin{cases} \forall \mu \in Or(x), \forall t'' \in \mathbf{R}, \exists t' \in I^{\sigma^{t''}(x)}, \\ \forall t \in \mathbf{T}_{\mu}^{\sigma^{t''}(x)} \cap [t', \infty), \{t + zT | z \in \mathbf{Z}\} \cap [t', \infty) \subset \mathbf{T}_{\mu}^{\sigma^{t''}(x)}, \end{cases} \tag{1.9}$$

$$\begin{cases} \forall \mu \in Or(x), \exists t' \in I^x, \forall t \geq t', \\ x(t) = \mu \Longrightarrow (x(t) = x(t+T) \text{ and } t - T \geq t' \Longrightarrow x(t) = x(t-T)), \end{cases} \tag{1.10}$$

$$\begin{cases} \forall \mu \in Or(x), \exists t' \in I^x, \forall t_1' \geq t', \forall t \geq t_1', x(t) = \mu \Longrightarrow \\ \Longrightarrow (x(t) = x(t+T) \text{ and } t - T \geq t_1' \Longrightarrow x(t) = x(t-T)), \end{cases} \tag{1.11}$$

$$\begin{cases} \forall \mu \in Or(x), \forall t'' \in \mathbf{R}, \exists t' \in I^{\sigma^{t''}(x)}, \\ \forall t \geq t', \sigma^{t''}(x)(t) = \mu \Longrightarrow (\sigma^{t''}(x)(t) = \sigma^{t''}(x)(t+T) \text{ and} \\ \text{and } t - T \geq t' \Longrightarrow \sigma^{t''}(x)(t) = \sigma^{t''}(x)(t-T)). \end{cases} \tag{1.12}$$

Proof. a) The proof of the implications

$$(1.1) \Longrightarrow (1.2) \Longrightarrow (1.3) \Longrightarrow (1.4) \Longrightarrow (1.5) \Longrightarrow (1.6)$$

follows from the proof of Theorem 23.

(1.6)$\Longrightarrow$(1.1) Let $\mu \in \widehat{Or}(\widehat{x}), k \in \widehat{\mathbf{T}}^{\widehat{x}}_{\mu}$ and $z \in \mathbf{Z}$ arbitrary such that $k + zp \geq -1$. Hypothesis (1.6) written for $k'' = 0$ gives

$$\begin{cases} \forall k_1 \in \mathbf{N}_{\_}, \widehat{x}(k_1) = \mu \Longrightarrow \\ \Longrightarrow (\widehat{x}(k_1) = \widehat{x}(k_1 + p) \text{ and } k_1 - p \geq -1 \Longrightarrow \widehat{x}(k_1) = \widehat{x}(k_1 - p)). \end{cases} \tag{1.13}$$

Case $z < 0$,

$$\mu = \widehat{x}(k) \overset{(1.13)}{=} \widehat{x}(k-p) \overset{(1.13)}{=} \widehat{x}(k-2p) \overset{(1.13)}{=} \ldots \overset{(1.13)}{=} \widehat{x}(k+zp);$$

Case $z = 0$,

$$\mu = \widehat{x}(k+zp);$$

Case $z > 0$,

$$\mu = \widehat{x}(k) \overset{(1.13)}{=} \widehat{x}(k+p) \overset{(1.13)}{=} \widehat{x}(k+2p) \overset{(1.13)}{=} \ldots \overset{(1.13)}{=} \widehat{x}(k+zp).$$

We infer in all the three cases that $k + zp \in \widehat{\mathbf{T}}^{\widehat{x}}_{\mu}$.

b) The proof of the following implications

$$(1.7) \Longrightarrow (1.8) \Longrightarrow (1.9) \Longrightarrow (1.10) \Longrightarrow (1.11) \Longrightarrow (1.12)$$

follows from the proof of Theorem 23, page 47.

(1.12)$\Longrightarrow$(1.7) Let $\mu \in Or(x)$. Statement (1.12) written for t'' sufficiently small in order that $\sigma^{t''}(x) = x$ gives the existence of $t' \in I^x$ with

$$\begin{cases} \forall t_1 \geq t', x(t_1) = \mu \Longrightarrow \\ \Longrightarrow (x(t_1) = x(t_1 + T) \text{ and } t_1 - T \geq t' \Longrightarrow x(t_1) = x(t_1 - T)). \end{cases} \tag{1.14}$$

From Lemma 2, page 223 we have $\mathbf{T}^x_{\mu} \cap [t', \infty) \neq \varnothing$. We take $t \in \mathbf{T}^x_{\mu} \cap [t', \infty)$ and $z \in \mathbf{Z}$ arbitrary with $t + zT \geq t'$ and we have the following possibilities.

Case $z < 0$,

$$\mu = x(t) \overset{(1.14)}{=} x(t-T) \overset{(1.14)}{=} x(t-2T) \overset{(1.14)}{=} \ldots \overset{(1.14)}{=} x(t+zT);$$

Case $z = 0$,

$$\mu = x(t+zT);$$

Case $z > 0$,

$$\mu = x(t) \overset{(1.14)}{=} x(t+T) \overset{(1.14)}{=} x(t+2T) \overset{(1.14)}{=} \ldots \overset{(1.14)}{=} x(t+zT).$$

We have obtained in all these cases that $t + zT \in \mathbf{T}^x_{\mu}$. We infer the truth of (1.7). □

2. The Second Group of Periodicity Properties

Remark 125. *The properties (2.1),...,(2.3) and (2.4),...,(2.6) from this group have occurred for the first time as* $(4.1)_{page\ 59}$,...,$(4.3)_{page\ 59}$ *and* $(4.4)_{page\ 59}$,..., $(4.6)_{page\ 59}$ *in Theorem 25, page 59. These properties refer to the signals themselves, and not to their values.*

Theorem 86. *The signals* $\widehat{x} \in \widehat{S}^{(n)}, x \in S^{(n)}$ *are given.*

a) The following properties are equivalent, for any $p \geq 1$, *with any of (1.1),..., (1.6):*

$$\forall k \in \mathbf{N}_{\_}, \widehat{x}(k) = \widehat{x}(k+p), \tag{2.1}$$

$$\forall k' \in \mathbf{N}_{\_}, \forall k \geq k', \widehat{x}(k) = \widehat{x}(k+p), \tag{2.2}$$

$$\forall k'' \in \mathbf{N}, \forall k \in \mathbf{N}_{\_}, \widehat{\sigma}^{k''}(\widehat{x})(k) = \widehat{\sigma}^{k''}(\widehat{x})(k+p). \tag{2.3}$$

b) For any $T > 0$, *the following properties are equivalent with any of (1.7),..., (1.12):*

$$\exists t' \in I^x, \forall t \geq t', x(t) = x(t+T), \tag{2.4}$$

$$\exists t' \in I^x, \forall t_1' \geq t', \forall t \geq t_1', x(t) = x(t+T), \tag{2.5}$$

$$\forall t'' \in \mathbf{R}, \exists t' \in I^{\sigma^{t''}(x)}, \forall t \geq t', \sigma^{t''}(x)(t) = \sigma^{t''}(x)(t+T). \tag{2.6}$$

Proof. a) The proof of (2.1)$\Longrightarrow$(2.2)$\Longrightarrow$(2.3) follows from the proof of Theorem 25.

(1.1)$\Longrightarrow$(2.1) Let $k \in \mathbf{N}_{\_}$ arbitrary, fixed and we choose $\mu \in \widehat{Or}(\widehat{x})$ with the property $\widehat{x}(k) = \mu$. We infer

$$k+p \in \{k+zp | z \in \mathbf{Z}\} \cap \mathbf{N}_{\_} \overset{(1.1)}{\subset} \widehat{\mathbf{T}}_{\mu}^{\widehat{x}},$$

thus $\widehat{x}(k+p) = \mu = \widehat{x}(k)$.

(2.3)$\Longrightarrow$(1.1) Let $\mu \in \widehat{Or}(\widehat{x}), k \in \widehat{\mathbf{T}}_{\mu}^{\widehat{x}}$ and $z \in \mathbf{Z}$ arbitrary with $k+zp \geq -1$. We apply (2.3) written for $k'' = 0$,

$$\forall k_1 \in \mathbf{N}_{\_}, \widehat{x}(k_1) = \widehat{x}(k_1+p) \tag{2.7}$$

and we have the following cases:

Case $z > 0$,

$$\mu = \widehat{x}(k) \overset{(2.7)}{=} \widehat{x}(k+p) \overset{(2.7)}{=} \widehat{x}(k+2p) \overset{(2.7)}{=}$$
$$\overset{(2.7)}{=} \ldots \overset{(2.7)}{=} \widehat{x}(k+zp);$$

Case $z = 0$,

$$\mu = \widehat{x}(k) = \widehat{x}(k+zp);$$

Case $z < 0$,

$$\widehat{x}(k+zp) \overset{(2.7)}{=} \widehat{x}(k+(z+1)p) \overset{(2.7)}{=}$$
$$\overset{(2.7)}{=} \widehat{x}(k+(z+2)p) \overset{(2.7)}{=} \ldots \overset{(2.7)}{=} \widehat{x}(k) = \mu.$$

In all these cases we have obtained that $k+zp \in \widehat{\mathbf{T}}_{\mu}^{\widehat{x}}$.

b) The proof of the implications (2.4)$\Longrightarrow$(2.5)$\Longrightarrow$(2.6) follows from the proof of Theorem 25, page 59.

(1.7)$\Longrightarrow$(2.4) We suppose that $Or(x) = \{\mu^1, ..., \mu^s\}$ and from (1.7) we have the existence $\forall i \in \{1, ..., s\}$ of $t_i' \in I^x$ with

$$\forall t_1 \in \mathbf{T}^x_{\mu^i} \cap [t_i', \infty), \{t_1 + zT | z \in \mathbf{Z}\} \cap [t_i', \infty) \subset \mathbf{T}^x_{\mu^i} \tag{2.8}$$

fulfilled. With the notation $t' = \max\{t_1', ..., t_s'\}$, we get the truth of $t' \in I^x$,

$$\forall t \in \mathbf{T}^x_{\mu^i} \cap [t', \infty), \{t + zT | z \in \mathbf{Z}\} \cap [t', \infty) \subset \mathbf{T}^x_{\mu^i} \tag{2.9}$$

for any $i \in \{1, ..., s\}$, see Lemma 3, page 224. Let now $t \geq t'$ arbitrary. Some $i \in \{1, ..., s\}$ exists such that $x(t) = \mu^i$, for which we can write

$$t + T \in \{t + zT | z \in \mathbf{Z}\} \cap [t', \infty) \overset{(2.9)}{\subset} \mathbf{T}^x_{\mu^i},$$

in other words $x(t+T) = \mu^i = x(t)$.

(2.6)$\Longrightarrow$(1.7) We take in (2.6) $t'' \in \mathbf{R}$ sufficiently small so that $\sigma^{t''}(x) = x$ and the existence of $t' \in I^x$ results with

$$\forall t_1 \geq t', x(t_1) = x(t_1 + T). \tag{2.10}$$

Let $\mu \in Or(x)$ arbitrary. We have from Lemma 2, page 223 that $\mathbf{T}^x_\mu \cap [t', \infty) \neq \emptyset$ and we take $t \in \mathbf{T}^x_\mu \cap [t', \infty), z \in \mathbf{Z}$ arbitrary such that $t + zT \geq t'$. The following possibilities exist:

Case $z > 0$,

$$\mu = x(t) \overset{(2.10)}{=} x(t+T) \overset{(2.10)}{=} x(t+2T) \overset{(2.10)}{=} ... \overset{(2.10)}{=} x(t+zT);$$

Case $z = 0$,

$$\mu = x(t) = x(t+zT);$$

Case $z < 0$,

$$x(t+zT) \overset{(2.10)}{=} x(t+(z+1)T) \overset{(2.10)}{=} x(t+(z+2)T) \overset{(2.10)}{=} ... \overset{(2.10)}{=} x(t) = \mu.$$

In all these cases, the satisfaction of $t + zT \in \mathbf{T}^x_\mu$ is proved. □

Remark 126. *A signal is periodic with the period p, T if and only if all the points of its orbit are periodic with the period p, T :*

$$(1.1)_{page\,134} \Longleftrightarrow \forall \mu \in \widehat{Or}(\widehat{x}), (1.1)_{page\,110},$$

$$(1.7)_{page\,134} \Longleftrightarrow \forall \mu \in Or(x), (1.7)_{page\,110}.$$

Remark 127. *Statements (1.1),...,(1.6) and (1.7),...,(1.12) refer to left-and-right time shifts, and statements (2.1),...,(2.3) and (2.4),...,(2.6) refer to right time shifts only.*

Theorem 87. *Let the periodic signals $\widehat{x} \in \widehat{S}^{(n)}, x \in S^{(n)}$. We have $\widehat{\omega}(\widehat{x}) = \widehat{Or}(\widehat{x}), \omega(x) = Or(x)$.*

Proof. In order to prove the real time statement, we suppose that $T > 0, t' \in I^x$ exist such that

$$\forall \mu \in Or(x), \forall t \in \mathbf{T}^x_\mu \cap [t', \infty), \{t + zT | z \in \mathbf{Z}\} \cap [t', \infty) \subset \mathbf{T}^x_\mu$$

is true and let $\mu \in Or(x)$ arbitrary. The fact that $\mathbf{T}^x_\mu$ is superiorly unbounded shows that $\mu \in \omega(x)$, wherefrom the conclusion that $Or(x) \subset \omega(x)$. As far as the inclusion $\omega(x) \subset Or(x)$ is always true, we infer that $\omega(x) = Or(x)$. □

3. The Accessibility of the Orbit

Theorem 88. *a) If* $\widehat{x} \in \widehat{S}^{(n)}$*, then*

$$\widehat{P}^{\widehat{x}} \neq \varnothing \Longrightarrow \forall k' \in \mathbf{N}_{\_}, \widehat{Or}(\widehat{x}) = \{\widehat{x}(k) | k \geq k'\}. \tag{3.1}$$

b) For $x \in S^{(n)}$ *we have*

$$P^x \neq \varnothing \Longrightarrow \forall t' \in \mathbf{R}, Or(x) = \{x(t) | t \geq t'\}. \tag{3.2}$$

Proof. a) We suppose that $\widehat{P}^{\widehat{x}} \neq \varnothing$. The fact that $\widehat{\omega}(\widehat{x}) \overset{Theorem\, 87, page\, 137}{=} \widehat{Or}(\widehat{x}) = \{\widehat{x}(k) | k \geq -1\}$ shows, from Theorem 7, page 11 that $\forall k' \geq -1, \widehat{\omega}(\widehat{x}) = \{\widehat{x}(k) | k \geq k'\}$. The conclusion is that $\forall k' \in \mathbf{N}_{\_}, \widehat{Or}(\widehat{x}) = \widehat{\omega}(\widehat{x}) = \{\widehat{x}(k) | k \geq k'\}$.

b) The hypothesis states the periodicity of $x, P^x \neq \varnothing$. On the other hand $t' \in I^x$ exists with the property $\omega(x) \overset{Theorem\, 87, page\, 137}{=} Or(x) = \{x(t) | t \geq t'\}$. Theorem 7 shows that $\forall t'' \geq t', \omega(x) = \{x(t) | t \geq t''\}$ and the conclusion is $\forall t'' \geq t', Or(x) = \omega(x) = \{x(t) | t \geq t''\}$. Let now an arbitrary $t'' < t'$. We infer $\{x(t) | t \geq t''\} = \{x(t) | t \geq t'\} = Or(x) = \omega(x)$. Statement b) is proved. □

Theorem 89. *a) We suppose that* $\widehat{x}$ *is periodic, with the period* $p \geq 1$ *:*

$$\forall k \in \mathbf{N}_{\_}, \widehat{x}(k) = \widehat{x}(k+p). \tag{3.3}$$

Then

$$\forall k \in \mathbf{N}_{\_}, \widehat{Or}(\widehat{x}) = \{\widehat{x}(i) | i \in \{k, k+1, ..., k+p-1\}\}. \tag{3.4}$$

b) If x is periodic with the period $T > 0 : t' \in I^x$ *exists with*

$$\forall t \geq t', x(t) = x(t+T), \tag{3.5}$$

then

$$\forall t \geq t', Or(x) = \{x(\xi) | \xi \in [t, t+T)\}. \tag{3.6}$$

Proof. We apply Theorem 47, page 93 with $\widehat{\omega}(\widehat{x}) = \widehat{Or}(\widehat{x})$, $k' = -1$ and $\omega(x) = Or(x)$, $t' \in I^x$. □

Remark 128. *The previous theorem states the property that, in the case of the periodic signals, all the points of the orbit are accessible in a time interval with the length of a period.*

4. The Limit of Periodicity

Theorem 90. *If* $\widehat{x}$ *is periodic, then*

$$\forall \mu \in \widehat{Or}(\widehat{x}), \widehat{L}^{\widehat{x}} = \widehat{L}^{\widehat{x}}_{\mu} = \mathbf{N}_{\_}. \tag{4.1}$$

Proof. The fact that the periodicity of $\widehat{x}$ implies $\widehat{L^x} = \mathbf{N}\_$ is obvious and the fact that $\forall\mu \in \widehat{Or}(\widehat{x}), \widehat{L^x} \subset \widehat{L^x_\mu}$ results from Theorem 49, page 95, where we put $\widehat{Or}(\widehat{x}) = \widehat{\omega}(\widehat{x})$. For any $\mu \in \widehat{Or}(\widehat{x})$, we infer

$$\mathbf{N}\_ = \widehat{L^x} \subset \widehat{L^x_\mu} \subset \mathbf{N}\_.$$

□

Example 29. *Let $\mu, \mu', \mu'' \in \mathbf{B}^n$ distinct and $x \in S^{(n)}$ defined this way:*

$$x(t) = \mu \cdot \chi_{(-\infty,0)}(t) \oplus \mu'' \cdot \chi_{[0,1)}(t) \oplus \mu' \cdot \chi_{[1,2)}(t) \oplus \mu'' \cdot \chi_{[2,4)}(t) \oplus \mu' \cdot \chi_{[4,5)}(t)$$

$$\oplus \mu \cdot \chi_{[5,6)}(t) \oplus \mu'' \cdot \chi_{[6,7)}(t) \oplus \mu' \cdot \chi_{[7,8)}(t) \oplus \mu'' \cdot \chi_{[8,10)}(t) \oplus \mu' \cdot \chi_{[10,11)}(t)$$

$$\oplus \mu \cdot \chi_{[11,12)}(t) \oplus \mu'' \cdot \chi_{[12,13)}(t) \oplus \mu' \cdot \chi_{[13,14)}(t) \oplus \mu'' \cdot \chi_{[14,16)}(t) \oplus \mu' \cdot \chi_{[16,17)}(t) \oplus \ldots$$

In this example $Or(x) = \{\mu, \mu', \mu''\}$; $L^x_\mu = [-1,\infty), L^x_{\mu'} = [-1,\infty), L^x_{\mu''} = [-2,\infty)$, $L^x = [-1,\infty)$; $P^x_\mu = \{6,12,18,\ldots\}, P^x_{\mu'} = \{3,6,9,\ldots\}, P^x_{\mu''} = \{6,12,18,\ldots\}, P^x = \{6,12,18,\ldots\}$. We notice the falsity of (4.1) in the real time case, expressed under the form $L^x \neq L^x_{\mu''}$.

Theorem 91. *The non constant signal x is given, together with $T > 0$ and we suppose that $t' \in I^x$ exists with the property that*

$$\forall t \geq t', x(t) = x(t+T). \tag{4.2}$$

Then $t'_0, t_0 \in \mathbf{R}$ exist, $t'_0 < t_0$ such that $\forall t'' \in [t'_0, t_0)$, we have $t'' \in I^x$,

$$\forall t \geq t'', x(t) = x(t+T) \tag{4.3}$$

and if $t'' \notin [t'_0, t_0)$, then at least one of $t'' \in I^x$, (4.3) is false. In other words $[t'_0, t_0) = I^x \cap L^x$.

Proof. The first proof. As x is not constant, $t_0 \in \mathbf{R}$ exists with $I^x = (-\infty, t_0)$. We suppose that $Or(x) = \{\mu^1, \ldots, \mu^s\}, s \geq 2$ and from the periodicity of x we get $\omega(x) = \{\mu^1, \ldots, \mu^s\}$. From Theorem 30, page 70 we infer the existence of $t'_1, \ldots, t'_s$ with $L^x_{\mu^1} = [t'_1, \infty), \ldots, L^x_{\mu^s} = [t'_s, \infty)$. The periodicity of $\mu^1, \ldots, \mu^s$ implies $I^x \cap L^x_{\mu^1} \neq \varnothing, \ldots, I^x \cap L^x_{\mu^s} \neq \varnothing$ and the eventual periodicity of x shows, from Theorem 49, page 95 that $L^x = L^x_{\mu^1} \cap \ldots \cap L^x_{\mu^s}$. It has resulted the fact that $t'_0 = \max\{t'_1, \ldots, t'_s\}$ satisfies

$$I^x \cap L^x = I^x \cap (L^x_{\mu^1} \cap \ldots \cap L^x_{\mu^s}) = (-\infty, t_0) \cap ([t'_1, \infty) \cap \ldots \cap [t'_s, \infty))$$

$$= (-\infty, t_0) \cap [t'_0, \infty) = [t'_0, t_0) \neq \varnothing.$$

□

Proof. The second proof. We define t_0 in the following way:

$$\forall t < t_0, x(t) = x(-\infty+0), \tag{4.4}$$

$$x(t_0) \neq x(-\infty+0) \tag{4.5}$$

and this is possible since x is not constant. From (4.4), (4.5) we have $I^x = (-\infty, t_0)$ and since $t' \in I^x$, we infer that $t' < t_0$. We have from (4.2):

$$\forall t \in [t', t_0), x(-\infty+0) = x(t) = x(t+T), \tag{4.6}$$

thus

$$\forall t \in [t'+T, t_0+T), x(t) = x(-\infty+0), \tag{4.7}$$

$$x(t_0+T) = x(t_0) \overset{(4.5)}{\neq} x(-\infty+0). \tag{4.8}$$

We can see that $(-\infty, t_0) \cup [t'+T, t_0+T) \subset \mathbf{T}^x_{x(-\infty+0)}, t_0, t_0+T \notin \mathbf{T}^x_{x(-\infty+0)}$, where $t_0 < t'+T$ is the only possibility, thus

$$t' < t_0 < t'+T < t_0+T$$

is true. Then $t'_0 \leq t'$ exists such that $t'_0 + T > t_0$ and

$$\forall t \in [t'_0+T, t_0+T), x(t) = x(-\infty+0), \tag{4.9}$$

$$x(t'_0+T-0) \neq x(-\infty+0). \tag{4.10}$$

We take some arbitrary $t'' \in [t'_0, t_0)$, some arbitrary $t \in \mathbf{R}$ and we have the following possibilities.

a) Case $t'' \in [t', t_0)$

a.1) Case $t \leq t''$, when $x(t) = x(-\infty+0)$,

a.2) Case $t \geq t''$, when $t \geq t'$ and $x(t) \overset{(4.2)}{=} x(t+T)$;

b) Case $t'' \in [t'_0, t')$

b.1) Case $t \leq t''$, when $x(t) = x(-\infty+0)$,

b.2) Case $t \geq t''$

b.2.1) Case $t \in [t'', t')$, when

$$t'_0+T \leq t''+T \leq t+T < t'+T < t_0+T$$

and $x(t) = x(-\infty+0) \overset{(4.9)}{=} x(t+T)$,

b.2.2) Case $t \geq t'$, when $x(t) \overset{(4.2)}{=} x(t+T)$.

In all these cases $t'' \in I^x$ and (4.3) hold.

We suppose now, against all reason, that $t'' \in I^x$, (4.3) hold and $t'' \notin [t'_0, t_0)$. The following possibilities exist.

i) Case $t'' < t'_0$

Some $\varepsilon_1 > 0$ exists such that

$$\forall t \in (t'_0+T-\varepsilon_1, t'_0+T), x(t) = x(t'_0+T-0) \tag{4.11}$$

and let $\varepsilon \in (0, \min\{t'_0 - t'', \varepsilon_1\})$, for which

$$(t'_0-\varepsilon, t'_0) \subset [t'', \infty), \tag{4.12}$$

$$(t'_0+T-\varepsilon, t'_0+T) \subset (t'_0+T-\varepsilon_1, t'_0+T). \tag{4.13}$$

We take an arbitrary $t \in (t'_0-\varepsilon, t'_0)$. We have the contradiction

$$x(-\infty+0) = x(t) \overset{(4.3),(4.12)}{=} x(t+T) \overset{(4.11),(4.13)}{=} x(t'_0+T-0) \overset{(4.10)}{\neq} x(-\infty+0).$$

ii) Case $t'' \geq t_0$

$$x(-\infty+0) = x(t'') = x(t_0) \overset{(4.5)}{\neq} x(-\infty+0),$$

contradiction. □

Corollary 5. *Let x be not constant, $Or(x) = \{\mu^1, \mu^2, ..., \mu^s\}, s \geq 2$, with $\mu^1 = x(-\infty + 0)$ and we suppose that x has the period $T > 0$. We consider the statements*

$$\forall t < t_0, x(t) = \mu^1, \tag{4.14}$$

$$x(t_0) \neq \mu^1, \tag{4.15}$$

$$\forall t \in [t_0' + T, t_0 + T), x(t) = \mu^1, \tag{4.16}$$

$$x(t_0' + T - 0) = \mu^2, \tag{4.17}$$

$$x(t_1 - 0) = \mu^2, \tag{4.18}$$

$$[t_1, t_0 + T) \cap \mathbf{T}^x_{\mu^2} = \varnothing, \tag{4.19}$$

$$x(t_2 - 0) = \mu^3, \tag{4.20}$$

$$[t_2, t_0 + T) \cap \mathbf{T}^x_{\mu^3} = \varnothing, \tag{4.21}$$

$$...$$

$$x(t_{s-1} - 0) = \mu^s, \tag{4.22}$$

$$[t_{s-1}, t_0 + T) \cap \mathbf{T}^x_{\mu^s} = \varnothing \tag{4.23}$$

where (4.14), (4.15) define t_0, (4.16), (4.17) define $t_0' < t_0$, (4.18), (4.19) define $t_1 \in (t_0, t_0 + T)$, (4.20), (4.21) define $t_2 \in (t_0, t_1)$,..., (4.22), (4.23) define $t_{s-1} \in (t_0, t_{s-2})$. Then $t_0' = t_1 - T$, $[t_0', t_0) = I^x \cap L^x = I^x \cap L^x_{\mu^1}$, $[t_1 - T, t_0) = I^x \cap L^x_{\mu^2}$, $[t_2 - T, t_0) = I^x \cap L^x_{\mu^3}$, ..., $[t_{s-1} - T, t_0) = I^x \cap L^x_{\mu^s}$ and we have

$$[t_0', t_0) = [t_1 - T, t_0) = [t_1 - T, t_0) \cap [t_2 - T, t_0) \cap ... \cap [t_{s-1} - T, t_0). \tag{4.24}$$

Proof. Statements (4.14), (4.15) coincide with $(4.4)_{page\ 139}$, $(4.5)_{page\ 139}$ and statements (4.16), (4.17) result from $(4.9)_{page\ 140}$, $(4.10)_{page\ 140}$ in the proof of Theorem 91, page 139. By comparing (4.17) with (4.18) and (4.16) with (4.19) we obtain $t_0' = t_1 - T$. From Theorem 91 we have that $[t_0', t_0) = I^x \cap L^x$ and from Corollary 3, page 116 we have $[t_0', t_0) = I^x \cap L^x_{\mu^1}$.

The periodicity of x with the period T implies the periodicity of its values $\mu^1, \mu^2, ..., \mu^s$ with the period T and the bounds of the initial times=limits of periodicity are given by $[t_1 - T, t_0), [t_1 - T, t_0), [t_2 - T, t_0), ..., [t_{s-1} - T, t_0)$, where (4.18), (4.19); (4.20), (4.21);... (4.22), (4.23) result from $(4.24)_{page\ 116}$, $(4.25)_{page\ 116}$ of Corollary 3. The truth of (4.24) results from the remark that $t_1 = \max\{t_1, t_2, ..., t_{s-1}\}$. □

Remark 129. *Notice that previously*

$$I^x \cap L^x = I^x \cap L^x_{\mu^1} = I^x \cap L^x_{\mu^2} = [t_0', t_0).$$

5. A property of Constancy

Theorem 92. *Let the signals $\widehat{x}, x$.*

a) If the statement

$$\forall k \in \mathbf{N}_{\_}, \widehat{x}(k) = \widehat{x}(k+p) \tag{5.1}$$

is true for $p = 1$, then $\mu \in \widehat{Or}(\widehat{x})$ exists such that

$$\forall k \in \mathbf{N}_{\_}, \widehat{x}(k) = \mu \tag{5.2}$$

and (5.1) is true for any $p \geq 1$.

b) We suppose that $t_0 \in \mathbf{R}, h > 0$ exist such that x is of the form

$$\begin{aligned} x(t) = x(-\infty+0) \cdot \chi_{(-\infty,t_0)}(t) \oplus x(t_0) \cdot \chi_{[t_0,t_0+h)}(t) \oplus \ldots \\ \ldots \oplus x(t_0+kh) \cdot \chi_{[t_0+kh,t_0+(k+1)h)}(t) \oplus \ldots \end{aligned} \tag{5.3}$$

If the statement

$$\forall t \geq t', x(t) = x(t+T) \tag{5.4}$$

is true for some $t' \in I^x$, $T \in (0,h) \cup (h,2h) \cup \ldots \cup (qh,(q+1)h) \cup \ldots$, then some $\mu \in Or(x)$ exists such that

$$\forall t \in \mathbf{R}, x(t) = \mu \tag{5.5}$$

and (5.4) is true for any $t' \in \mathbf{R}$, $T > 0$.

c) We presume that (5.3) is true under the form

$$\begin{aligned} x(t) = \widehat{x}(-1) \cdot \chi_{(-\infty,t_0)}(t) \oplus \widehat{x}(0) \cdot \chi_{[t_0,t_0+h)}(t) \oplus \ldots \\ \ldots \oplus \widehat{x}(k) \cdot \chi_{[t_0+kh,t_0+(k+1)h)}(t) \oplus \ldots \end{aligned} \tag{5.6}$$

Then:

c.1) the fulfillment of (5.1) for $p = 1$ implies that $\mu \in \widehat{Or}(\widehat{x}) = Or(x)$ exists such that (5.2), (5.5) are true, (5.1) holds for any $p \geq 1$ and (5.4) holds for any $t' \in \mathbf{R}$ and any $T > 0$;

c.2) the satisfaction of the statement (5.4) for some $t' \in I^x, T \in (0,h) \cup (h,2h) \cup \ldots \cup (qh,(q+1)h) \cup \ldots$ implies the existence of $\mu \in \widehat{Or}(\widehat{x}) = Or(x)$ such that (5.2), (5.5) are true, (5.1) holds for any $p \geq 1$ and (5.4) holds for any $t' \in \mathbf{R}$ and any $T > 0$.

Proof. a) From (5.1) written for $p = 1$ we get the existence of $\mu = \widehat{x}(-1)$ such that (5.2) is true. Moreover, as far as $\widehat{x}$ is the constant function, (5.1) holds for any $p \geq 1$.

b) We suppose against all reason that x is not constant, thus $t_0' \in \mathbf{R}$ exists such that $I^x = (-\infty, t_0')$. The hypothesis states the existence of $t' \in I^x$ with the property that (5.4) is true for $T \in (0,h) \cup (h,2h) \cup \ldots \cup (qh,(q+1)h) \cup \ldots$ In these conditions by using Theorem 51, page 96 we infer the existence of $\mu \in \omega(x)$ such that

$$\forall t \geq t', x(t) = \mu,$$

$$\forall t \leq t', x(t) = x(-\infty+0)$$

are true. We have obtained that $\mu = x(-\infty+0)$, contradiction with our supposition that x is not constant. Property (5.5) holds. As in this situation $I^x = L^x = \mathbf{R}$ and $P^x = (0,\infty)$ are true, b) is proved.

c) This is a consequence of a) and b). □

6. Discussion on Constancy

Remark 130. *Theorem 69, page 117 (concerning the periodic points) and Theorem 92, page 142 (concerning the periodic signals) express essentially the same idea, namely that in the situation when $\widehat{x}, x$ are related by*

$$x(t) = \widehat{x}(-1) \cdot \chi_{(-\infty,t_0)}(t) \oplus \widehat{x}(0) \cdot \chi_{[t_0,t_0+h)}(t) \oplus \ldots$$
$$\ldots \oplus \widehat{x}(k) \cdot \chi_{[t_0+kh,t_0+(k+1)h)}(t) \oplus \ldots$$

any of a)

$$\forall k \in \widehat{\mathbf{T}}^{\widehat{x}}_{\mu}, \{k + zp|z \in \mathbf{Z}\} \cap \mathbf{N}_{\_} \subset \widehat{\mathbf{T}}^{\widehat{x}}_{\mu}$$

or

$$\forall k \in \mathbf{N}_{\_}, \widehat{x}(k) = \widehat{x}(k+p)$$

true for $p = 1$,
b)

$$\exists t' \in I^x, \forall t \in \mathbf{T}^x_{\mu} \cap [t', \infty), \{t + zT|z \in \mathbf{Z}\} \cap [t', \infty) \subset \mathbf{T}^x_{\mu}$$

or

$$\exists t' \in I^x, \forall t \geq t', x(t) = x(t+T)$$

true for $T \in (0,h) \cup (h,2h) \cup \ldots \cup (qh,(q+1)h) \cup \ldots$
implies the truth of

$$\forall k \in \mathbf{N}_{\_}, \widehat{x}(k) = \mu,$$
$$\forall t \in \mathbf{R}, x(t) = \mu$$

thus $\widehat{x}, x$ are equal with the same constant μ. The validity $\forall p \geq 1$, of

$$\forall \mu \in \widehat{Or}(\widehat{x}), \forall k \in \widehat{\mathbf{T}}^{\widehat{x}}_{\mu}, \{k + zp|z \in \mathbf{Z}\} \cap \mathbf{N}_{\_} \subset \widehat{\mathbf{T}}^{\widehat{x}}_{\mu}$$

$$\overset{\text{Theorem 86, page 136}}{\Longleftrightarrow} \forall k \in \mathbf{N}_{\_}, \widehat{x}(k) = \widehat{x}(k+p)$$

and $\forall T > 0$, of

$$\exists t' \in I^x, \forall \mu \in Or(x), \forall t \in \mathbf{T}^x_{\mu} \cap [t', \infty), \{t + zT|z \in \mathbf{Z}\} \cap [t', \infty) \subset \mathbf{T}^x_{\mu}$$

$$\overset{\text{Theorem 86, page 136}}{\Longleftrightarrow} \exists t' \in I^x, \forall t \geq t', x(t) = x(t+T)$$

shows the fact that the common conclusion of Theorem 69, page 117, and Theorem 92, page 142, is not surprising.[1]

[1]Note that at $(1.7)_{page\ 134}$ we have the order of the quantifiers $\forall \mu \in Or(x), \exists t' \in I^x$ but in the proof $(1.7)_{page\ 134} \Longrightarrow (2.4)_{page\ 136}$ from Theorem 86, page 136 we could make use of $\exists t' \in I^x, \forall \mu \in Or(x)$, thus the previous argument is correct. We shall refer again to the possibility of changing the order of some quantifiers in stating periodicity properties in Section 14 of this chapter.

7. Discrete Time vs Real Time

Theorem 93. *We presume that $\widehat{x}, x$ satisfy*

$$x(t) = \widehat{x}(-1) \cdot \chi_{(-\infty,t_0)}(t) \oplus \widehat{x}(0) \cdot \chi_{[t_0,t_0+h)}(t) \oplus ... \\ ... \oplus \widehat{x}(k) \cdot \chi_{[t_0+kh,t_0+(k+1)h)}(t) \oplus ... \tag{7.1}$$

for some $t_0 \in \mathbf{R}$ and $h > 0$. Then the existence of $p \geq 1$ such that

$$\forall k \in \mathbf{N}_{\_}, \widehat{x}(k) = \widehat{x}(k+p) \tag{7.2}$$

implies that, for $T = ph$ we have $\exists t' \in I^x$,

$$\forall t \geq t', x(t) = x(t+T). \tag{7.3}$$

Proof. The hypothesis states that $t_0 \in \mathbf{R}, h > 0$ and $p \geq 1$ exist such that (7.1), (7.2) hold and we denote $t' = t_0 - h$. We have

$$x(-\infty+0) = \widehat{x}(-1), \tag{7.4}$$

$$\forall t \leq t', x(t) = x(-\infty+0). \tag{7.5}$$

We fix some arbitrary $t \geq t'$. Then $k \in \mathbf{N}_{\_}$ exists with $t \in [t_0 + kh, t_0 + (k+1)h)$ wherefrom, for $T = ph$ we infer

$$t + T \in [t_0 + kh + T, t_0 + (k+1)h + T) = [t_0 + (k+p)h, t_0 + (k+p+1)h).$$

We finally get

$$x(t) = \widehat{x}(k) \overset{(7.2)}{=} \widehat{x}(k+p) = x(t+T)$$

and, by taking into account (7.5) also, we infer that $t' \in I^x$ exists such that (7.3) is true. □

Theorem 94. *If $\widehat{x}, x$ are not constant and*
i) $t_0 \in \mathbf{R}$, $h > 0$ exist such that $\widehat{x}, x$ fulfill (7.1),
ii) $T > 0$, $t' \in I^x$ exist such that x fulfills (7.3)
then $\frac{T}{h} \in \{1, 2, 3, ...\}$ and $k' \in \mathbf{N}_{\_}$ exists making

$$\forall k \geq k', \widehat{x}(k) = \widehat{x}(k+p), \tag{7.6}$$

$$\forall k \in \{-1, 0, ..., k'\}, \widehat{x}(k) = \widehat{x}(k') \tag{7.7}$$

true for $p = \frac{T}{h}$.

Proof. We presume that $t_0 \in \mathbf{R}$, $h > 0$ exist such that (7.1) holds. We have also the existence of $T > 0$, $t' \in I^x$ such that (7.3) is true.

If $T \in (0,h) \cup (h,2h) \cup ... \cup (qh,(q+1)h) \cup ...$ then $\widehat{x}, x$ are both constant from Theorem 92 b), page 142, contradiction with the hypothesis. We suppose at this moment that $T \in \{h, 2h, 3h, ...\}$ and let $p = \frac{T}{h}$, $p \geq 1$. Let $k_1 \in \mathbf{Z}$ be the number that fulfills $t' \in [t_0 + k_1 h, t_0 + (k_1+1)h)$ and we define $k' = \max\{k_1, -1\}$. We have the following possibilities.

Case $k_1 \leq -2$.

We have $k' = -1$. Let an arbitrary $k \geq k'$ for which an arbitrary, fixed $t \in [t_0 + kh, t_0 + (k+1)h)$ fulfills

$$t + T \in [t_0 + kh + T, t_0 + (k+1)h + T) = [t_0 + (k+p)h, t_0 + (k+p+1)h)$$

and moreover

$$t \geq t_0 + kh \geq t_0 + k'h = t_0 - h > t'. \tag{7.8}$$

We can write that

$$\widehat{x}(k) = x(t) \overset{(7.3),(7.8)}{=} x(t+T) = x(t_0 + (k+p)h) = \widehat{x}(k+p),$$

thus (7.6) is true and (7.7) is also true.

Case $k_1 \geq -1$.

In this situation we have $k' = k_1$. Let us take some arbitrary, fixed $k \geq k'$. We can write

$$t_0 + kh \leq t' + (k - k_1)h < t_0 + (k+1)h, \tag{7.9}$$

$$t_0 + (k+p)h \leq t' + (k - k_1 + p)h < t_0 + (k+p+1)h, \tag{7.10}$$

where

$$t' + (k - k_1)h \geq t'. \tag{7.11}$$

We infer

$$\widehat{x}(k) = x(t_0 + kh) \overset{(7.9)}{=} x(t' + (k - k_1)h) \overset{(7.3),(7.11)}{=} x(t' + (k - k_1)h + T)$$

$$= x(t' + (k - k_1 + p)h) \overset{(7.10)}{=} x(t_0 + (k+p)h) = \widehat{x}(k+p).$$

We have proved the truth of (7.6) and the truth of (7.7) results from the fact that

$$\forall t \leq t', \forall k \in \{-1, 0, ..., k'\}, \widehat{x}(k) = x(t) = x(-\infty + 0).$$

□

Example 30. *We define $\widehat{x} \in \widehat{S}^{(1)}$ by*

$$\forall k \in \mathbf{N}_{\_}, \widehat{x}(k) = \begin{cases} 1, if\ k \in \{2,4,6,8,...\} \\ 0, otherwise \end{cases}$$

and $x \in S^{(1)}$ respectively by

$$x(t) = \widehat{x}(-1) \cdot \chi_{(-\infty,-4)}(t) \oplus \widehat{x}(0) \cdot \chi_{[-4,-2)}(t) \oplus$$

$$\oplus \widehat{x}(1) \cdot \chi_{[-2,0)}(t) \oplus \widehat{x}(2) \cdot \chi_{[0,2)}(t) \oplus ...$$

We have $I^x = (-\infty, 0), L^x = [-2, \infty)$,

$$\forall t \leq -2, x(t) = 0,$$

$$\forall t \geq -2, x(t) = x(t+4),$$

$$\forall k \geq 1, \widehat{x}(k) = \widehat{x}(k+2),$$

$$\forall k \in \{-1, 0, 1\}, \widehat{x}(k) = 0$$

thus $(7.3)_{page\ 144}$ is fulfilled with $T = 4, t' = -2$ and (7.6), (7.7) are true with $p = 2, k' = 1$. Furthermore, in this example $h = 2$.

Remark 131. *In Theorem 94, the conjunction of (7.6) with (7.7) gives a special case of eventual periodicity.*

Remark 132. *Theorem 93 and Theorem 94 represent the periodic version of Theorem 52, page 98 and Theorem 53, page 99 that refer to eventual periodicity.*

8. Sums, Differences and Multiples of Periods

Theorem 95. *Let the signals $\widehat{x}, x$.*

a) We suppose that $\widehat{x}$ has the periods $p, p' \geq 1$,

$$\forall k \in \mathbf{N}_{\_}, \widehat{x}(k) = \widehat{x}(k+p), \tag{8.1}$$

$$\forall k \in \mathbf{N}_{\_}, \widehat{x}(k) = \widehat{x}(k+p'). \tag{8.2}$$

Then $p+p' \geq 1$, $\widehat{x}$ has the period $p+p'$,

$$\forall k \in \mathbf{N}_{\_}, \widehat{x}(k) = \widehat{x}(k+p+p') \tag{8.3}$$

and if $p > p'$, then $p-p' \geq 1$ and $\widehat{x}$ has the period $p-p'$,

$$\forall k \in \mathbf{N}_{\_}, \widehat{x}(k) = \widehat{x}(k+p-p'). \tag{8.4}$$

b) Let $T, T' > 0$, $t' \in I^x$ arbitrary with

$$\forall t \geq t', x(t) = x(t+T), \tag{8.5}$$

$$\forall t \geq t', x(t) = x(t+T') \tag{8.6}$$

fulfilled. We have on one hand that $T+T' > 0$ and

$$\forall t \geq t', x(t) = x(t+T+T'), \tag{8.7}$$

and on the other hand that $T > T'$ implies $T-T' > 0$ and

$$\forall t \geq t', x(t) = x(t+T-T'). \tag{8.8}$$

Proof. This is a special case of Theorem 54, page 100 with the limit of periodicity $k' = -1$ at a) and $t' \in I^x$ at b). □

Theorem 96. *We consider the signals $\widehat{x}, x$.*

a) Let $p, k_1 \geq 1$. We have that $p' = k_1 p$ fulfills $p' \geq 1$ and

$$\forall k \in \mathbf{N}_{\_}, \widehat{x}(k) = \widehat{x}(k+p) \tag{8.9}$$

implies

$$\forall k \in \mathbf{N}_{\_}, \widehat{x}(k) = \widehat{x}(k+p'). \tag{8.10}$$

b) We suppose that $T > 0$, $t' \in I^x$, $k_1 \geq 1$ are given. We infer that $T' = k_1 T$ fulfills $T' > 0$ and

$$\forall t \geq t', x(t) = x(t+T) \tag{8.11}$$

implies

$$\forall t \geq t', x(t) = x(t+T'). \tag{8.12}$$

Proof. This is a consequence of Theorem 95, the first assertion from a), b) that refers to the addition. □

Remark 133. *We can express the statements of Theorem 96 in an equivalent way under the form: if* $p \in \widehat{P^{\widehat{x}}}$ *then* $\{p, 2p, 3p, ...\} \subset \widehat{P^{\widehat{x}}}$ *and if* $T \in P^x$ *then* $\{T, 2T, 3T, ...\} \subset P^x$.

9. The Set of the Periods

Theorem 97. *a) We suppose that for* $\widehat{x} \in \widehat{S}^{(n)}$, *the set* $\widehat{P^{\widehat{x}}}$ *is non empty. Some* $\widetilde{p} \geq 1$ *exists then with the property*

$$\widehat{P^{\widehat{x}}} = \{\widetilde{p}, 2\widetilde{p}, 3\widetilde{p}, ...\}. \tag{9.1}$$

b) Let $x \in S^{(n)}$ *be not constant and we suppose that the set* P^x *is not empty. Then* $\widetilde{T} > 0$ *exists such that*

$$P^x = \{\widetilde{T}, 2\widetilde{T}, 3\widetilde{T}, ...\}. \tag{9.2}$$

Proof. This is a special case of Theorem 56, page 101 with $\widehat{L^{\widehat{x}}} = \mathbf{N}\_$ at a) and $I^x \cap L^x \neq \varnothing$ at b). □

Remark 134. *If in Theorem 97, item a)* $\widehat{x}$ *is constant, then* $\widetilde{p} = 1$ *and* $\widehat{P^{\widehat{x}}} = \{1, 2, 3, ...\}$, *thus (9.1) is still true. Unlike this situation, if x is constant, then (9.2) is false and we get* $P^x = (0, \infty)$ *instead.*

Theorem 98. *We suppose that the relation between* $\widehat{x}$ *and x is given by*

$$\begin{aligned} x(t) = \widehat{x}(-1) \cdot \chi_{(-\infty, t_0)}(t) \oplus \widehat{x}(0) \cdot \chi_{[t_0, t_0+h)}(t) \oplus \\ \oplus \widehat{x}(1) \cdot \chi_{[t_0+h, t_0+2h)}(t) \oplus ... \oplus \widehat{x}(k) \cdot \chi_{[t_0+kh, t_0+(k+1)h)}(t) \oplus ... \end{aligned} \tag{9.3}$$

where $t_0 \in \mathbf{R}$ *and* $h > 0$. *If* $\widehat{x}$, *x are periodic, two possibilities exist:*

a) $\widehat{x}$, *x are both constant,* $\widehat{P^{\widehat{x}}} = \{1, 2, 3, ...\}$ *and* $P^x = (0, \infty)$;

b) $\widehat{x}$, *x are both non-constant,* $\min \widehat{P^{\widehat{x}}} = p > 1$ *is the prime period of* $\widehat{x}$ *and* $\min P^x = T = ph$ *is the prime period of x.*

Proof. The proof is analogue with the proof of Theorem 74, page 121.

b) Theorem 93, page 144 shows that $p \in \widehat{P^{\widehat{x}}} \Longrightarrow T = ph \in P^x$ and from Theorem 94, page 144 we get that $T \in P^x \Longrightarrow p = \frac{T}{h} \in \widehat{P^{\widehat{x}}}$. We suppose, see Theorem 97, that $\widehat{P^{\widehat{x}}} = \{\widetilde{p}, 2\widetilde{p}, 3\widetilde{p}, ...\}$ and $P^x = \{\widetilde{T}, 2\widetilde{T}, 3\widetilde{T}, ...\}$. Then $\widetilde{T} = \widetilde{p}h$. □

10. Necessity Conditions of Periodicity

Theorem 99. *Let* $\widehat{x} \in \widehat{S}^{(n)}$ *with* $\widehat{Or}(\widehat{x}) = \{\mu^1, ..., \mu^s\}, s \geq 2$ *be periodic with the period* $p \geq 1$. *We have the existence, for any* $i \in \{1, ..., s\}$, *of* $n_1^i, n_2^i, ..., n_{k_i}^i \in \{-1, 0, ..., p-2\}$, $k_i \geq 1$, *such that*

$$\widehat{\mathbf{T}}^{\widehat{x}}_{\mu^i} = \bigcup_{k \in \mathbf{N}} \{n_1^i + kp, n_2^i + kp, ..., n_{k_i}^i + kp\}. \tag{10.1}$$

Proof. If $\widehat{x}$ is periodic with the period p, then we can apply Theorem 75, page 121 for any $i \in \{1, ..., s\}$, since any μ^i is periodic with the period p. □

Theorem 100. *Let $x \in S^{(n)}$ non constant with $Or(x) = \{\mu^1, ..., \mu^s\}$ and we denote $\mu^1 = x(-\infty + 0)$. We suppose that x is periodic with the period $T > 0$. Then t_0 and $a_1^i, b_1^i, a_2^i, b_2^i, ..., a_{k_i}^i, b_{k_i}^i \in \mathbf{R}$ exist, $k_i \geq 1$, $i \in \{1, ..., s\}$ such that*

$$\forall t < t_0, x(t) = \mu^1, \tag{10.2}$$

$$x(t_0) \neq \mu^1, \tag{10.3}$$

$$t_0 < a_1^1 < b_1^1 < a_2^1 < b_2^1 < ... < a_{k_1}^1 < b_{k_1}^1 = t_0 + T, \tag{10.4}$$

$$[a_1^1, b_1^1) \cup [a_2^1, b_2^1) \cup ... \cup [a_{k_1}^1, b_{k_1}^1) = \mathbf{T}_{\mu^1}^x \cap [t_0, t_0 + T), \tag{10.5}$$

$$\mathbf{T}_{\mu^1}^x = \begin{array}{c} (-\infty, t_0) \cup \bigcup_{k \in \mathbf{N}} ([a_1^1 + kT, b_1^1 + kT) \cup [a_2^1 + kT, b_2^1 + kT) \cup ... \\ ... \cup [a_{k_1}^1 + kT, b_{k_1}^1 + kT)) \end{array} \tag{10.6}$$

and for any $i \in \{2, ..., s\}$,

$$t_0 \leq a_1^i < b_1^i < a_2^i < b_2^i < ... < a_{k_i}^i < b_{k_i}^i < t_0 + T, \tag{10.7}$$

$$[a_1^i, b_1^i) \cup [a_2^i, b_2^i) \cup ... \cup [a_{k_i}^i, b_{k_i}^i) = \mathbf{T}_{\mu^i}^x \cap [t_0, t_0 + T), \tag{10.8}$$

$$\mathbf{T}_{\mu^i}^x = \bigcup_{k \in \mathbf{N}} ([a_1^i + kT, b_1^i + kT) \cup [a_2^i + kT, b_2^i + kT) \cup ... \cup [a_{k_1}^i + kT, b_{k_1}^i + kT)) \tag{10.9}$$

are fulfilled.

Proof. As x is non constant, periodic with the period T, all of $\mu^1, ..., \mu^s$ are periodic, with the period T. Theorem 76, page 122 shows then the existence of $t_0, a_1^1, b_1^1, a_2^1, b_2^1, ..., a_{k_1}^1, b_{k_1}^1 \in \mathbf{R}$, $k_1 \geq 1$ such that (10.2),...,(10.6) are true. The existence of $t_0, a_1^i, b_1^i, a_2^i, b_2^i, ..., a_{k_i}^i, b_{k_i}^i \in \mathbf{R}$, $k_i \geq 1$ for $i \in \{2, ..., s\}$ such that (10.2), (10.3), (10.7),..., (10.9) hold results from Theorem 77, page 123. □

Example 31. *The periodic signal $x \in S^{(1)}$,*

$$x(t) = \chi_{(-\infty,0)}(t) \oplus \chi_{[1,2)}(t) \oplus \chi_{[3,5)}(t) \oplus \chi_{[6,7)}(t) \oplus \chi_{[8,10)}(t) \oplus \chi_{[11,12)}(t) \oplus ...$$

fulfills $\mu^1 = 1, \mu^2 = 0, k_1 = k_2 = 2, T = 5, t_0 = a_1^2 = 0, a_1^1 = b_1^2 = 1, b_1^1 = a_2^2 = 2, a_2^1 = b_2^2 = 3, b_2^1 = t_0 + 5 = 5$.

11. Sufficiency Conditions of Periodicity

Theorem 101. *Let $\widehat{x} \in \widehat{S}^{(n)}$, $\widehat{Or}(\widehat{x}) = \{\mu^1, ..., \mu^s\}$ and $p \geq 1$. We suppose that $\forall i \in \{1, ..., s\}$, $n_1^i, n_2^i, ..., n_{k_i}^i \in \{-1, 0, ..., p-2\}, k_i \geq 1$ exist such that*

$$\widehat{\mathbf{T}}_{\mu^i}^{\widehat{x}} = \bigcup_{k \in \mathbf{N}} \{n_1^i + kp, n_2^i + kp, ..., n_{k_i}^i + kp\}. \tag{11.1}$$

Then $\forall i \in \{1, ..., s\}$,

$$\forall k \in \widehat{\mathbf{T}}_{\mu^i}^{\widehat{x}}, \{k + zp | z \in \mathbf{Z}\} \cap \mathbf{N}_{\_} \subset \widehat{\mathbf{T}}_{\mu^i}^{\widehat{x}}. \tag{11.2}$$

Proof. If $\forall i \in \{1,...,s\}, n_1^i, n_2^i, ..., n_{k_i}^i \in \{-1,0,...,p-2\}, k_i \geq 1$ exist such that (11.1) holds, we infer from Theorem 78, page 124 that all of $\mu^1,...,\mu^s$ are periodic with the period p, i.e. $\widehat{x}$ is periodic with the period p. □

Theorem 102. *The signal $x \in S^{(n)}$ is given, such that $Or(x) = \{\mu^1,...,\mu^s\}, s \geq 2$ and we suppose that the initial value of x is μ^1. We ask that $T > 0$ and the points $t_0, a_1^i, b_1^i, a_2^i, b_2^i, ..., a_{k_i}^i, b_{k_i}^i \in \mathbf{R}$, $k_i \geq 1$ exist, $i \in \{1,...,s\}$ with*

$$t_0 < a_1^1 < b_1^1 < a_2^1 < b_2^1 < ... < a_{k_1}^1 < b_{k_1}^1 = t_0 + T, \tag{11.3}$$

$$\mathbf{T}_{\mu^1}^x = \begin{array}{l} (-\infty, t_0) \cup \bigcup\limits_{k \in \mathbf{N}} ([a_1^1 + kT, b_1^1 + kT) \cup [a_2^1 + kT, b_2^1 + kT) \cup ... \\ \quad ... \cup [a_{k_1}^1 + kT, b_{k_1}^1 + kT)) \end{array} \tag{11.4}$$

and for any $i \in \{2,...,s\}$,

$$b_{k_i}^i - T < t_0 \leq a_1^i < b_1^i < a_2^i < b_2^i < ... < a_{k_i}^i < b_{k_i}^i, \tag{11.5}$$

$$\mathbf{T}_{\mu^i}^x = \bigcup_{k \in \mathbf{N}} ([a_1^i + kT, b_1^i + kT) \cup [a_2^i + kT, b_2^i + kT) \cup ... \cup [a_{k_1}^i + kT, b_{k_1}^i + kT)). \tag{11.6}$$

In such conditions, for any $t' \in [a_{k_1}^1 - T, t_0)$, we have $t' \in I^x$ and

$$\forall i \in \{1,...,s\}, \forall t \in \mathbf{T}_{\mu^i}^x \cap [t', \infty), \{t + zT | z \in \mathbf{Z}\} \cap [t', \infty) \subset \mathbf{T}_{\mu^i}^x. \tag{11.7}$$

Proof. We suppose that T and $t_0, a_1^1, b_1^1, a_2^1, b_2^1, ..., a_{k_1}^1, b_{k_1}^1$, $k_1 \geq 1$ exist such that (11.3), (11.4) are true and let $t' \in [a_{k_1}^1 - T, t_0)$ arbitrary, fixed. From Theorem 79, page 124, we have that for an arbitrary $t_1' \in [a_{k_1}^1 - T, t_0)$, the statements $t_1' \in I^x$,

$$\forall t \in \mathbf{T}_{\mu^1}^x \cap [t_1', \infty), \{t + zT | z \in \mathbf{Z}\} \cap [t_1', \infty) \subset \mathbf{T}_{\mu^1}^x \tag{11.8}$$

hold. We suppose furthermore that $\forall i \in \{2,...,s\}, a_1^i, b_1^i, a_2^i, b_2^i, ..., a_{k_i}^i, b_{k_i}^i$, $k_i \geq 1$ exist with the property that (11.5), (11.6) are true. As far as

$$\forall t < t_0, x(t) = \mu^1,$$

$$x(t_0) \overset{(11.3),(11.4)}{\neq} \mu^1,$$

Theorem 80, page 125 shows that for any $t_i' \in [b_{k_i}^i - T, t_0)$, the statements $t_i' \in I^x$,

$$\forall t \in \mathbf{T}_{\mu^i}^x \cap [t_i', \infty), \{t + zT | z \in \mathbf{Z}\} \cap [t_i', \infty) \subset \mathbf{T}_{\mu^i}^x \tag{11.9}$$

hold also, i.e. $\mu^1,...,\mu^s$ are all periodic with the same period T. Corollary 5, page 141, states basically that

$$[a_{k_1}^1 - T, t_0) = [a_{k_1}^1 - T, t_0) \cap [b_{k_2}^2 - T, t_0) \cap ... \cap [b_{k_s}^s - T, t_0),$$

i.e. in (11.8),(11.9) we can make the choice $t_i' = t', i \in \{1,...,s\}$. In such circumstances $t' \in I^x$ and (11.7) holds. □

12. A Special Case

Theorem 103. *Let $\widehat{x} \in \widehat{S}^{(n)}$ with $\widehat{Or}(\widehat{x}) = \{\mu^1, ..., \mu^s\}$ and $p \geq 1$. We suppose that $\forall i \in \{1, ..., s\}, n^i \in \{-1, 0, ..., p-2\}$ exists such that*

$$\widehat{\mathbf{T}}^{\widehat{x}}_{\mu^i} = \{n^i, n^i + p, n^i + 2p, ...\}. \tag{12.1}$$

a) We have $\forall i \in \{1, ..., s\}$,

$$\forall k \in \widehat{\mathbf{T}}^{\widehat{x}}_{\mu^i}, \{k + zp | z \in \mathbf{Z}\} \cap \mathbf{N}_{\_} \subset \widehat{\mathbf{T}}^{\widehat{x}}_{\mu^i}. \tag{12.2}$$

b) Any $p' \geq 1$ making $\forall i \in \{1, ..., s\}$,

$$\forall k \in \widehat{\mathbf{T}}^{\widehat{x}}_{\mu^i}, \{k + zp' | z \in \mathbf{Z}\} \cap \mathbf{N}_{\_} \subset \widehat{\mathbf{T}}^{\widehat{x}}_{\mu^i} \tag{12.3}$$

true fulfills $p' \in \{p, 2p, 3p, ...\}$, i.e. p is the prime period of $\widehat{x}$.

Proof. The theorem is a special case of Theorem 63, page 104. In particular item a) is a special case of Theorem 101, page 148 with $k_i = 1, i = \overline{1,s}$. □

Theorem 104. *Let $x \in S^{(n)}$ with $Or(x) = \{\mu^1, ..., \mu^s\}$, $s \geq 2$, $T > 0$ and the points $t_0, t_1, ..., t_{s-1} \in \mathbf{R}$ with the following properties:*

$$t_0 < t_1 < ... < t_{s-1} < t_0 + T, \tag{12.4}$$

$$\mathbf{T}^x_{\mu^1} = (-\infty, t_0) \cup [t_{s-1}, t_0 + T) \cup [t_{s-1} + T, t_0 + 2T) \cup [t_{s-1} + 2T, t_0 + 3T) \cup ... \tag{12.5}$$

$$\mathbf{T}^x_{\mu^2} = [t_0, t_1) \cup [t_0 + T, t_1 + T) \cup [t_0 + 2T, t_1 + 2T) \cup ... \tag{12.6}$$

$$...$$

$$\mathbf{T}^x_{\mu^s} = [t_{s-2}, t_{s-1}) \cup [t_{s-2} + T, t_{s-1} + T) \cup [t_{s-2} + 2T, t_{s-1} + 2T) \cup ... \tag{12.7}$$

a) For any $t' \in [t_{s-1} - T, t_0)$, the statements $t' \in I^x$,

$$\forall i \in \{1, ..., s\}, \forall t \in \mathbf{T}^x_{\mu^i} \cap [t', \infty), \{t + zT | z \in \mathbf{Z}\} \cap [t', \infty) \subset \mathbf{T}^x_{\mu^i} \tag{12.8}$$

are true.

b) Let $t'' \in [t_{s-1} - T, t_0)$ arbitrary. For any $T' > 0$ such that

$$\forall i \in \{1, ..., s\}, \forall t \in \mathbf{T}^x_{\mu^i} \cap [t'', \infty), \{t + zT' | z \in \mathbf{Z}\} \cap [t'', \infty) \subset \mathbf{T}^x_{\mu^i}, \tag{12.9}$$

we have $T' \in \{T, 2T, 3T, ...\}$.

Proof. The theorem is a special case of Theorem 64, page 104. In particular item a) is a special case of Theorem 102, page 149. By definition we have $t_0 = a_1^2, t_1 = a_1^3 = b_1^2, ..., t_{s-2} = a_1^s = b_1^{s-1}, t_{s-1} = a_1^1 = b_1^s, t_0 + T = b_1^1$. □

13. Periodicity vs Eventual Periodicity

Theorem 105. *a) If* $\widehat{x}$ *is periodic, then for any* $\widetilde{k} \in \mathbf{N}$, $\widehat{\sigma}^{\widetilde{k}}(\widehat{x})$ *is periodic and* $\widehat{P}^{\widehat{x}} = \widehat{P}^{\widehat{\sigma}^{\widetilde{k}}(\widehat{x})}$.

b) We suppose that x is periodic. For arbitrary $\widetilde{t} \in \mathbf{R}$, *we have that* $\sigma^{\widetilde{t}}(x)$ *is periodic and* $P^x = P^{\sigma^{\widetilde{t}}(x)}$.

Proof. a) We suppose that $\widehat{P}^{\widehat{x}} \neq \varnothing$ and let $\widetilde{k} \in \mathbf{N}$ arbitrary.

The first proof. For any $\mu \in \widehat{Or}(\widehat{x}) = \widehat{\omega}(\widehat{x}) \overset{Theorem\,5,page\,9}{=} \widehat{\omega}(\widehat{\sigma}^{\widetilde{k}}(\widehat{x})) = \widehat{Or}(\widehat{\sigma}^{\widetilde{k}}(\widehat{x}))$ we have $\widehat{P}^{\widehat{x}}_{\mu} \overset{Theorem\,84,page\,128}{=} \widehat{P}^{\widehat{\sigma}^{\widetilde{k}}(\widehat{x})}_{\mu}$, thus

$$\widehat{P}^{\widehat{x}} \overset{Theorem\,58,page\,102}{=} \bigcap_{\mu \in \widehat{\omega}(\widehat{x})} \widehat{P}^{\widehat{x}}_{\mu} = \bigcap_{\mu \in \widehat{\omega}(\widehat{\sigma}^{\widetilde{k}}(\widehat{x}))} \widehat{P}^{\widehat{\sigma}^{\widetilde{k}}(\widehat{x})}_{\mu} \overset{Theorem\,58,page\,102}{=} \widehat{P}^{\widehat{\sigma}^{\widetilde{k}}(\widehat{x})}.$$

The second proof. We prove $\widehat{P}^{\widehat{x}} \subset \widehat{P}^{\widehat{\sigma}^{\widetilde{k}}(\widehat{x})}$. We take an arbitrary $p \in \widehat{P}^{\widehat{x}}$, meaning that

$$\forall k \in \mathbf{N}_{\_}, \widehat{x}(k) = \widehat{x}(k+p) \tag{13.1}$$

holds and we show that

$$\forall k \in \mathbf{N}_{\_}, \widehat{\sigma}^{\widetilde{k}}(\widehat{x})(k) = \widehat{\sigma}^{\widetilde{k}}(\widehat{x})(k+p) \tag{13.2}$$

is true. Indeed, for any $k \in \mathbf{N}_{\_}$ we get

$$\widehat{\sigma}^{\widetilde{k}}(\widehat{x})(k) = \widehat{x}(k+\widetilde{k}) \overset{(13.1)}{=} \widehat{x}(k+\widetilde{k}+p) = \widehat{\sigma}^{\widetilde{k}}(\widehat{x})(k+p).$$

We prove $\widehat{P}^{\widehat{\sigma}^{\widetilde{k}}(\widehat{x})} \subset \widehat{P}^{\widehat{x}}$. Let $p \in \widehat{P}^{\widehat{x}}$, thus (13.1) is true. We suppose against all reason that $\widehat{P}^{\widehat{\sigma}^{\widetilde{k}}(\widehat{x})} \subset \widehat{P}^{\widehat{x}}$ is false, i.e. some $p' \in \widehat{P}^{\widehat{\sigma}^{\widetilde{k}}(\widehat{x})} \setminus \widehat{P}^{\widehat{x}}$ exists. This means, by rewriting (13.2) under the form

$$\forall k \geq \widetilde{k}-1, \widehat{x}(k) = \widehat{x}(k+p'), \tag{13.3}$$

that

$$\exists k_1 \in \{-1, 0, ..., \widetilde{k}-2\}, \widehat{x}(k_1) \neq \widehat{x}(k_1+p'). \tag{13.4}$$

Let $\overline{k} \in \mathbf{N}$ with the property that $k_1 + \overline{k}p \geq \widetilde{k}-1$. We infer:

$$\widehat{x}(k_1) \overset{(13.1)}{=} \widehat{x}(k_1+\overline{k}p) \overset{(13.3)}{=} \widehat{x}(k_1+\overline{k}p+p') \overset{(13.1)}{=} \widehat{x}(k_1+p'). \tag{13.5}$$

Statements (13.4) and (13.5) are contradictory.

b) We suppose that $P^x \neq \varnothing$ and we take $\widetilde{t} \in \mathbf{R}$ arbitrarily.

We prove $P^x \subset P^{\sigma^{\widetilde{t}}(x)}$. Let $T \in P^x$ arbitrary, for which $t' \in I^x$ exists such that

$$\forall t \geq t', x(t) = x(t+T) \tag{13.6}$$

holds. We must prove the existence of $t'' \in I^{\sigma^{\widetilde{t}}(x)}$ making

$$\forall t \geq t'', \sigma^{\widetilde{t}}(x)(t) = \sigma^{\widetilde{t}}(x)(t+T) \tag{13.7}$$

true. If x is constant, then $\sigma^{\widetilde{t}}(x)=x$ and the statements $t''\in I^{\sigma^{\widetilde{t}}(x)}$, (13.7) are trivially true for any $t''\in R$ so that we shall suppose from now that x is not constant. Some $t_0\in\mathbf{R}$ exists with

$$\forall t<t_0, x(t)=x(-\infty+0), \tag{13.8}$$

$$x(t_0)\neq x(-\infty+0). \tag{13.9}$$

From (13.8), (13.9) we get $I^x=(-\infty,t_0)$ and since $t'\in I^x$, we infer $t'<t_0$. Two possibilities exist.

Case $\widetilde{t}\leq t_0$

In this case $\sigma^{\widetilde{t}}(x)=x$ and the statements $t''\in I^{\sigma^{\widetilde{t}}(x)}$, (13.7) are true again for $t''=t'$.

Case $\widetilde{t}>t_0$

Some $\varepsilon>0$ exists with the property $\forall\xi\in(\widetilde{t}-\varepsilon,\widetilde{t}), x(\xi)=x(\widetilde{t}-0)$. We infer from here that $\widetilde{t}-\varepsilon\geq t_0>t'$ and for $t''\in(\widetilde{t}-\varepsilon,\widetilde{t})$ arbitrary, fixed we have

$$\sigma^{\widetilde{t}}(x)(t)=\begin{cases} x(t), t\geq t'',\\ x(\widetilde{t}-0), t<\widetilde{t}.\end{cases} \tag{13.10}$$

We notice that $t''\in I^{\sigma^{\widetilde{t}}(x)}$ and, on the other hand, we can write for any $t\geq t''$ that

$$\sigma^{\widetilde{t}}(x)(t)\overset{(13.10)}{=}x(t)\overset{(13.6)}{=}x(t+T)\overset{(13.10)}{=}\sigma^{\widetilde{t}}(x)(t+T),$$

wherefrom the truth of (13.7).

We prove $P^{\sigma^{\widetilde{t}}(x)}\subset P^x$. Let $T\in P^x$ arbitrary, thus $t'\in I^x$ exists such that (13.6) takes place. We suppose against all reason that $P^{\sigma^{\widetilde{t}}(x)}\subset P^x$ is false, i.e. some $T'\in P^{\sigma^{\widetilde{t}}(x)}\setminus P^x$ exists. This means the existence of $t''\in I^{\sigma^{\widetilde{t}}(x)}$ with

$$\forall t\geq t'', \sigma^{\widetilde{t}}(x)(t)=\sigma^{\widetilde{t}}(x)(t+T'), \tag{13.11}$$

$$\forall t'''\in I^x, \exists t_1\geq t''', x(t_1)\neq x(t_1+T'). \tag{13.12}$$

Let $t'''\in I^x\cap[t',\infty)$ arbitrary and we take $\overline{k}\in\mathbf{N}$ such that for $t_1\geq t'''$ whose existence is stated in (13.12), we have $t_1+\overline{k}T\geq\max\{t'',\widetilde{t}\}$. We get

$$\begin{aligned} x(t_1)&\overset{(13.6)}{=}x(t_1+\overline{k}T)=\sigma^{\widetilde{t}}(x)(t_1+\overline{k}T)\\ &\overset{(13.11)}{=}\sigma^{\widetilde{t}}(x)(t_1+\overline{k}T+T')=x(t_1+\overline{k}T+T')\overset{(13.6)}{=}x(t_1+T'),\end{aligned} \tag{13.13}$$

and (13.12), (13.13) are contradictory. □

14. Changing the Order of the Quantifiers

Theorem 106. *a) The statements*

$$\exists p\geq 1, \forall\mu\in\widehat{Or}(\widehat{x}), \forall k\in\widehat{\mathbf{T}}_{\mu}^{\widehat{x}}, \{k+zp|z\in\mathbf{Z}\}\cap\mathbf{N}_{\_}\subset\widehat{\mathbf{T}}_{\mu}^{\widehat{x}}, \tag{14.1}$$

$$\forall\mu\in\widehat{Or}(\widehat{x}), \exists p\geq 1, \forall k\in\widehat{\mathbf{T}}_{\mu}^{\widehat{x}}, \{k+zp|z\in\mathbf{Z}\}\cap\mathbf{N}_{\_}\subset\widehat{\mathbf{T}}_{\mu}^{\widehat{x}} \tag{14.2}$$

are equivalent.

b) The real time statements

$$\begin{cases} \exists T>0, \exists t' \in I^x, \forall \mu \in Or(x), \\ \forall t \in \mathbf{T}^x_\mu \cap [t',\infty), \{t+zT | z\in \mathbf{Z}\} \cap [t',\infty) \subset \mathbf{T}^x_\mu, \end{cases} \tag{14.3}$$

$$\begin{cases} \exists T>0, \forall \mu \in Or(x), \exists t' \in I^x, \\ \forall t \in \mathbf{T}^x_\mu \cap [t',\infty), \{t+zT | z\in \mathbf{Z}\} \cap [t',\infty) \subset \mathbf{T}^x_\mu, \end{cases} \tag{14.4}$$

$$\begin{cases} \exists t' \in I^x, \forall \mu \in Or(x), \exists T>0, \\ \forall t \in \mathbf{T}^x_\mu \cap [t',\infty), \{t+zT | z\in \mathbf{Z}\} \cap [t',\infty) \subset \mathbf{T}^x_\mu, \end{cases} \tag{14.5}$$

$$\begin{cases} \forall \mu \in Or(x), \exists T>0, \exists t' \in I^x, \\ \forall t \in \mathbf{T}^x_\mu \cap [t',\infty), \{t+zT | z\in \mathbf{Z}\} \cap [t',\infty) \subset \mathbf{T}^x_\mu \end{cases} \tag{14.6}$$

are also equivalent.

Proof. a) Since (14.1)$\Longrightarrow$(14.2) is obvious, we prove (14.2)$\Longrightarrow$(14.1). We denote $\widehat{Or}(\widehat{x}) = \{\mu^1,...,\mu^s\}$. From (14.2) we get that for any $i \in \{1,...,s\}$, some $p_i \geq 1$ exists such that

$$\forall k \in \widehat{\mathbf{T}}^{\widehat{x}}_{\mu^i}, \{k+zp_i | z\in \mathbf{Z}\} \cap \mathbf{N}_\_ \subset \widehat{\mathbf{T}}^{\widehat{x}}_{\mu^i}.$$

The number $p = p_1 \cdot ... \cdot p_s \geq 1$ fulfills the property that $\forall i \in \{1,...,s\}$, we have

$$\forall k \in \widehat{\mathbf{T}}^{\widehat{x}}_{\mu^i}, \{k+zp | z\in \mathbf{Z}\} \cap \mathbf{N}_\_ \subset \{k+zp_i | z\in \mathbf{Z}\} \cap \mathbf{N}_\_ \subset \widehat{\mathbf{T}}^{\widehat{x}}_{\mu^i}, \tag{14.7}$$

i.e. (14.1) is true.

b) The implications (14.3)$\Longrightarrow$(14.4)$\Longrightarrow$(14.6) and (14.3)$\Longrightarrow$(14.5)$\Longrightarrow$(14.6) are obvious.

The implication (14.6)$\Longrightarrow$(14.3) has no proof so far. □

Remark 135. *In the previous theorem, where the proof of the implication (14.6)$\Longrightarrow$(14.3) is missing, we address the problem of changing the order of the quantifiers in stating periodicity properties of the signals. The importance of this aspect is given by the fact that we are tempted to define the periodic signals by (14.2), (14.6) (all the points of $\widehat{Or}(\widehat{x})$, $Or(x)$ are periodic) and to use (14.1), (14.3) instead.*

15. Other Issues

Remark 136. *Remarks 100 and 101, page 106 from the eventually periodic signals, as well as Section 14., page 107 from the same chapter may be restated in the case of the periodic signals also, where they are still interesting.*

Chapter 9

Examples

We sketch in this chapter some constructions that either weaken, in discrete time and real time, the periodicity of the points to eventual periodicity, or change their sets of periods.

1. Discrete Time, Periodic Points

Remark 137. *In the following, the signal $\widehat{x} \in \widehat{S^{(n)}}$ and the periodic point $\mu \in \widehat{Or}(\widehat{x})$ with the period $\widetilde{p} \geq 1$ are considered. The sets $\widehat{\mathbf{T}}_{\mu}^{\widehat{x}}$ and $\widehat{P}_{\mu}^{\widehat{x}}$ are given by*

$$\widehat{\mathbf{T}}_{\mu}^{\widehat{x}} = \{k_0, k_0 + \widetilde{p}, k_0 + 2\widetilde{p}, ...\}, k_0 \in \{-1, 0, ..., \widetilde{p} - 2\},$$

$$\widehat{P}_{\mu}^{\widehat{x}} = \{\widetilde{p}, 2\widetilde{p}, 3\widetilde{p}, ...\}.$$

This simple form of $\widehat{\mathbf{T}}_{\mu}^{\widehat{x}}$, corresponding to the special case from Theorem 81, page 126 does not restrict the generality of the exposure.

Example 32. *Let $\mu' \in \mathbf{B}^n, \mu' \neq \mu$ and the time instant $k' \in \widehat{\mathbf{T}}_{\mu}^{\widehat{x}}$. We define $\widehat{y} \in \widehat{S^{(n)}}$ by $\forall k \in \mathbf{N}_{\_}$,*

$$\widehat{y}(k) = \begin{cases} \widehat{x}(k), k \neq k', \\ \mu', k = k'. \end{cases}$$

We say that $\widehat{y}$ is obtained by removing from $\widehat{x}$ the instant k' of periodicity of μ and we interpret the fact that $\widehat{y}(k') = \mu' \neq \mu = \widehat{x}(k')$ instead of $\widehat{y}(k') = \mu = \widehat{x}(k')$ as representing an error, or a perturbation of the periodicity of μ. We have: after removing from $\widehat{x}$ the instant k' of periodicity of μ we loose the periodicity of μ, but we still have eventual periodicity with the same sets of periods $\widehat{P}_{\mu}^{\widehat{y}} = \widehat{P}_{\mu}^{\widehat{x}}$ and with the limit of periodicity $k' + 1$.

Example 33. *The previous example is continued by taking the points $\mu^1, ..., \mu^s \in \mathbf{B}^n$ that are not necessarily distinct, but they differ from μ: $\mu^1 \neq \mu, ..., \mu^s \neq \mu$ and also the distinct time instants $k^1, ..., k^s \in \widehat{\mathbf{T}}_{\mu}^{\widehat{x}}$. The signal $\widehat{y} \in \widehat{S^{(n)}}$ is defined in the following way: $\forall k \in \mathbf{N}_{\_}$,*

$$\widehat{y}(k) = \begin{cases} \widehat{x}(k), k \notin \{k^1, ..., k^s\}, \\ \mu^j, \exists j \in \{1, ..., s\}, k = k^j. \end{cases}$$

We use to say that $\widehat{y}$ is obtained by the removal from $\widehat{x}$ of the instants $k^1, ..., k^s$ of periodicity of μ. Then $\widehat{P}_{\mu}^{\widehat{y}} = \widehat{P}_{\mu}^{\widehat{x}}$ again, where μ is an eventually periodic point of $\widehat{y}$ and $1 + \max\{k^1, ..., k^s\}$ is its limit of periodicity.

Example 34. *We give the countable version of the construction. The sequence* $\mu^j \in \mathbf{B}^n, j \in \mathbf{N}_-$ *is considered with* $\forall j \in \mathbf{N}_-, \mu^j \neq \mu$ *and also the sequence* $k^j \in \widehat{\mathbf{T}}_{\mu}^{\widehat{x}}, j \in \mathbf{N}_-$ *of distinct time instants. We define* $\widehat{y} \in \widehat{S}^{(n)}$*:* $\forall k \in \mathbf{N}_-$,

$$\widehat{y}(k) = \begin{cases} \widehat{x}(k), \forall j \in \mathbf{N}_-, k \neq k^j, \\ \mu^j, \exists j \in \mathbf{N}_-, k = k^j, \end{cases}$$

thus $\widehat{y}$ *is obtained by removing from* $\widehat{x}$ *the instants* (k^j) *of periodicity of* μ. *We get several possibilities that result from this construction, we give here only one of these possibilities:*

$$\mu^j = \mu' \in \mathbf{B}^n \setminus \widehat{Or}(\widehat{x}), j \in \mathbf{N}_-,$$

$$(k^{-1}, k^0, k^1, ...) = (k_0, k_0 + 2\widetilde{p}, k_0 + 4\widetilde{p}, ...)$$

and the periodic point μ *gives birth to two periodic points,* μ *and* μ'*, with*

$$\widehat{P}_{\mu}^{\widehat{y}} = \widehat{P}_{\mu'}^{\widehat{y}} = \{2\widetilde{p}, 4\widetilde{p}, 6\widetilde{p}, ...\},$$

$$\widehat{\mathbf{T}}_{\mu}^{\widehat{y}} = \{k_0 + \widetilde{p}, k_0 + 3\widetilde{p}, k_0 + 5\widetilde{p}, ...\},$$

$$\widehat{\mathbf{T}}_{\mu'}^{\widehat{y}} = \{k_0, k_0 + 2\widetilde{p}, k_0 + 4\widetilde{p}, ...\}.$$

Example 35. *For the time instant* $k' \in \mathbf{N}_- \setminus \widehat{\mathbf{T}}_{\mu}^{\widehat{x}}$*, we define* $\widehat{y} \in \widehat{S}^{(n)}$ *like that:* $\forall k \in \mathbf{N}_-$,

$$\widehat{y}(k) = \begin{cases} \widehat{x}(k), k \neq k', \\ \mu, k = k'. \end{cases}$$

We say that $\widehat{y}$ *is obtained by adding to* $\widehat{x}$ *the instant* k' *of equality with* μ. *We notice that after adding to* $\widehat{x}$ *the instant* k' *of equality with* μ *we loose the periodicity of* μ*, but we still have eventual periodicity with the same sets of periods and with the limit of periodicity* $k' + 1$.

Example 36. *We take now the distinct time instants* $k^1, ..., k^s \in \mathbf{N}_- \setminus \widehat{\mathbf{T}}_{\mu}^{\widehat{x}}$ *and we define* $\widehat{y} \in \widehat{S}^{(n)}$ *by:* $\forall k \in \mathbf{N}_-$,

$$\widehat{y}(k) = \begin{cases} \widehat{x}(k), k \notin \{k^1, ..., k^s\}, \\ \mu, \exists j \in \{1, ..., s\}, k = k^j. \end{cases}$$

In this situation we say that $\widehat{y}$ *is obtained from* $\widehat{x}$ *by addition of the instants* $k^1, ..., k^s$ *of equality with* μ. *We get the same sets of periods but eventual periodicity of* μ *only, with the limit of periodicity* $1 + \max\{k^1, ..., k^s\}$.

Example 37. *We consider the sequence of distinct time instants* $k^j \in \mathbf{N}_- \setminus \widehat{\mathbf{T}}_{\mu}^{\widehat{x}}, j \in \mathbf{N}_-$. *We define* $\widehat{y} \in \widehat{S}^{(n)}$ *by* $\forall k \in \mathbf{N}_-$,

$$\widehat{y}(k) = \begin{cases} \widehat{x}(k), \forall j \in \mathbf{N}_-, k \neq k^j, \\ \mu, \exists j \in \mathbf{N}_-, k = k^j \end{cases}$$

meaning that we have constructed $\widehat{y}$ *by addition to* $\widehat{x}$ *of the instants* (k^j) *of equality with* μ. *Several possibilities exist in this construction, we point out the following two situations only, given by*

$$(k^{-1}, k^0, k^1, ...) = (k_0 + 2, k_0 + 2 + \widetilde{p}, k_0 + 2 + 2\widetilde{p}, ...), k_0 + 2 \in \{-1, 0, ..., \widetilde{p} - 2\},$$

when, after the passage from $\widehat{x}$ to $\widehat{y}$, the point μ is still periodic and

i) it keeps its prime period if $\widetilde{p}=5, k_0=-1$; during an interval with the length of a period, by adding instants of equality with μ, instead of one occurrence of μ at k_0+k5, we have two occurrences, at k_0+k5 and $k_0+2+k5, k\in \mathbf{N}$;

ii) it doubles its prime period if $\widetilde{p}=4, k_0=-1$, with one occurrence of μ only during an interval with the length of a period, at $k_0+k2, k\in \mathbf{N}$.

2. Real Time, Periodic Points

Remark 138. *In this section $x\in S^{(n)}$, $t_0,t_1\in \mathbf{R}$, $\widetilde{T}>0$ and the periodic point $\mu\in Or(x)$ are given, such that*

$$t_0<t_1<t_0+\widetilde{T},$$

$$\mathbf{T}^x_\mu=(-\infty,t_0)\cup[t_1,t_0+\widetilde{T})\cup[t_1+\widetilde{T},t_0+2\widetilde{T})\cup[t_1+2\widetilde{T},t_0+3\widetilde{T})\cup...$$

$$P^x_\mu=\{\widetilde{T},2\widetilde{T},3\widetilde{T},...\}.$$

We notice that $\mu=x(-\infty+0)$ and we are in the special case of periodicity from Theorem 82, page 126 but a different choice of μ or of $\mathbf{T}^x_\mu$ does not change things significantly.

Definition 21. *We define the function $t\longmapsto \underline{t}$ that associates to each real number $t\in\mathbf{R}$ an interval $\underline{t}\subset\mathbf{R}$ in the following way:*

$$\underline{t}=\begin{cases} I^x, \text{ if } t\in I^x,\\ [a,b), \text{ if } t\in[a,b), [a,b)\subset \mathbf{T}^x_{x(t)}, x(a-0)\neq x(t), x(b)\neq x(t).\end{cases}$$

Remark 139. *We notice that $\underline{t}$ is the greatest interval that contains t and where x has the constant value $x(t)$.*

Remark 140. *The definition of $\underline{t}$ is possible since $card(\omega(x))\geq 2$; the fact that $\omega(x)$ has at least two elements is inferred from the form of $\mathbf{T}^x_\mu$.*

Example 38. *Let now $\mu'\in\mathbf{B}^n$ with $\mu'\neq\mu$ and let also the time instant $t''\in\mathbf{T}^x_\mu$. We define the signal*

$$y(t)=\begin{cases} x(t), t\notin \underline{t''},\\ \mu', t\in\underline{t''}.\end{cases}$$

We say that y is obtained by removing from x the interval $\underline{t''}$ of periodicity of μ. After the removal of $\underline{t''}$, the periodicity of μ is lost, but eventual periodicity still holds; the set of the periods is the same $P^y_\mu=P^x_\mu$ and the limit of periodicity is $\sup\underline{t''}$.

Example 39. *The points $\mu^1,...,\mu^s\in\mathbf{B}^n$ are taken and they are not required to be distinct, but we ask that they are distinct from $\mu: \mu^1\neq\mu,...,\mu^s\neq\mu$ and we also take the time instants $t'_1,...,t'_s\in\mathbf{T}^x_\mu$ with the property that the intervals $\underline{t'_1},...,\underline{t'_s}$ are disjoint. We define*

$$y(t)=\begin{cases} x(t), t\notin\underline{t'_1}\cup...\cup\underline{t'_s},\\ \mu^j, \exists j\in\{1,...,s\}, t\in\underline{t'_j}.\end{cases}$$

Obviously the phenomenon is the same, μ is not periodic any longer, but it is eventually periodic with $P^y_\mu=P^x_\mu$ and the limit of periodicity is $max\{\sup\underline{t'_1},...,\sup\underline{t'_s}\}$.

Example 40. *We consider the sequence* $\mu^j \in \mathbf{B}^n \setminus \{\mu\}, j \in \mathbf{N}$ *and also the time instants* $t'_j \in \mathbf{T}^x_\mu, j \in \mathbf{N}$ *having the property that* $\underline{t'_0}, \underline{t'_1}, \underline{t'_2}, \ldots$ *are disjoint. We define*

$$y(t) = \begin{cases} x(t), t \notin \underline{t'_0} \cup \underline{t'_1} \cup \underline{t'_2} \cup \ldots \\ \mu^j, \exists j \in \mathbf{N}, t \in \underline{t'_j}. \end{cases}$$

Similarly with Example 34, page 156 several possibilities may occur, for example the periodic point μ *gives birth to the periodic points* μ, μ', μ'' *with*

$$P^y_\mu = P^y_{\mu'} = P^y_{\mu''} = \{3\widetilde{T}, 6\widetilde{T}, 9\widetilde{T}, \ldots\},$$

$$\mathbf{T}^y_\mu = (-\infty, t_0) \cup [t_1 + 2\widetilde{T}, t_0 + 3\widetilde{T}) \cup [t_1 + 5\widetilde{T}, t_0 + 6\widetilde{T}) \cup [t_1 + 8\widetilde{T}, t_0 + 9\widetilde{T}) \cup \ldots$$

$$\mathbf{T}^y_{\mu'} = [t_1, t_0 + \widetilde{T}) \cup [t_1 + 3\widetilde{T}, t_0 + 4\widetilde{T}) \cup [t_1 + 6\widetilde{T}, t_0 + 7\widetilde{T}) \cup \ldots$$

$$\mathbf{T}^y_{\mu''} = [t_1 + \widetilde{T}, t_0 + 2\widetilde{T}) \cup [t_1 + 4\widetilde{T}, t_0 + 5\widetilde{T}) \cup [t_1 + 7\widetilde{T}, t_0 + 8\widetilde{T}) \cup \ldots$$

Example 41. *Let us increase now the initial time instant* t_0 *of* x *to* $t_0 + \varepsilon$*, where* $\varepsilon \in (0, t_1 - t_0)$*. We have:*

$$\mathbf{T}^y_\mu = (-\infty, t_0 + \varepsilon) \cup [t_1, t_0 + \widetilde{T}) \cup [t_1 + \widetilde{T}, t_0 + 2\widetilde{T}) \cup [t_1 + 2\widetilde{T}, t_0 + 3\widetilde{T}) \cup \ldots$$

The periodicity of $\mu \in Or(x)$ *has become the eventual periodicity of* $\mu \in Or(y)$*, the two sets of periods are equal* $P^y_\mu = P^x_\mu$ *and the new limit of periodicity is* $t_0 + \varepsilon$*.*

If this increase with $\varepsilon \in (0, t_1 - t_0)$ *is applied at the time instant* $t_0 + k\widetilde{T}$*, then*

$$\mathbf{T}^y_\mu = (-\infty, t_0) \cup [t_1, t_0 + \widetilde{T}) \cup \ldots \cup [t_1 + (k-1)\widetilde{T}, t_0 + k\widetilde{T} + \varepsilon) \cup [t_1 + k\widetilde{T}, t_0 + (k+1)\widetilde{T}) \cup \ldots,$$

$P^y_\mu = P^x_\mu$ *and the new limit of periodicity is* $t_0 + k\widetilde{T} + \varepsilon$*. In this construction we say that we have added the intervals* $[t_0, t_0 + \varepsilon), [t_0 + k\widetilde{T}, t_0 + k\widetilde{T} + \varepsilon)$ *of equality with* μ*.*

Example 42. *Let* $\varepsilon > 0$ *be arbitrary and we decrease the initial time instant from* t_0 *to* $t_0 - \varepsilon$*. Then*

$$\mathbf{T}^y_\mu = (-\infty, t_0 - \varepsilon) \cup [t_1, t_0 + \widetilde{T}) \cup [t_1 + \widetilde{T}, t_0 + 2\widetilde{T}) \cup [t_1 + 2\widetilde{T}, t_0 + 3\widetilde{T}) \cup \ldots$$

$\mu \in Or(y)$ *is eventually periodic,* $P^y_\mu = P^x_\mu$ *and the new limit of periodicity is* t_0*. If we decrease however* $t_0 + k\widetilde{T}$ *to* $t_0 + k\widetilde{T} - \varepsilon$*, with* $\varepsilon \in (0, t_0 - t_1 + \widetilde{T})$*, we see that*

$$\mathbf{T}^y_\mu = (-\infty, t_0) \cup [t_1, t_0 + \widetilde{T}) \cup \ldots \cup [t_1 + (k-1)\widetilde{T}, t_0 + k\widetilde{T} - \varepsilon) \cup [t_1 + k\widetilde{T}, t_0 + (k+1)\widetilde{T}) \cup \ldots$$

$\mu \in Or(y)$ *is eventually periodic,* $P^y_\mu = P^x_\mu$ *and the new limit of periodicity is* $t_0 + k\widetilde{T}$*. In this example we have removed from* x *the time intervals* $[t_0 - \varepsilon, t_0), [t_0 + k\widetilde{T} - \varepsilon, t_0 + k\widetilde{T})$ *of equality with* μ*.*

Chapter 10

Computation Functions

In order to introduce the flows, we need to present first the computation functions. The computation functions show when and how the Boolean functions are computed, in discrete and in real time. They give (subject to the additional request of progressiveness) the so called unbounded delay model of computation of the Boolean functions. Their definition is the purpose of Section 1 of the chapter and in Section 2 we make a discussion.

The eventual periodicity and the periodicity of the computation functions are defined in Section 3, by analogy with the signals. Section 4 gives, in case of eventual periodicity and periodicity, the bounds of the initial time and of the limit of periodicity.

In the last three sections we define the progressiveness property and we show that the forgetful function keeps the (eventual) periodicity and the progressiveness of the computation functions.

1. The Definition of the Computation Functions

Definition 22. *The* ***discrete time computation functions*** *are by definition the sequences* $\alpha : \mathbf{N} \to \mathbf{B}^n$. *Their set is denoted by* $\widehat{\Pi}'_n$. *In general, we write* α^k *instead of* $\alpha(k), k \in \mathbf{N}$.

The ***real time computation functions*** $\rho : \mathbf{R} \to \mathbf{B}^n$ *are by definition the functions of the form*

$$\rho(t) = \alpha^0 \cdot \chi_{\{t_0\}}(t) \oplus \alpha^1 \cdot \chi_{\{t_1\}}(t) \oplus ... \oplus \alpha^k \cdot \chi_{\{t_k\}}(t) \oplus ... \qquad (1.1)$$

where $(t_k) \in Seq$. *Their set is denoted with* Π'_n.

Definition 23. *The* ***initial value*** *of* $\alpha \in \widehat{\Pi}'_n$ *is* $\alpha^0 \in \mathbf{B}^n$ *and the* ***initial value*** *of* $\rho \in \Pi'_n$ *is by definition* $(0,...,0) \in \mathbf{B}^n$.

Definition 24. *The initial time instant of* α *is not defined and the* ***initial time (instant)*** *of* ρ *is any number* $t' \in \mathbf{R}$ *that fulfills*

$$\forall t \leq t', \rho(t) = (0,...,0).$$

The set of the initial time instants of ρ *is denoted by* I^{ρ}.

Remark 141. *Similarly with Theorem 2, page 5, for any* $\rho \in \Pi'_n$, *two possibilities exist:*

a) ρ *is non null and* $t_0 \in \mathbf{R}$ *exists with* $I^{\rho} = (-\infty, t_0)$ *(see* $(3.7)_{page\,5}$*);*

b) ρ *is null and* $I^{\rho} = \mathbf{R}$ *(see* $(3.9)_{page\,5}$*).*

Notation 14. *We use the notation*

$$\widehat{Seq'} = \{(k_j)|k_j \in \mathbf{N}, j \in \mathbf{N} \text{ and } k_0 < k_1 < k_2 < ...\}.$$

Definition 25. *The* ***orbits*** *of* $\alpha \in \widehat{\Pi}'_n$, $\rho \in \Pi'_n$ *are defined in the following way:*

$$\widehat{Or}(\alpha) = \{\alpha^k | k \in \mathbf{N}\},$$

$$Or(\rho) = \{\rho(t) | t \in \mathbf{R}\}.$$

Definition 26. *The* $\omega-$***limit set*** *of* $\alpha \in \widehat{\Pi}'_n$ *is defined by*

$$\widehat{\omega}(\alpha) = \{\mu | \mu \in \mathbf{B}^n, \exists (k_j) \in \widehat{Seq'}, \forall j \in \mathbf{N}, \alpha^{k_j} = \mu\}$$

and the $\omega-$***limit set*** *of* $\rho \in \Pi'_n$ *is defined by*

$$\omega(\rho) = \{\mu | \mu \in \mathbf{B}^n, \exists (t_j) \in Seq, \forall j \in \mathbf{N}, \rho(t_j) = \mu\}.$$

Example 43. $\alpha^k = (0,...,0), k \in \mathbf{N}$ *and* $\rho(t) = (0,...,0)$, *respectively* $\alpha^k = (1,...,1), k \in \mathbf{N}$ *and*

$$\rho(t) = (1,...,1) \cdot \chi_{\{0\}}(t) \oplus (1,...,1) \cdot \chi_{\{1\}}(t) \oplus (1,...,1) \cdot \chi_{\{2\}}(t) \oplus ...$$

are typical examples of computation functions.

Remark 142. *Notice the truth of* $\forall \rho \in \Pi'_n, (0,...,0) \in \omega(\rho)$, *while the discrete time statement* $\forall \alpha \in \widehat{\Pi}'_n, (0,...,0) \in \widehat{\omega}(\alpha)$ *is false.*

2. Discussion

Remark 143. *The meaning of the computation functions* $\alpha \in \widehat{\Pi}'_n$, $\rho \in \Pi'_n$, *subject to the additional property of progressiveness that will be stated later in this chapter, is that of showing when -in discrete time and in real time- and how the Boolean functions are computed. The manner of computing that will result, called* ***asynchronous***, *gives the so called* ***unbounded delay model*** *of computation of the Boolean functions.*

Remark 144. *The definitions and the notations referring to* $\alpha \in \widehat{\Pi}'_n$, $\rho \in \Pi'_n$ *are analogue with those of the signals* $\widehat{x} \in \widehat{S}^{(n)}, x \in S^{(n)}$. *Differences exist between the signals and the computation functions:* α *is not defined in* -1, *while* ρ *is equal with* $(0,...,0)$ *on all of* $\mathbf{R}$ *except for some time instants* (t_k) *where it can take non null values.*

Remark 145. *In the situation when probably all the definitions related with the signals have appropriate definitions related with the computation functions, we shall avoid restating many of such concepts referring to the computation functions when they are not necessary. For example the final value of* ρ *may be defined and, if it exists, it is equal with* $(0,...,0)$, *like the initial value, but this fact has no special meaning in our exposure.*

Remark 146. *Taking the risk of being boring, we shall give the necessary definitions for the computation functions during the exposure, even if the signals and the computation functions are similar.*

Remark 147. *The more general class of functions that the real time signals and the real time computation functions belong to has been called by the author in previous works the class of the **differentiable** functions. A function $h : \mathbf{R} \to \mathbf{B}^n$ is by definition differentiable if it is of the form*

$$h(t) = ... \oplus h(t_{-1}) \cdot \chi_{\{t_{-1}\}}(t) \oplus h(\tfrac{t_{-1}+t_0}{2}) \cdot \chi_{(t_{-1},t_0)}(t)$$
$$\oplus h(t_0) \cdot \chi_{\{t_0\}}(t) \oplus h(\tfrac{t_0+t_1}{2}) \cdot \chi_{(t_0,t_1)}(t) \oplus h(t_1) \cdot \chi_{\{t_1\}}(t) \oplus ...$$

where the upper and lower unbounded sequence $t_z \in \mathbf{R}, z \in \mathbf{Z}$ is strictly increasing. The definitions of $h(-\infty+0), h(\infty-0), I^h, F^h, h(t-0), h(t+0), Or(h), \omega(h), ...$ are obvious now. The signals x are the differentiable functions that fulfill $x(t) = x(t+0)$ (right continuity) and the existence of the initial value, while the computation functions ρ are the differentiable functions that fulfill $\rho(t-0) = \rho(t+0) = (0,...,0)$ and the existence of the initial value.

Remark 148. *Working with $\widehat{\Pi}'_n$ and Π'_n might prove to be too general for most applications and this will become obvious after introducing progressiveness. The author needed however $\widehat{\Pi}'_n$ and Π'_n in the study of the symmetry of the asynchronous systems*[1]*, where certain types of symmetry require this generality.*

3. Eventual Periodicity and Periodicity

Definition 27. *The computation function $\alpha \in \widehat{\Pi}'_n$ is said to be **eventually periodic** with the **period** $p \geq 1$ and the **limit of periodicity** $k' \in \mathbf{N}$ if we have*

$$\forall k \geq k', \alpha^k = \alpha^{k+p} \tag{3.1}$$

*and the computation function $\rho \in \Pi'_n$ is called **eventually periodic** with the **period** $T > 0$ and the **limit of periodicity** $t' \in \mathbf{R}$ if*

$$\forall t \geq t', \rho(t) = \rho(t+T). \tag{3.2}$$

Definition 28. *In the special case when (3.1) is true for $k' = 0$, α is called **periodic** (with the **period** $p \geq 1$) and in the special case when (3.2) holds for some $t' \in I^\rho$, the computation function ρ is called **periodic** (with the **period** T).*

Notation 15. *We use the notations*

$$\widehat{P}^\alpha = \{p | p \geq 1, \exists k' \in \mathbf{N}, \text{(3.1) holds}\},$$

$$P^\rho = \{T | T > 0, \exists t' \in \mathbf{R}, \text{(3.2) holds}\}.$$

Notation 16. *We also denote*

$$\widehat{L}^\alpha = \{k' | k' \in \mathbf{N}, \exists p \geq 1, \text{(3.1) holds}\},$$

$$L^\rho = \{t' | t' \in \mathbf{R}, \exists T > 0, \text{(3.2) holds}\}.$$

[1] Our paper Defining the symmetry of the universal semi-regular autonomous asynchronous systems, Symmetry, Nr. 4, 2012, pp. 116-128, where $\widehat{\Pi}'_n$ and Π'_n occur with different notations.

Definition 29. *The numbers*

$$\widetilde{p} = \min \widehat{P}^{\alpha},$$

$$\widetilde{T} = \min P^{\rho}$$

are called **prime periods** *(of* α, ρ*) and the numbers*

$$\widetilde{k'} = \min \widehat{L}^{\alpha},$$

$$\widetilde{t'} = \min L^{\rho}$$

are called **prime limits of periodicity** *(of* α, ρ*).*

Theorem 107. *a) We suppose that* α *is eventually periodic, with the period* $p \geq 1$ *and the limit of periodicity* $k' \in \mathbf{N}$*. Then*

$$\forall k \geq k', \widehat{\omega}(\alpha) = \{\alpha^{k+i} | i \in \{0, 1, ..., p-1\}\}. \tag{3.3}$$

b) If $\rho \in \Pi'_n$ *is eventually periodic with the period* $T > 0$ *and the limit of periodicity* $t' \in \mathbf{R}$*, then*

$$\forall t \geq t', \omega(\rho) = \{\rho(\xi) | \xi \in [t, t+T)\}. \tag{3.4}$$

Proof. a) The hypothesis states that

$$\forall k \geq k', \alpha^k = \alpha^{k+p} \tag{3.5}$$

is true. We show first that

$$\forall k \geq k', \widehat{\omega}(\alpha) \subset \{\alpha^k, \alpha^{k+1}, ..., \alpha^{k+p-1}\}. \tag{3.6}$$

Let $k \geq k'$ and $\mu \in \widehat{\omega}(\alpha)$ be arbitrary, fixed thus $k_1 \geq k'$ exists such that $\mu = \alpha^{k_1}$. Such a choice of k_1 is possible because the set $\{k'' | k'' \in \mathbf{N}, \alpha^{k''} = \mu\}$ is infinite. If $z' \in \mathbf{Z}$ fulfills $k_1 + z'p \in \{k, k+1, ..., k+p-1\}$, then we have

$$\mu = \alpha^{k_1} = \alpha^{k_1 + z'p} \in \{\alpha^k, \alpha^{k+1}, ..., \alpha^{k+p-1}\},$$

thus, as μ was arbitrary, (3.6) is true. On the other hand the inclusion

$$\forall k \geq k', \widehat{\omega}(\alpha) \supset \{\alpha^k, \alpha^{k+1}, ..., \alpha^{k+p-1}\} \tag{3.7}$$

is true due to (3.5). Statement (3.3) holds.

b) The proof runs through the same steps like the proof of a). From the hypothesis we have

$$\forall t \geq t', \rho(t) = \rho(t+T). \tag{3.8}$$

We show the truth of

$$\forall t \geq t', \omega(\rho) \subset \{\rho(\xi) | \xi \in [t, t+T)\}. \tag{3.9}$$

Let $t \geq t'$ and $\mu \in \omega(\rho)$ arbitrary, fixed, thus $t_1 \geq t'$ exists with $\mu = \rho(t_1)$. This choice of t_1 is possible since the set $\{t'' | t'' \in \mathbf{R}, \rho(t'') = \mu\}$ is superiorly unbounded. If $z' \in \mathbf{Z}$ satisfies $t_1 + z'T \in [t, t+T)$, then we can write

$$\mu = \rho(t_1) = \rho(t_1 + z'T) \in \{\rho(\xi) | \xi \in [t, t+T)\}$$

and, as μ was arbitrary, (3.9) holds. On the other hand the inclusion

$$\forall t \geq t', \omega(\rho) \supset \{\rho(\xi) | \xi \in [t, t+T)\} \tag{3.10}$$

is true from (3.8). Property (3.4) is proved. □

Remark 149. *Theorem 107 is a version of Theorem 47, page 93. We can obviously state a version of Theorem 89, page 138, if we replace in Theorem 107 eventual periodicity with periodicity and* $\widehat{\omega}(\alpha), \omega(\rho)$ *with* $\widehat{Or}(\alpha), Or(\rho)$.

4. The Bounds of the Initial Time and of the Limit of Periodicity

Theorem 108. *For* $\rho \in \Pi'_n$ *non null,* $T > 0$ *and* $t'_0 \in \mathbf{R}$, *the following properties are equivalent:*

a) the statement

$$\forall t'' \in (t'_0, \infty), \forall t \geq t'', \rho(t) = \rho(t+T) \tag{4.1}$$

is true and the statement

$$\exists t'' \leq t'_0, \forall t \geq t'', \rho(t) = \rho(t+T) \tag{4.2}$$

is false;

b) the statements

$$\forall t > t'_0, \rho(t) = \rho(t+T), \tag{4.3}$$

$$\rho(t'_0) \neq \rho(t'_0+T) \tag{4.4}$$

hold;

c) $L^{\rho} = (t'_0, \infty)$.

Proof. a)$\Longrightarrow$b) Let $t > t'_0$ arbitrary, fixed. We take $t'' \in (t'_0, t]$, for which (4.1) gives $\rho(t) = \rho(t+T)$, thus (4.3) is true. On the other hand the falsity of (4.2) means that

$$\forall t'' \leq t'_0, \exists t \geq t'', \rho(t) \neq \rho(t+T) \tag{4.5}$$

and we take $t'' = t'_0$; we have

$$\exists t \geq t'_0, \rho(t) \neq \rho(t+T). \tag{4.6}$$

By comparing (4.1) with (4.6) we infer (4.4).

b)$\Longrightarrow$a) Let $t'' > t'_0$ and $t \geq t''$, both arbitrary and fixed. As $t > t'_0$, (4.3) implies $\rho(t) = \rho(t+T)$, thus (4.1) holds.

On the other hand (4.4) implies (4.6), thus (4.5) is true and finally (4.2) is false.

b)$\Longleftrightarrow$c) is obvious. □

Theorem 109. *The non null computation function* ρ *is considered, having the property that* $T > 0, t' \in \mathbf{R}$ *exist with*

$$\forall t \geq t', \rho(t) = \rho(t+T). \tag{4.7}$$

Then $t'_0 < t'$ *exists such that* $L^{\rho} = (t'_0, \infty)$.

Proof. The hypothesis that ρ is non null shows the existence of $t_0 \in \mathbf{R}$ such that $\rho(t_0) \neq (0,...,0)$ and $\forall t < t_0, \rho(t) = (0,...,0)$. We infer from here, since $(0,...,0) = \rho(t_0 - T) \neq \rho(t_0)$, that $L^\rho \subset (t_0 - T, \infty)$, in other words we have the existence of $t_0' > t_0 - T$ with $L^\rho = [t_0', \infty)$ or of $t_0' \geq t_0 - T$ with $L^\rho = (t_0', \infty)$.

Case $t_0' \in (t_0 - T, t'], L^\rho = [t_0', \infty)$

In this situation, the next statements hold:

$$\forall t \geq t_0', \rho(t) = \rho(t+T), \tag{4.8}$$

$$\exists \varepsilon > 0, \forall t \in (t_0' - \varepsilon, t_0'), \rho(t) = (0,...,0), \tag{4.9}$$

$$\forall \delta > 0, \exists \xi \geq t_0' - \delta, \rho(\xi) \neq \rho(\xi + T), \tag{4.10}$$

$$\exists \varepsilon' > 0, \forall t \in (t_0' + T - \varepsilon', t_0' + T), \rho(t) = (0,...,0). \tag{4.11}$$

We fix an arbitrary $\delta \in (0, \max\{\varepsilon, \varepsilon'\})$. From (4.10) we infer

$$\exists \xi \geq t_0' - \delta, \rho(\xi) \neq \rho(\xi + T) \tag{4.12}$$

and this, together with (4.8), gives

$$\exists \xi \in [t_0' - \delta, t_0'), \rho(\xi) \neq \rho(\xi + T). \tag{4.13}$$

We can see that

$$t_0' - \varepsilon < t_0' - \delta \leq \xi < t_0', \tag{4.14}$$

$$t_0' + T - \varepsilon' < t_0' + T - \delta \leq \xi + T < t_0' + T \tag{4.15}$$

hold, thus

$$(0,...,0) \overset{(4.9),(4.14)}{=} \rho(\xi) \overset{(4.13)}{\neq} \rho(\xi + T) \overset{(4.11),(4.15)}{=} (0,...,0),$$

contradiction.

We have proved that $t_0' \in [t_0 - T, t')$, $L^\rho = (t_0', \infty)$ is the only possibility. □

Remark 150. *We suppose that $\rho \in \Pi_n'$ is non null. Its property of eventual periodicity is independent on the choice of the limit of periodicity t' in some interval (t_0', ∞); outside this interval, any choice of t' makes the property of eventual periodicity of ρ be false.*

Theorem 110. *The non null function $\rho \in \Pi_n'$ is given, together with $T > 0$ and $t' \in I^\rho$ with the property that*

$$\forall t \geq t', \rho(t) = \rho(t+T) \tag{4.16}$$

holds. Then $t_0', t_0 \in \mathbf{R}$ exist, $t_0' < t_0$ such that $\forall t'' \in (t_0', t_0)$, we have $t'' \in I^\rho$,

$$\forall t \geq t'', \rho(t) = \rho(t+T) \tag{4.17}$$

and for any $t'' \notin (t_0', t_0)$, at least one of $t'' \in I^\rho$, (4.17) is false. In other words $I^\rho \cap L^\rho = (t_0', t_0)$.

Proof. The fact that $t_0', t_0 \in \mathbf{R}$ exist such that $I^\rho = (-\infty, t_0)$ and $L^\rho = (t_0', \infty)$ results from Remark 141, page 159 and Theorem 109. From the hypothesis of the Theorem we infer $I^\rho \cap L^\rho \neq \varnothing$. □

Remark 151. *The property of periodicity of the non null function $\rho \in \Pi_n'$ is independent on the choice of the initial time t'= limit of periodicity in some interval (t_0', t_0); outside this interval, any choice of t' makes the periodicity property of ρ be false.*

5. The Forgetful Function

Definition 30. *The* ***discrete time*** $\widehat{\sigma}^{k'} : \widehat{\Pi}'_n \to \widehat{\Pi}'_n$, $k' \in \mathbf{N}$ *and the* ***continuous time*** $\sigma^{t'} : \Pi'_n \to \Pi'_n$ ***forgetful functions***, $t' \in \mathbf{R}$, *are defined by:* $\forall \alpha \in \widehat{\Pi}'_n, \forall k \in \mathbf{N}$,

$$(\widehat{\sigma}^{k'}(\alpha))^k = \alpha^{k+k'}$$

and $\forall \rho \in \Pi'_n, \forall t \in \mathbf{R}$,

$$\sigma^{t'}(\rho)(t) = \rho(t) \cdot \chi_{[t',\infty)}(t).$$

Remark 152. *Definition 30 was given by analogy with Definition 8, page 7. Even if the left limit function was defined so far for real time signals only. We can accept that, for any* $\rho \in \Pi'_n$ *and any* t, $\rho(t-0) \in \mathbf{B}^n$ *is the one that formally satisfies* $(2.1)_{page\,3}$:

$$\exists \varepsilon > 0, \forall \xi \in (t-\varepsilon, t), \rho(\xi) = \rho(t-0);$$

then $\forall t \in \mathbf{R}, \rho(t-0) = (0,...,0)$ *and*

$$\sigma^{t'}(\rho)(t) = \begin{cases} \rho(t), t \geq t', \\ \rho(t'-0), t < t' \end{cases} = \begin{cases} \rho(t), t \geq t', \\ (0,...,0), t < t' \end{cases} = \rho(t) \cdot \chi_{[t',\infty)}(t)$$

indeed.

Theorem 111. *We consider the computation functions* $\alpha \in \widehat{\Pi}'_n$ *and* $\rho \in \Pi'_n$.

a) If α *is periodic, then for any* $k' \in \mathbf{N}$, $\beta = \widehat{\sigma}^{k'}(\alpha)$ *is periodic and* $\widehat{P}^{\alpha} = \widehat{P}^{\beta}$.

b) If ρ *is periodic then, for arbitrary* $t' \in \mathbf{R}$, *the computation function* $\rho' = \sigma^{t'}(\rho)$ *is periodic and* $P^{\rho} = P^{\rho'}$.

Proof. a) The proof is simple and it coincides with the proof of Theorem 105 a), page 151.

b) We suppose that $P^{\rho} \neq \varnothing$ and we fix an arbitrary $t' \in \mathbf{R}$.

We prove $P^{\rho} \subset P^{\rho'}$. For an arbitrary $T \in P^{\rho}$, $t'_1 \in I^{\rho}$ exists with

$$\forall t \geq t'_1, \rho(t) = \rho(t+T) \tag{5.1}$$

true and we must prove the existence of $t''_1 \in I^{\rho'}$ with

$$\forall t \geq t''_1, \rho'(t) = \rho'(t+T). \tag{5.2}$$

If ρ is null the statement is obvious, because $\sigma^{t'}(\rho)$ is null. We suppose now that ρ is not null and that it is of the form

$$\rho(t) = \rho(t_0) \cdot \chi_{\{t_0\}}(t) \oplus \rho(t_1) \cdot \chi_{\{t_1\}}(t) \oplus \rho(t_2) \cdot \chi_{\{t_2\}}(t) \oplus ... \tag{5.3}$$

where $(t_k) \in Seq$ and, without loosing the generality, we suppose that $\rho(t_0) \neq (0,...,0)$. This means that $I^{\rho} = (-\infty, t_0)$ and in (5.1) we have $t'_1 < t_0$. If $t' \leq t_0$, then $\rho' = \rho$ and (5.2) is true with $t''_1 = t'_1$. We take now $t' > t_0$, meaning the existence of $k \geq 1$ with $t' \in (t_{k-1}, t_k]$ and

$$\rho'(t) = \rho(t_k) \cdot \chi_{\{t_k\}}(t) \oplus \rho(t_{k+1}) \cdot \chi_{\{t_{k+1}\}}(t) \oplus \rho(t_{k+2}) \cdot \chi_{\{t_{k+2}\}}(t) \oplus ... \tag{5.4}$$

where we can suppose that $\rho(t_k) \neq (0,...,0)$[2]. We notice that $I^{\rho'} = (-\infty, t_k)$. Because the set $\{t | t \in [t_k, t_k + T), \rho(t) \neq (0,...,0)\}$ is finite, from Lemma 11, page 231, we can define

$$t_0' = \max\{t | t \in [t_k, t_k + T), \rho(t) \neq (0,...,0)\} - T, \tag{5.5}$$

thus we have $t_0' < t_k \leq t_0' + T$. Obviously,

$$\forall t \in (t_0' + T, t_k + T), \rho(t) = (0,...,0). \tag{5.6}$$

We claim that $t_0' \geq t_0$. Indeed, let us suppose against all reason that $t_0' < t_0$, meaning that

$$t_k \leq t_0' + T < t_0 + T < t_k + T \tag{5.7}$$

is true. But

$$(0,...,0) \neq \rho(t_0) \overset{(5.1)}{=} \rho(t_0 + T) \overset{(5.6),(5.7)}{=} (0,...,0)$$

represents a contradiction. We infer $t_0' \geq t_0$ and let $t_1'' \in (t_0', t_k)$ arbitrary. For any $t \in (t_0', t_k)$ we have

$$t_1' < t_0 \leq t_0' < t < t_k \leq t_0' + T < t + T < t_k + T,$$

$$\rho(t + T) \overset{(5.6)}{=} (0,...,0)$$

thus $\forall t \in (t_0', t_k)$,

$$\rho(t) = \rho(t + T) = (0,...,0) = \rho'(t)$$

wherefrom

$$\forall t > t_0', \rho'(t) = \rho(t). \tag{5.8}$$

Let $t \geq t_1''$ arbitrary. We have

$$\rho'(t) \overset{(5.8)}{=} \rho(t) \overset{(5.1)}{=} \rho(t + T) \overset{(5.8)}{=} \rho'(t + T).$$

We prove $P^{\rho'} \subset P^{\rho}$. We take an arbitrary $T \in P^{\rho}$. This implies the existence of $t_1' \in I^{\rho}$ with the property that (5.1) is true. We suppose against all reason that the inclusion $P^{\rho'} \subset P^{\rho}$ is not true, i.e. $T' \in P^{\rho'} \setminus P^{\rho}$ and $t_1'' \in I^{\rho'}$ exist such that

$$\forall t \geq t_1'', \rho'(t) = \rho'(t + T'), \tag{5.9}$$

$$\forall t''' \in I^{\rho}, \exists t_1 \geq t''', \rho(t_1) \neq \rho(t_1 + T') \tag{5.10}$$

hold. Let $t''' \in I^{\rho} \cap [t_1', \infty)$ arbitrary and we take $\bar{k} \in \mathbf{N}$ such that for $t_1 \geq t'''$ whose existence is stated in (5.10), we have $t_1 + \bar{k}T \geq \max\{t_1'', t'\}$. We infer

$$\begin{aligned} \rho(t_1) &\overset{(5.1)}{=} \rho(t_1 + \bar{k}T) = \rho'(t_1 + \bar{k}T) \\ &\overset{(5.9)}{=} \rho'(t_1 + \bar{k}T + T') = \rho(t_1 + \bar{k}T + T') \overset{(5.1)}{=} \rho(t_1 + T') \end{aligned} \tag{5.11}$$

and we see that (5.10), (5.11) are contradictory. □

Remark 153. *The previous theorem represents the same property like the one from Theorem 105, page 151.*

---

[2] ρ is non null and periodic implies $\omega(\rho) \neq \{(0,...,0)\}$. On the other hand, analoguely with Theorem 5, page 9, we have $\omega(\rho) = \omega(\sigma^{t'}(\rho))$. Thus we can suppose that $\rho(t_k) = \rho'(t_k) = \sigma^{t'}(\rho)(t_k) \neq (0,...,0)$.

6. Progressiveness

Theorem 112. *Let the computation functions* $\alpha \in \widehat{\Pi}'_n$ *and* $\rho \in \Pi'_n$ *and we denote*

$$I_k = \{i | i \in \{1,...,n\}, \alpha_i^k = 1\}, k \in \mathbf{N},$$

$$J_t = \{i | i \in \{1,...,n\}, \rho_i(t) = 1\}, t \in \mathbf{R}.$$

a) The following properties are equivalent:

$$\forall i \in \{1,...,n\}, \exists \mu \in \widehat{\omega}(\alpha), \mu_i = 1, \tag{6.1}$$

$$\forall i \in \{1,...,n\}, \textit{the set } \{k | k \in \mathbf{N}, \alpha_i^k = 1\} \textit{ is infinite,} \tag{6.2}$$

$$\forall k \in \mathbf{N}, \exists N \in \mathbf{N}, I_k \cup I_{k+1} \cup ... \cup I_{k+N} = \{1,...,n\}. \tag{6.3}$$

b) The following properties are also equivalent:

$$\forall i \in \{1,...,n\}, \exists \mu \in \omega(\rho), \mu_i = 1, \tag{6.4}$$

$$\begin{gathered}\forall i \in \{1,...,n\}, \textit{the set } \{t | t \in \mathbf{R}, \rho_i(t) = 1\} \\ \textit{is unbounded from above,}\end{gathered} \tag{6.5}$$

$$\forall t \in \mathbf{R}, \exists \delta > 0, \bigcup_{\xi \in [t, t+\delta]} J_\xi = \{1,...,n\}. \tag{6.6}$$

c) We suppose that the relation between ρ *and* α *is given by the existence of* $(t_k) \in Seq$ *such that*

$$\rho(t) = \alpha^0 \cdot \chi_{\{t_0\}}(t) \oplus \alpha^1 \cdot \chi_{\{t_1\}}(t) \oplus ... \oplus \alpha^k \cdot \chi_{\{t_k\}}(t) \oplus ... \tag{6.7}$$

Then any of (6.1),..., (6.3) is equivalent with any of (6.4),..., (6.6).

Proof. (6.1)$\Longrightarrow$(6.2) Let $i \in \{1,...,n\}$ be arbitrary, fixed. A sequence $(k_j) \in \widehat{Seq'}$ exists such that $\forall j \in \mathbf{N}, \alpha^{k_j} = \mu$ and $\alpha_i^{k_j} = \mu_i = 1$. It has resulted that the sets $\{k_j | j \in \mathbf{N}\} \subset \{k | k \in \mathbf{N}, \alpha_i^k = 1\}$ are both infinite.

(6.2)$\Longrightarrow$(6.3) Let $k \in \mathbf{N}$ arbitrary, fixed. As the sets $\{k' | k' \in \mathbf{N}, \alpha_i^{k'} = 1\}$, $i = \overline{1,n}$ are infinite, the intersections $\{k' | k' \in \mathbf{N}, \alpha_1^{k'} = 1\} \cap \{k, k+1, k+2, ...\}$,..., $\{k' | k' \in \mathbf{N}, \alpha_n^{k'} = 1\} \cap \{k, k+1, k+2, ...\}$ are not empty and let $k_1 \geq k$ such that $\alpha_1^{k_1} = 1, ..., k_n \geq k$ such that $\alpha_n^{k_n} = 1$. The number $k'' = \max\{k_1, ..., k_n\}$ fulfills $I_k \cup I_{k+1} \cup ... \cup I_{k''} = \{1,...,n\}$.

(6.3)$\Longrightarrow$(6.1) Let us suppose against all reason that (6.1) is not true, i.e.

$$\exists i \in \{1,...,n\}, \forall \mu \in \widehat{\omega}(\alpha), \mu_i = 0. \tag{6.8}$$

Let $k_1 \in \mathbf{N}$ with the property that

$$\widehat{\omega}(\alpha) = \{\alpha^k | k \geq k_1\}, \tag{6.9}$$

thus we infer that

$$\forall k \geq k_1, \alpha_i^k = 0. \tag{6.10}$$

The property (6.3) states the existence of N with

$$I_{k_1} \cup I_{k_1+1} \cup ... \cup I_{k_1+N} = \{1,...,n\}. \tag{6.11}$$

As (6.10) shows that $i \notin I_{k_1} \cup I_{k_1+1} \cup ... \cup I_{k_1+N}$, the equations (6.10) and (6.11) are contradictory. Statement (6.1) holds.

((6.1) and (6.7))$\Longrightarrow$(6.4)[3] $\forall i \in \{1,...,n\}$, the sequence $(k_j) \in \widehat{Seq'}$ exists such that $\forall j \in \mathbf{N}, \alpha^{k_j} = \mu$ and $\alpha_i^{k_j} = \mu_i = 1$. In this situation the sequence $(t_{k_j}) \in Seq$ fulfills $\forall j \in \mathbf{N}, \rho(t_{k_j}) = \alpha^{k_j} = \mu$ and $\rho_i(t_{k_j}) = \alpha_i^{k_j} = \mu_i = 1$ and obviously $\mu \in \omega(\rho)$. □

Definition 31. *The discrete time computation function $\alpha \in \widehat{\Pi}'_n$ is called **progressive** if one of the previous properties (6.1),..., (6.3) is true. The set of the discrete time progressive computation functions is denoted by $\widehat{\Pi}_n$.*

*The real time computation function $\rho \in \Pi'_n$ is called **progressive** if one of (6.4),..., (6.6) is true. The set of the real time progressive computation functions is denoted by Π_n.*

Theorem 113. *We consider the computation functions $\alpha \in \widehat{\Pi}'_n$ and $\rho \in \Pi'_n$.*

a) We presume that $\widehat{\omega}(\alpha) = \{\mu^1, ..., \mu^p\}$ with $p \geq 1$. Then $p = 1$ implies

$$\alpha \in \widehat{\Pi}_n \Longleftrightarrow \mu^1 = (1, ..., 1)$$

and if $p > 1$ then

$$\alpha \in \widehat{\Pi}_n \Longleftrightarrow \mu^1 \cup ... \cup \mu^p = (1, ..., 1).$$

Previously, $\cup$ is the notation of the union of the binary $n-$tuples, made coordinatewise.

b) We presume that $\omega(\rho) = \{\mu^1, ..., \mu^p\}$ where $p \geq 1$. If $p = 1$ then $\omega(\rho) = \{(0, ..., 0)\}$ and $\rho \notin \Pi_n$; if $p > 1$ then

$$\rho \in \Pi_n \Longleftrightarrow \mu^1 \cup ... \cup \mu^p = (1, ..., 1).$$

c) If $(t_k) \in Seq$ exists such that (6.7) holds, then

$$\alpha \in \widehat{\Pi}_n \Longleftrightarrow \rho \in \Pi_n.$$

Proof. a) If $\exists \mu^1 \in \mathbf{B}^n, \widehat{\omega}(\alpha) = \{\mu^1\}$, then from (6.1) we infer that $\alpha \in \widehat{\Pi}_n \Leftrightarrow \mu^1 = (1, ..., 1)$. For $\widehat{\omega}(\alpha) = \{\mu^1, ..., \mu^p\}, p > 1$, (6.1) is equivalent with $\mu^1 \cup ... \cup \mu^p = (1, ..., 1)$.

b) In this case $(0, ..., 0) \in \omega(\rho)$ and (6.4) is false for $p = 1$. For $p > 1$, the condition that $\rho \in \Pi_n$ holds as expressed by (6.4) is $\mu^1 \cup ... \cup \mu^p = (1, ..., 1)$, like previously.

c) The following equivalencies hold:

$$\alpha \in \widehat{\Pi}_n \Longleftrightarrow \forall i \in \{1, ..., n\}, \exists \mu \in \widehat{\omega}(\alpha), \mu_i = 1$$

$$\Longleftrightarrow \forall i \in \{1, ..., n\}, \exists (k_j) \in \widehat{Seq'}, \alpha_i^{k_j} = 1$$

$$\Longleftrightarrow \forall i \in \{1, ..., n\}, \exists (k_j) \in \widehat{Seq'}, \rho_i(t_{k_j}) = 1$$

$$\Longleftrightarrow \forall i \in \{1, ..., n\}, \exists \mu \in \omega(\rho), \mu_i = 1 \Longleftrightarrow \rho \in \Pi_n.$$

□

Example 44. *The constant sequence $\alpha \in \widehat{\Pi}'_n$ equal with $(1, ..., 1)$ fulfills $\widehat{\omega}(\alpha) = \{(1, ..., 1)\}$, thus $\alpha \in \widehat{\Pi}_n$ from the first statement of Theorem 113 a).*

[3] We must prove that (6.7)$\Longrightarrow$((6.1)$\Longrightarrow$(6.4)), but this is equivalent with ((6.1) and (6.7))$\Longrightarrow$(6.4).

Example 45. *The sequence* $\alpha \in \widehat{\Pi}_n'$ *given* $\forall k \in \mathbf{N}, k = k_1 n + j$, *where* $k_1 \in \mathbf{N}, j \in \{0, 1, ..., n-1\}$, *by*

$$\alpha^k = (0, 0, ..., \underset{j+1}{1}, ..., 0),$$

fulfills $\widehat{\omega}(\alpha) = \{(1, 0, ..., 0), (0, 1, ..., 0), ..., (0, 0, ..., 1)\}$ *thus* $\alpha \in \widehat{\Pi}_n$ *from the second statement of Theorem 113 a).*

Theorem 114. *a) Let the computation function* $\alpha \in \widehat{\Pi}_n'$. *The following equivalence holds:*

$$\alpha \in \widehat{\Pi}_n \Longleftrightarrow \widehat{\sigma}^1(\alpha) \in \widehat{\Pi}_n.$$

b) The computation function $\rho \in \Pi_n'$ *and* $t' \in \mathbf{R}$ *are given. The following equivalence holds:*

$$\rho \in \Pi_n \Longleftrightarrow \sigma^{t'}(\rho) \in \Pi_n.$$

Proof. a) For any $i \in \{1, ..., n\}$, the sets $\{k | k \geq 0, \alpha_i^k = 1\}, \{k | k \geq 1, \alpha_i^k = 1\}$ are simultaneously finite or infinite.

b) We suppose that ρ is of the form

$$\rho(t) = \rho(t_0) \cdot \chi_{\{t_0\}}(t) \oplus \rho(t_1) \cdot \chi_{\{t_1\}}(t) \oplus ... \oplus \rho(t_k) \cdot \chi_{\{t_k\}}(t) \oplus ... \quad (6.12)$$

with $(t_k) \in Seq$. We denote with $k' \geq 0$ the rank of the sequence (t_k) that is defined by

$$\sigma^{t'}(\rho)(t) = \rho(t_{k'}) \cdot \chi_{\{t_{k'}\}}(t) \oplus \rho(t_{k'+1}) \cdot \chi_{\{t_{k'+1}\}}(t) \oplus ...$$

For any $i \in \{1, ..., n\}$, the sets $\{k | k \geq 0, \rho_i(t_k) = 1\}, \{k | k \geq k', \rho_i(t_k) = 1\}$ are simultaneously finite or infinite. □

Remark 154. *From Theorem 114 a) we get the following conclusion. For* $\alpha \in \widehat{\Pi}_n'$, *we have the equivalence:* $\forall k \in \mathbf{N}$,

$$\alpha \in \widehat{\Pi}_n \Longleftrightarrow \widehat{\sigma}^k(\alpha) \in \widehat{\Pi}_n.$$

Remark 155. *For* $\lambda^1, ..., \lambda^{k_1} \in \mathbf{B}^n, k_1 \in \mathbf{N}$ *and* $\alpha \in \widehat{\Pi}_n$, *we define* $\beta : \mathbf{N} \longrightarrow \mathbf{B}^n$ *in the following way:*

$$\forall k \in \mathbf{N}, \beta^k = \begin{cases} \lambda^k, if\ k \in \{0, ..., k_1\}, \\ \alpha^{k-k_1-1}, if\ k \geq k_1 + 1. \end{cases}$$

Since $\widehat{\sigma}^{k_1+1}(\beta) = \alpha$ *and* $\alpha \in \widehat{\Pi}_n$, *we infer* $\beta \in \widehat{\Pi}_n$.

7. Progressive Periodic Computation Functions

Theorem 115. *a) We presume that* $p \geq 1$ *exists such that the computation function* $\alpha \in \widehat{\Pi}_n'$ *fulfills*

$$\forall k \in \mathbf{N}, \alpha^k = \alpha^{k+p}. \quad (7.1)$$

If $p = 1$ *then*

$$\alpha \in \widehat{\Pi}_n \Longleftrightarrow \forall k \in \mathbf{N}, \alpha^k = (1, ..., 1) \quad (7.2)$$

and if $p > 1$ then

$$\alpha \in \widehat{\Pi}_n \Longleftrightarrow \forall k \in \mathbf{N}, \alpha^k \cup \ldots \cup \alpha^{k+p-1} = (1, \ldots, 1). \tag{7.3}$$

b) Let the computation function $\rho \in \Pi'_n$,

$$\rho(t) = \rho(t_0) \cdot \chi_{\{t_0\}}(t) \oplus \rho(t_1) \cdot \chi_{\{t_1\}}(t) \oplus \ldots \oplus \rho(t_k) \cdot \chi_{\{t_k\}}(t) \oplus \ldots \tag{7.4}$$

where $(t_k) \in Seq$ and we presume that $T > 0, t' \in I^\rho$ exist such that

$$\forall t \geq t', \rho(t) = \rho(t+T). \tag{7.5}$$

Then

$$\rho \in \Pi_n \Longleftrightarrow \forall t \geq t', \bigcup_{\xi \in [t, t+T)} \rho(\xi) = (1, \ldots, 1). \tag{7.6}$$

Proof. a) If α is periodic with the period p, then from Theorem 107, page 162, written for $k' = 0$, we have $\forall k \in \mathbf{N}$,

$$\widehat{\omega}(\alpha) = \{\alpha^k, \alpha^{k+1}, \ldots, \alpha^{k+p-1}\}.$$

Let us fix such an arbitrary k. For $p = 1$ we have $\widehat{\omega}(\alpha) = \{\alpha^k\}$ and Theorem 113, page 168 shows that

$$\alpha \in \widehat{\Pi}_n \Longleftrightarrow \alpha^k = (1, \ldots, 1).$$

For $p > 1$ from Theorem 113 we get

$$\alpha \in \widehat{\Pi}_n \Longleftrightarrow \alpha^k \cup \alpha^{k+1} \cup \ldots \cup \alpha^{k+p-1} = (1, \ldots, 1).$$

□

Example 46. *The computation function α from Example 44, page 168 is periodic with $p = 1$ and the computation function*

$$\rho(t) = (1, \ldots, 1) \cdot \chi_{\{0\}}(t) \oplus (1, \ldots, 1) \cdot \chi_{\{1\}}(t) \oplus \ldots \oplus (1, \ldots, 1) \cdot \chi_{\{k\}}(t) \oplus \ldots$$

is periodic with the period $T = 1$.

Example 47. *The computation function α from Example 45, page 169 is periodic with the period $p = n$ and the computation function*

$$\rho(t) = (1,0,\ldots,0) \cdot \chi_{\{0\}}(t) \oplus (0,1,\ldots,0) \cdot \chi_{\{1\}}(t) \oplus \ldots \oplus (0,0,\ldots,1) \cdot \chi_{\{n-1\}}(t)$$
$$\oplus (1,0,\ldots,0) \cdot \chi_{\{n\}}(t) \oplus (0,1,\ldots,0) \cdot \chi_{\{n+1\}}(t) \oplus \ldots \oplus (0,0,\ldots,1) \cdot \chi_{\{2n-1\}}(t)$$
$$\oplus (1,0,\ldots,0) \cdot \chi_{\{2n\}}(t) \oplus \ldots$$

is periodic with $T = n$.

Corollary 6. *a) If $\alpha \in \widehat{\Pi}'_n$ is progressive and periodic then, for any $k' \in \mathbf{N}$, we have that $\widehat{\sigma}^{k'}(\alpha)$ is progressive, periodic and $\widehat{P}^\alpha = \widehat{P}^{\widehat{\sigma}^{k'}(\alpha)}$.*

b) If $\rho \in \Pi'_n$ is progressive, periodic and $t' \in \mathbf{R}$ is arbitrary, then $\sigma^{t'}(\rho)$ is progressive, periodic and $P^\rho = P^{\sigma^{t'}(\rho)}$.

Proof. From Theorem 114, page 169 we have that the forgetful function brings progressive computation functions in progressive computation functions and Theorem 111, page 165 states that the forgetful function brings periodic computation functions in periodic computation functions with the same set of periods. The corollary represents the conjunction of the previous statements. □

Chapter 11

Flows

The flows are signals that model the evolution of the electrical signals when it is generated by the asynchronous iterations of a Boolean function. They are defined in Section 1 and in Section 2 we show that they can be identified with some equations.

In Section 3 we compare the consistency, the composition and the causality properties of the flows with the appropriate defining properties of the dynamical systems from [4].

We show in Section 4 how the equivalence of the computation functions can be defined and what this property implies.

1. Flows

Definition 32. *For the function* $\Phi : \mathbf{B}^n \longrightarrow \mathbf{B}^n$ *and* $\lambda \in \mathbf{B}^n$, *we define* $\Phi^\lambda : \mathbf{B}^n \longrightarrow \mathbf{B}^n$ *by* $\forall \mu \in \mathbf{B}^n$,

$$\Phi^\lambda(\mu) = (\overline{\lambda_1} \cdot \mu_1 \oplus \lambda_1 \cdot \Phi_1(\mu), \ldots, \overline{\lambda_n} \cdot \mu_n \oplus \lambda_n \cdot \Phi_n(\mu)).$$

Definition 33. *Let* $\alpha^0, \ldots, \alpha^k, \alpha^{k+1} \in \mathbf{B}^n, k \geq 0$. *We define iteratively the function* $\Phi^{\alpha^0 \ldots \alpha^k \alpha^{k+1}} : \mathbf{B}^n \longrightarrow \mathbf{B}^n$ *by* $\forall \mu \in \mathbf{B}^n$,

$$\Phi^{\alpha^0 \ldots \alpha^k \alpha^{k+1}}(\mu) = \Phi^{\alpha^{k+1}}(\Phi^{\alpha^0 \ldots \alpha^k}(\mu)).$$

Definition 34. *a) The function* $\mathbf{B}^n \times \mathbf{N}_{\_} \times \widehat{\Pi}_n \ni (\mu, k, \alpha) \longmapsto \widehat{\Phi}^\alpha(\mu, k) \in \mathbf{B}^n$ *defined by* $\forall k \in \mathbf{N}_{\_}$,

$$\widehat{\Phi}^\alpha(\mu, k) = \begin{cases} \mu, if\ k = -1, \\ \Phi^{\alpha^0 \ldots \alpha^k}(\mu), if\ k \geq 0 \end{cases}$$

is called (discrete time) ***evolution function****, or* ***(state) transition function****, or* ***next state function****.* $\mathbf{B}^n$ *is called* ***state space*** *(or* ***phase space****),* μ *is called the* ***initial (value of the) state*** *and* α *is the* ***computation function****. The value*

$$\widehat{x}(k) = \widehat{\Phi}^\alpha(\mu, k)$$

is the ***state*** $\widehat{x}(k)$ ***resulted at the time instant*** k ***from the initial (value of the) state*** μ ***under the (action of the) computation function*** α.

b) We define the function $\mathbf{B}^n \times \mathbf{R} \times \Pi_n \ni (\mu,t,\rho) \longmapsto \Phi^\rho(\mu,t) \in \mathbf{B}^n$ *in the following way. Let* $\forall t \in \mathbf{R}$,

$$\rho(t) = \alpha^0 \cdot \chi_{\{t_0\}}(t) \oplus \alpha^1 \cdot \chi_{\{t_1\}}(t) \oplus \ldots \oplus \alpha^k \cdot \chi_{\{t_k\}}(t) \oplus \ldots \tag{1.1}$$

where $\alpha \in \widehat{\Pi}_n$ *and* $(t_k) \in Seq$. *Then*

$$\Phi^\rho(\mu,t) = \widehat{\Phi}^\alpha(\mu,-1) \cdot \chi_{(-\infty,t_0)}(t) \oplus \widehat{\Phi}^\alpha(\mu,0) \cdot \chi_{[t_0,t_1)}(t) \oplus \ldots$$
$$\ldots \oplus \widehat{\Phi}^\alpha(\mu,k) \cdot \chi_{[t_k,t_{k+1})}(t) \oplus \ldots$$

is called (real time) ***evolution function****, (state)* ***transition function*** *or* ***next state function****,* $\mathbf{B}^n$ *is the* ***state space*** *(or* ***phase space****),* μ *is the* ***initial (value of the) state*** *and* ρ *is the* ***computation function****. The value*

$$x(t) = \Phi^\rho(\mu,t)$$

is the ***state resulted at the time instant*** t ***from the initial (value of the) state*** μ ***under the (action of the) computation function*** ρ.

Definition 35. *a) We fix in the argument of the discrete time evolution function* $\mu \in \mathbf{B}^n$ *and* $\alpha \in \widehat{\Pi}_n$. *The signal* $\widehat{\Phi}^\alpha(\mu,\cdot) \in \widehat{S}^{(n)}$ *is called (discrete time)* ***flow (through*** μ*,* ***under*** α*) and, more general, if previously* $\alpha \in \widehat{\Pi}'_n$*, then* $\widehat{\Phi}^\alpha(\mu,\cdot)$ *is called* ***semi-flow****.*

b) We fix in the argument of the real time evolution function $\mu \in \mathbf{B}^n$ *and* $\rho \in \Pi_n$. *The signal* $\Phi^\rho(\mu,\cdot) \in S^{(n)}$ *is called (real time)* ***flow (through*** μ*,* ***under*** ρ*). More general, if previously* $\rho \in \Pi'_n$*, then* $\Phi^\rho(\mu,\cdot)$ *is called* ***semi-flow****.*

Remark 156. *The function* Φ *applied to the argument* μ *is computed on all its coordinates:* $\Phi(\mu) = (\Phi_1(\mu),\Phi_2(\mu),...,\Phi_n(\mu))$. *The function* Φ^λ *applied to* μ *computes these coordinates* Φ_i *of* Φ *for which* $\lambda_i = 1$ *and it does not compute these coordinates* Φ_i *for which* $\lambda_i = 0$ *:* $\forall i \in \{1,2,...,n\}$,

$$\Phi_i^\lambda(\mu) = \begin{cases} \Phi_i(\mu), \lambda_i = 1, \\ \mu_i, \lambda_i = 0. \end{cases}$$

Unlike the usual computations from the dynamical systems theory that happen synchronously on all the coordinates: $\Phi(\mu)$, $(\Phi \circ \Phi)(\mu)$, $(\Phi \circ \Phi \circ \Phi)(\mu)$,... *here things happen on some coordinates only, as shown in Definitions 32, 33, 34. The asynchronous flows represent a generalization of the computations from the dynamical systems theory, since the constant sequence* $\alpha^k = (1,...,1) \in \mathbf{B}^n, k \in \mathbf{N}$ *belongs to* $\widehat{\Pi}_n$*, see Example 44, page 168 and it gives for any* $\mu \in \mathbf{B}^n$ *that* $\Phi^{\alpha^0}(\mu) = \Phi(\mu)$, $\Phi^{\alpha^0\alpha^1}(\mu) = (\Phi \circ \Phi)(\mu)$, $\Phi^{\alpha^0\alpha^1\alpha^2}(\mu) = (\Phi \circ \Phi \circ \Phi)(\mu)$,...

Another interpretation of the flows $\widehat{\Phi}^\alpha(\mu,\cdot)$, $\Phi^\rho(\mu,\cdot)$ *coming from the dynamical systems theory is that they represent systems with variable structure: instead of a function* Φ *that iterates, we have several functions* $\Phi^{\alpha^k}, k \in \mathbf{N}$ *that iterate. This is a special case of variable structure, since the functions* Φ^{α^k} *have all the origin in* Φ.

Remark 157. *We give the meaning of progressiveness:* $\alpha \in \widehat{\Pi}_n, \rho \in \Pi_n$ *show that* $\widehat{\Phi}^\alpha(\mu,\cdot)$, $\Phi^\rho(\mu,\cdot)$ *compute each coordinate* $\Phi_i, i = \overline{1,n}$ *infinitely many times as* $k \to \infty, t \to \infty$. *In electrical engineering, this corresponds to the so called* ***unbounded delay model*** *of computation of the Boolean functions, stating basically that each coordinate* i *of* Φ *is computed independently on the other coordinates, in finite time.*

Remark 158. *In the following we shall always suppose that the progressiveness requirement on α, ρ is fulfilled, thus we shall work with flows.*

Example 48. $\Phi : \mathbf{B}^2 \to \mathbf{B}^2$ *is defined by* $\forall \mu \in \mathbf{B}^2$,

$$\Phi(\mu_1, \mu_2) = (\overline{\mu_1}, \overline{\mu_2})$$

and $\alpha \in \widehat{\Pi}_2$ *is the following one:*

$$\alpha = \underset{0}{(1,0)}, \underset{1}{(0,1)}, \underset{2}{(1,0)}, \underset{3}{(0,1)}, \underset{4}{(1,0)}, \ldots$$

We obtain

$$\widehat{\Phi}^{\alpha}(\mu, \cdot) = \underset{-1}{(\mu_1, \mu_2)}, \underset{0}{(\overline{\mu_1}, \mu_2)}, \underset{1}{(\overline{\mu_1}, \overline{\mu_2})}, \underset{2}{(\mu_1, \overline{\mu_2})}, \underset{3}{(\mu_1, \mu_2)}, \underset{4}{(\overline{\mu_1}, \mu_2)}, \ldots$$

To be noticed that the periodicity of α *has implied the periodicity of* $\widehat{\Phi}^{\alpha}(\mu, \cdot)$ *and that* $p = 4$ *is their common period.*

Notation 17. *The set of the initial time instants of* $\Phi^{\rho}(\mu, \cdot)$ *has the notation*

$$I^{\rho}_{\mu} = I^{\Phi^{\rho}(\mu, \cdot)}.$$

Notation 18. *The orbits of* $\widehat{\Phi}^{\alpha}(\mu, \cdot), \Phi^{\rho}(\mu, \cdot)$ *have the following notations*

$$\widehat{Or}^{\alpha}(\mu) = \widehat{Or}(\widehat{\Phi}^{\alpha}(\mu, \cdot)),$$

$$Or^{\rho}(\mu) = Or(\Phi^{\rho}(\mu, \cdot)).$$

Notation 19. *The omega limit sets of* $\widehat{\Phi}^{\alpha}(\mu, \cdot)$, $\Phi^{\rho}(\mu, \cdot)$ *are denoted by* $\widehat{\omega}^{\alpha}(\mu)$, $\omega^{\rho}(\mu)$:

$$\widehat{\omega}^{\alpha}(\mu) = \widehat{\omega}(\widehat{\Phi}^{\alpha}(\mu, \cdot)),$$

$$\omega^{\rho}(\mu) = \omega(\Phi^{\rho}(\mu, \cdot)).$$

Notation 20. *For* $\mu' \in \mathbf{B}^n$ *and* μ, α, ρ *like before, we denote*

$$\widehat{\mathbf{T}}^{\alpha}_{\mu,\mu'} = \widehat{\mathbf{T}}^{\widehat{\Phi}^{\alpha}(\mu, \cdot)}_{\mu'},$$

$$\mathbf{T}^{\rho}_{\mu,\mu'} = \mathbf{T}^{\Phi^{\rho}(\mu, \cdot)}_{\mu'}.$$

Remark 159. *The purpose of the Notations 17, 18, 19, 20 is that of eliminating* Φ, *as it is understood from the context.*

2. The Defining Equations of the Flows

Theorem 116. *a) $\mu \in \mathbf{B}^n$ and $\alpha \in \widehat{\Pi}_n$ are given, together with the statements:*

$$\begin{cases} \widehat{x}(-1) = \mu, \\ \forall k \in \mathbf{N}, \widehat{x}(k) = \Phi^{\alpha^k}(\widehat{x}(k-1)), \end{cases} \tag{2.1}$$

$$\forall k \in \mathbf{N}_{\_}, \widehat{x}(k) = \widehat{\Phi}^{\alpha}(\mu, k). \tag{2.2}$$

The equivalence (2.1)$\Longleftrightarrow$(2.2) holds.

b) Let $\mu \in \mathbf{B}^n$ and $\rho \in \Pi_n$. The following statements:

$$\begin{cases} x(-\infty+0) = \mu, \\ \forall t \in \mathbf{R}, x(t) = \Phi^{\rho(t)}(x(t-0)), \end{cases} \tag{2.3}$$

$$\forall t \in \mathbf{R}, x(t) = \Phi^{\rho}(\mu, t) \tag{2.4}$$

are equivalent.

Proof. b) We put ρ under the form

$$\rho(t) = \alpha^0 \cdot \chi_{\{t_0\}}(t) \oplus \alpha^1 \cdot \chi_{\{t_1\}}(t) \oplus \ldots \oplus \alpha^k \cdot \chi_{\{t_k\}}(t) \oplus \ldots \tag{2.5}$$

where $\alpha \in \widehat{\Pi}_n$ and $(t_k) \in Seq$. In (2.4) we have

$$\begin{aligned} x(t) = \mu \cdot \chi_{(-\infty,t_0)}(t) \oplus \Phi^{\alpha^0}(\mu) \cdot \chi_{[t_0,t_1)}(t) \oplus \Phi^{\alpha^0\alpha^1}(\mu) \cdot \chi_{[t_1,t_2)}(t) \oplus \ldots \\ \ldots \oplus \Phi^{\alpha^0\alpha^1\ldots\alpha^k}(\mu) \cdot \chi_{[t_k,t_{k+1})}(t) \oplus \ldots \end{aligned} \tag{2.6}$$

and we prove first the implication (2.3)$\Longrightarrow$(2.6). Some $\varepsilon > 0$ exists such that

$$\forall \xi \in (t_0 - \varepsilon, t_0), x(\xi) = x(t_0 - 0) = \mu$$

and on the other hand

$$\forall t < t_0, x(t) = \Phi^{\rho(t)}(x(t-0)) = \Phi^{(0,\ldots,0)}(x(t-0)) = x(t-0).$$

In such circumstances $\forall t < t_0, x(t) = \mu$, since the supposition that this is not true brings a contradiction. We have:

$t = t_0, x(t_0) = \Phi^{\rho(t_0)}(x(t_0 - 0)) = \Phi^{\alpha^0}(\mu);$
$t \in (t_0, t_1), x(t) = \Phi^{\rho(t)}(x(t-0)) = \Phi^{(0,\ldots,0)}(\Phi^{\alpha^0}(\mu)) = \Phi^{\alpha^0}(\mu);$
$t = t_1, x(t_1) = \Phi^{\rho(t_1)}(x(t_1 - 0)) = \Phi^{\alpha^1}(\Phi^{\alpha^0}(\mu)) = \Phi^{\alpha^0\alpha^1}(\mu);$
$t \in (t_1, t_2), x(t) = \Phi^{\rho(t)}(x(t-0)) = \Phi^{(0,\ldots,0)}(\Phi^{\alpha^0\alpha^1}(\mu)) = \Phi^{\alpha^0\alpha^1}(\mu);$

...

$t = t_k, x(t_k) = \Phi^{\rho(t_k)}(x(t_k - 0)) = \Phi^{\alpha^k}(\Phi^{\alpha^0\ldots\alpha^{k-1}}(\mu)) = \Phi^{\alpha^0\alpha^1\ldots\alpha^k}(\mu);$
$t \in (t_k, t_{k+1}), x(t) = \Phi^{\rho(t)}(x(t-0)) = \Phi^{(0,\ldots,0)}(\Phi^{\alpha^0\alpha^1\ldots\alpha^k}(\mu)) = \Phi^{\alpha^0\alpha^1\ldots\alpha^k}(\mu);$

...

We prove (2.6) $\Longrightarrow$ (2.3). Since $\forall t < t_0, x(t) = \mu$ we get the truth of the first equation from (2.3). We prove the second equation from (2.3) and we infer for any $t \in \mathbf{R}$ that

$$x(t-0) = \mu \cdot \chi_{(-\infty,t_0]}(t) \oplus \Phi^{\alpha^0}(\mu) \cdot \chi_{(t_0,t_1]}(t) \oplus \Phi^{\alpha^0\alpha^1}(\mu) \cdot \chi_{(t_1,t_2]}(t) \oplus \ldots$$
$$\ldots \oplus \Phi^{\alpha^0\alpha^1\ldots\alpha^k}(\mu) \cdot \chi_{(t_k,t_{k+1}]}(t) \oplus \ldots$$
$$= \mu \cdot \chi_{(-\infty,t_0)}(t) \oplus \mu \cdot \chi_{\{t_0\}}(t) \oplus \Phi^{\alpha^0}(\mu) \cdot \chi_{(t_0,t_1)}(t) \oplus \Phi^{\alpha^0}(\mu) \cdot \chi_{\{t_1\}}(t) \oplus$$
$$\oplus \Phi^{\alpha^0\alpha^1}(\mu) \cdot \chi_{(t_1,t_2)}(t) \oplus \Phi^{\alpha^0\alpha^1}(\mu) \cdot \chi_{\{t_2\}}(t) \oplus \ldots$$
$$\ldots \oplus \Phi^{\alpha^0\alpha^1\ldots\alpha^k}(\mu) \cdot \chi_{(t_k,t_{k+1})}(t) \oplus \Phi^{\alpha^0\alpha^1\ldots\alpha^k}(\mu) \cdot \chi_{\{t_{k+1}\}}(t) \oplus \ldots$$

thus

$$\Phi^{\rho(t)}(x(t-0)) = \Phi^{\rho(t)}(\mu \cdot \chi_{(-\infty,t_0)}(t) \oplus \mu \cdot \chi_{\{t_0\}}(t) \oplus$$
$$\oplus \Phi^{\alpha^0}(\mu) \cdot \chi_{(t_0,t_1)}(t) \oplus \Phi^{\alpha^0}(\mu) \cdot \chi_{\{t_1\}}(t) \oplus$$
$$\oplus \Phi^{\alpha^0\alpha^1}(\mu) \cdot \chi_{(t_1,t_2)}(t) \oplus \Phi^{\alpha^0\alpha^1}(\mu) \cdot \chi_{\{t_2\}}(t) \oplus \ldots$$
$$\ldots \oplus \Phi^{\alpha^0\alpha^1\ldots\alpha^k}(\mu) \cdot \chi_{(t_k,t_{k+1})}(t) \oplus \Phi^{\alpha^0\alpha^1\ldots\alpha^k}(\mu) \cdot \chi_{\{t_{k+1}\}}(t) \oplus \ldots)$$
$$= \Phi^{\rho(t)}(\mu) \cdot \chi_{(-\infty,t_0)}(t) \oplus \Phi^{\rho(t)}(\mu) \cdot \chi_{\{t_0\}}(t) \oplus$$
$$\oplus \Phi^{\rho(t)}(\Phi^{\alpha^0}(\mu)) \cdot \chi_{(t_0,t_1)}(t) \oplus \Phi^{\rho(t)}(\Phi^{\alpha^0}(\mu)) \cdot \chi_{\{t_1\}}(t) \oplus$$
$$\oplus \Phi^{\rho(t)}(\Phi^{\alpha^0\alpha^1}(\mu)) \cdot \chi_{(t_1,t_2)}(t) \oplus \Phi^{\rho(t)}(\Phi^{\alpha^0\alpha^1}(\mu)) \cdot \chi_{\{t_2\}}(t) \oplus \ldots$$
$$\ldots \oplus \Phi^{\rho(t)}(\Phi^{\alpha^0\alpha^1\ldots\alpha^k}(\mu)) \cdot \chi_{(t_k,t_{k+1})}(t) \oplus \Phi^{\rho(t)}(\Phi^{\alpha^0\alpha^1\ldots\alpha^k}(\mu)) \cdot \chi_{\{t_{k+1}\}}(t) \oplus \ldots$$
$$= \Phi^{(0,\ldots,0)}(\mu) \cdot \chi_{(-\infty,t_0)}(t) \oplus \Phi^{\alpha^0}(\mu) \cdot \chi_{\{t_0\}}(t) \oplus$$
$$\oplus \Phi^{(0,\ldots,0)}(\Phi^{\alpha^0}(\mu)) \cdot \chi_{(t_0,t_1)}(t) \oplus \Phi^{\alpha^1}(\Phi^{\alpha^0}(\mu)) \cdot \chi_{\{t_1\}}(t) \oplus$$
$$\oplus \Phi^{(0,\ldots,0)}(\Phi^{\alpha^0\alpha^1}(\mu)) \cdot \chi_{(t_1,t_2)}(t) \oplus \Phi^{\alpha^2}(\Phi^{\alpha^0\alpha^1}(\mu)) \cdot \chi_{\{t_2\}}(t) \oplus \ldots$$
$$\ldots \oplus \Phi^{(0,\ldots,0)}(\Phi^{\alpha^0\alpha^1\ldots\alpha^k}(\mu)) \cdot \chi_{(t_k,t_{k+1})}(t) \oplus \Phi^{\alpha^{k+1}}(\Phi^{\alpha^0\alpha^1\ldots\alpha^k}(\mu)) \cdot \chi_{\{t_{k+1}\}}(t) \oplus \ldots$$
$$= \mu \cdot \chi_{(-\infty,t_0)}(t) \oplus \Phi^{\alpha^0}(\mu) \cdot \chi_{\{t_0\}}(t) \oplus$$
$$\oplus \Phi^{\alpha^0}(\mu) \cdot \chi_{(t_0,t_1)}(t) \oplus \Phi^{\alpha^0\alpha^1}(\mu) \cdot \chi_{\{t_1\}}(t) \oplus$$
$$\oplus \Phi^{\alpha^0\alpha^1}(\mu) \cdot \chi_{(t_1,t_2)}(t) \oplus \Phi^{\alpha^0\alpha^1\alpha^2}(\mu) \cdot \chi_{\{t_2\}}(t) \oplus \ldots$$
$$\ldots \oplus \Phi^{\alpha^0\alpha^1\ldots\alpha^k}(\mu) \cdot \chi_{(t_k,t_{k+1})}(t) \oplus \Phi^{\alpha^0\alpha^1\ldots\alpha^k\alpha^{k+1}}(\mu) \cdot \chi_{\{t_{k+1}\}}(t) \oplus \ldots = x(t).$$

□

Definition 36. *The equations (2.1), as well as the equations (2.3) are called the **defining equations of the flows**.*

3. Consistency, Composition and Causality

Remark 160. *The properties stated in Theorems 117, 118 and 120 to follow are the adaptation to the present context of the properties of consistency, composition and causality of the transition function that are contained in the definition of a dynamical system from [4], page 11. At the same page, the authors state that the words 'dynamical', 'non-anticipatory' and 'causal' have approximately the same meaning, making us conclude that the property of causality to be introduced may be also called non-anticipation. We must add here the remark that in the cited work the systems had an input, unlike here where it is convenient to omit this aspect*[1]*, and consequently there causality referred to the input, unlike here where it refers to the computation function. The input controls the state and the computation function shows when and how the state is computed.*

We suppose in this section that a function $\Phi : \mathbf{B}^n \to \mathbf{B}^n$ is given, together with $\mu \in \mathbf{B}^n$, $\alpha \in \widehat{\Pi}_n$ and $\rho \in \Pi_n$. The relation between α and ρ is

$$\rho(t) = \alpha^0 \cdot \chi_{\{t_0\}}(t) \oplus \alpha^1 \cdot \chi_{\{t_1\}}(t) \oplus \ldots \oplus \alpha^k \cdot \chi_{\{t_k\}}(t) \oplus \ldots \tag{3.1}$$

where $(t_k) \in Seq$.

Theorem 117. *(Consistency)*

$$\widehat{\Phi}^{\alpha}(\mu, -1) = \mu, \tag{3.2}$$

$$\Phi^{\rho}(\mu, t_0 - 0) = \mu. \tag{3.3}$$

Proof. a) This follows from Definition 34, page 173.

b) Definition 34 shows that we have

$$\forall t < t_0, \Phi^{\rho}(\mu, t) = \mu,$$

wherefrom (3.3) follows. □

Theorem 118. *(Composition) a) $\forall k' \in \mathbf{N}, \forall k \in \mathbf{N}_{\_}$,*

$$\widehat{\sigma}^{k'}(\widehat{\Phi}^{\alpha}(\mu, \cdot))(k) = \widehat{\Phi}^{\widehat{\sigma}^{k'}(\alpha)}(\widehat{\Phi}^{\alpha}(\mu, k' - 1), k). \tag{3.4}$$

b) $\forall t' \in \mathbf{R}, \forall t \in \mathbf{R}$,

$$\sigma^{t'}(\Phi^{\rho}(\mu, \cdot))(t) = \Phi^{\sigma^{t'}(\rho)}(\Phi^{\rho}(\mu, t' - 0), t). \tag{3.5}$$

Proof. Let us notice first of all that $\alpha \in \widehat{\Pi}_n, \rho \in \Pi_n \Longrightarrow \widehat{\sigma}^{k'}(\alpha) \in \widehat{\Pi}_n, \sigma^{t'}(\rho) \in \Pi_n$ result from Theorem 114, page 169 and Remark 154, page 169, wherefrom the right members of equations (3.4), (3.5) make sense.

a) We have the following possibilities.

Case $k' = 0, k \in \mathbf{N}_{\_}$ arbitrary, when

$$\widehat{\sigma}^0(\widehat{\Phi}^{\alpha}(\mu, \cdot))(k) = \widehat{\Phi}^{\alpha}(\mu, k) = \widehat{\Phi}^{\widehat{\sigma}^0(\alpha)}(\mu, k) = \widehat{\Phi}^{\widehat{\sigma}^0(\alpha)}(\widehat{\Phi}^{\alpha}(\mu, -1), k).$$

[1]The systems without input (control) are called autonomous. In Section 5 of this chapter we shall show how the flows are defined when an input exists.

Case $k' \geq 1, k = -1$

$$\widehat{\sigma}^{k'}(\widehat{\Phi}^{\alpha}(\mu,\cdot))(-1) = \widehat{\Phi}^{\alpha}(\mu,k'-1) = \widehat{\Phi}^{\widehat{\sigma}^{k'}(\alpha)}(\widehat{\Phi}^{\alpha}(\mu,k'-1),-1).$$

Case $k' \geq 1, k \in \mathbf{N}$ arbitrary, for which

$$\widehat{\sigma}^{k'}(\widehat{\Phi}^{\alpha}(\mu,\cdot))(k) = \widehat{\Phi}^{\alpha}(\mu,k+k') = \Phi^{\alpha^0...\alpha^{k'}\alpha^{k'+1}...\alpha^{k'+k}}(\mu)$$

$$= \Phi^{\alpha^{k'}\alpha^{k'+1}...\alpha^{k'+k}}(\Phi^{\alpha^0...\alpha^{k'-1}}(\mu)) = \widehat{\Phi}^{\alpha^{k'},\alpha^{k'+1},\alpha^{k'+2},...}(\Phi^{\alpha^0...\alpha^{k'-1}}(\mu),k)$$

$$= \widehat{\Phi}^{\widehat{\sigma}^{k'}(\alpha)}(\Phi^{\alpha^0...\alpha^{k'-1}}(\mu),k) = \widehat{\Phi}^{\widehat{\sigma}^{k'}(\alpha)}(\widehat{\Phi}^{\alpha}(\mu,k'-1),k).$$

b) Equation (3.1) shows that we can put $\Phi^{\rho}(\mu,\cdot)$ under the form

$$\begin{aligned}\Phi^{\rho}(\mu,t) = \mu\cdot\chi_{(-\infty,t_0)}(t)\oplus\Phi^{\alpha^0}(\mu)\cdot\chi_{[t_0,t_1)}(t)\oplus\Phi^{\alpha^0\alpha^1}(\mu)\cdot\chi_{[t_1,t_2)}(t)\oplus...\\ ...\oplus\Phi^{\alpha^0...\alpha^k}(\mu)\cdot\chi_{[t_k,t_{k+1})}(t)\oplus...\end{aligned} \tag{3.6}$$

We take an arbitrary $t' \in \mathbf{R}$ and we have the following possibilities.

Case $t' \leq t_0$

In this situation

$$\sigma^{t'}(\rho)(t) = \rho(t),$$

$$\Phi^{\rho}(\mu,t'-0) = \mu,$$

thus

$$\sigma^{t'}(\Phi^{\rho}(\mu,\cdot))(t) = \Phi^{\rho}(\mu,t) = \Phi^{\sigma^{t'}(\rho)}(\Phi^{\rho}(\mu,t'-0),t).$$

Case $\exists k \in \mathbf{N}, t' \in (t_k,t_{k+1}]$

In this case we infer

$$\sigma^{t'}(\rho)(t) = \alpha^{k+1}\cdot\chi_{\{t_{k+1}\}}(t)\oplus\alpha^{k+2}\cdot\chi_{\{t_{k+2}\}}(t)\oplus...$$

$$\Phi^{\rho}(\mu,t'-0) = \Phi^{\alpha^0...\alpha^k}(\mu),$$

$$\sigma^{t'}(\Phi^{\rho}(\mu,\cdot))(t) = \Phi^{\alpha^0...\alpha^k}(\mu)\cdot\chi_{(-\infty,t_{k+1})}(t)\oplus\Phi^{\alpha^0...\alpha^k\alpha^{k+1}}(\mu)\cdot\chi_{[t_{k+1},t_{k+2})}(t)\oplus...$$

$$= \Phi^{\alpha^0...\alpha^k}(\mu)\cdot\chi_{(-\infty,t_{k+1})}(t)\oplus\Phi^{\alpha^{k+1}}(\Phi^{\alpha^0...\alpha^k}(\mu))\cdot\chi_{[t_{k+1},t_{k+2})}(t)\oplus...$$

$$= \Phi^{\alpha^{k+1}\cdot\chi_{\{t_{k+1}\}}\oplus\alpha^{k+2}\cdot\chi_{\{t_{k+2}\}}\oplus...}(\Phi^{\alpha^0...\alpha^k}(\mu),t)$$

$$= \Phi^{\sigma^{t'}(\rho)}(\Phi^{\alpha^0...\alpha^k}(\mu),t) = \Phi^{\sigma^{t'}(\rho)}(\Phi^{\rho}(\mu,t'-0),t).$$

□

Remark 161. *Theorem 118 b) written for arbitrary $t' \in \mathbf{R}$ in the manner: $\forall t \in \mathbf{R}$,*

$$\sigma^{t'}(\Phi^{\rho}(\mu,\cdot))(t) = \Phi^{\rho\cdot\chi_{[t',\infty)}}(\Phi^{\rho}(\mu,t'-0),t)$$

suggests the following version of this property.

Theorem 119. *(Composition) a) For arbitrary $k' \in \mathbf{N}\_$ we can write:* $\forall k \geq k'$,

$$\widehat{\Phi}^{\alpha}(\mu,k) = \widehat{\Phi}^{\widehat{\sigma}^{k'+1}(\alpha)}(\widehat{\Phi}^{\alpha}(\mu,k'),k-k'-1). \tag{3.7}$$

b) $\forall t' \in \mathbf{R}$ *we have:* $\forall t \geq t'$,

$$\Phi^{\rho}(\mu,t) = \Phi^{\rho\cdot\chi_{(t',\infty)}}(\Phi^{\rho}(\mu,t'),t). \tag{3.8}$$

Proof. a) We make the substitution $k = k' + p$, where $p \in \mathbf{N}$ and we prove (3.7) by induction on p. For $p = 0$, (3.7) becomes

$$\widehat{\Phi}^{\alpha}(\mu,k') = \widehat{\Phi}^{\widehat{\sigma}^{k'+1}(\alpha)}(\widehat{\Phi}^{\alpha}(\mu,k'),-1),$$

obvious. We suppose that

$$\widehat{\Phi}^{\alpha}(\mu,k'+p) = \widehat{\Phi}^{\widehat{\sigma}^{k'+1}(\alpha)}(\widehat{\Phi}^{\alpha}(\mu,k'),p-1) \tag{3.9}$$

is true and we infer that

$$\widehat{\Phi}^{\alpha}(\mu,k'+p+1) = \Phi^{\alpha^{k'+p+1}}(\widehat{\Phi}^{\alpha}(\mu,k'+p))$$

$$\overset{(3.9)}{=} \Phi^{\alpha^{k'+p+1}}(\widehat{\Phi}^{\alpha^{k'+1},\alpha^{k'+2},\ldots,\alpha^{k'+p},\ldots}(\widehat{\Phi}^{\alpha}(\mu,k'),p-1))$$

$$= \Phi^{\alpha^{k'+p+1}}(\Phi^{\alpha^{k'+1}\alpha^{k'+2}\ldots\alpha^{k'+p}}(\widehat{\Phi}^{\alpha}(\mu,k'))) = \Phi^{\alpha^{k'+1}\alpha^{k'+2}\ldots\alpha^{k'+p}\alpha^{k'+p+1}}(\widehat{\Phi}^{\alpha}(\mu,k'))$$

$$= \widehat{\Phi}^{\alpha^{k'+1},\alpha^{k'+2},\ldots,\alpha^{k'+p},\alpha^{k'+p+1},\ldots}(\widehat{\Phi}^{\alpha}(\mu,k'),p) = \widehat{\Phi}^{\widehat{\sigma}^{k'+1}(\alpha)}(\widehat{\Phi}^{\alpha}(\mu,k'),p).$$

b) Indeed, we shall suppose in the following that (3.1) is true. In the case $t' < t_0$, we have

$$\rho\cdot\chi_{(t',\infty)} = \rho,$$
$$\Phi^{\rho}(\mu,t') = \mu$$

and (3.8) is true under the form $\forall t \geq t', \Phi^{\rho}(\mu,t) = \Phi^{\rho}(\mu,t)$.

In the case $t' \in [t_k,t_{k+1}), k \in \mathbf{N}$,

$$\rho\cdot\chi_{(t',\infty)} = \alpha^{k+1}\cdot\chi_{\{t_{k+1}\}}(t) \oplus \alpha^{k+2}\cdot\chi_{\{t_{k+2}\}}(t) \oplus \ldots$$

$$\Phi^{\rho}(\mu,t') = \Phi^{\alpha^0\ldots\alpha^k}(\mu),$$

$\Phi^{\rho}(\mu,t)$ is given by (3.6) and

$$\Phi^{\rho\cdot\chi_{(t',\infty)}}(\Phi^{\rho}(\mu,t'),t) = \Phi^{\alpha^{k+1}\cdot\chi_{\{t_{k+1}\}}\oplus\alpha^{k+2}\cdot\chi_{\{t_{k+2}\}}\oplus\ldots}(\Phi^{\alpha^0\ldots\alpha^k}(\mu),t)$$

$$= \Phi^{\alpha^0\ldots\alpha^k}(\mu)\cdot\chi_{(-\infty,t_{k+1})}(t) \oplus \Phi^{\alpha^0\ldots\alpha^{k+1}}(\mu)\cdot\chi_{[t_{k+1},t_{k+2})}(t) \oplus \ldots$$

For $t \geq t'$, (3.8) is true. □

Theorem 120. *(Causality) For any $k \in \mathbf{N}$ and any $\alpha, \beta \in \widehat{\Pi}_n$ with*

$$\forall k' \in \{0,1,...,k\}, \alpha^{k'} = \beta^{k'}, \tag{3.10}$$

we have

$$\widehat{\Phi}^{a}(\mu,k) = \widehat{\Phi}^{\beta}(\mu,k). \tag{3.11}$$

b) Let $t' \in \mathbf{R}$ and $\rho, \rho' \in \Pi_n$ with the property that

$$\forall t \leq t', \rho(t) = \rho'(t). \tag{3.12}$$

Then

$$\Phi^{\rho}(\mu,t') = \Phi^{\rho'}(\mu,t'). \tag{3.13}$$

Proof. a) We infer

$$\widehat{\Phi}^{a}(\mu,k) = \Phi^{\alpha^0...\alpha^k}(\mu) = \Phi^{\beta^0...\beta^k}(\mu) = \widehat{\Phi}^{\beta}(\mu,k). \tag{3.14}$$

b) If $t' \in \mathbf{R}$ is such that

$$\forall t \leq t', \rho(t) = \rho'(t) = (0,...,0), \tag{3.15}$$

then

$$\Phi^{\rho}(\mu,t') = \mu = \Phi^{\rho'}(\mu,t').$$

Let $k \in \mathbf{N}$ be arbitrary and fixed. We suppose that $\alpha, \beta \in \widehat{\Pi}_n$ and $(t_j), (t'_j) \in Seq$ exist such that

$$\forall k' \in \{0,1,...,k\}, t_{k'} = t'_{k'},$$

$$\rho(t) = \alpha^0 \cdot \chi_{\{t_0\}}(t) \oplus ... \oplus \alpha^k \cdot \chi_{\{t_k\}}(t) \oplus \alpha^{k+1} \cdot \chi_{\{t_{k+1}\}}(t) \oplus \alpha^{k+2} \cdot \chi_{\{t_{k+2}\}}(t) \oplus ...$$

$$\rho'(t) = \alpha^0 \cdot \chi_{\{t_0\}}(t) \oplus ... \oplus \alpha^k \cdot \chi_{\{t_k\}}(t) \oplus \beta^{k+1} \cdot \chi_{\{t'_{k+1}\}}(t) \oplus \beta^{k+2} \cdot \chi_{\{t'_{k+2}\}}(t) \oplus ...$$

$\alpha^0 \neq (0,...,0)$ and $t' \in [t_k, t_{k+1}) \cap [t_k, t'_{k+1})$ hold. We get

$$\Phi^{\rho}(\mu,t') = \Phi^{\alpha^0...\alpha^k}(\mu) = \Phi^{\rho'}(\mu,t'). \tag{3.16}$$

□

4. Equivalent Progressive Computation Functions

Notation 21. *For $\alpha \in \widehat{\Pi}_n$, $\rho \in \Pi_n$ we denote with*

$$supp\ \alpha = \{k | k \in \mathbf{N}, \alpha^k \neq (0,...,0)\}$$

*the **support (set)** of α and with*

$$supp\ \rho = \{t | t \in \mathbf{R}, \rho(t) \neq (0,...,0)\}$$

*the **support (set)** of ρ.*

Notation 22. *Let supp* $\alpha = \{j_0, j_1, j_2, ...\}$ *with* $j_0 < j_1 < j_2 < ...$ *We denote with* $\varphi^\alpha : \mathbf{N} \longrightarrow$ *supp* α *the function* $\forall k \in \mathbf{N}, \varphi^\alpha(k) = j_k$.

Remark 162. *We notice that* $\varphi^\alpha(k)$ *gives the rank of the* $k-th$ *non null term of* α. *The function* φ^α *can be defined since the progressiveness of* α *implies that* α *has infinitely many non null terms.*

Example 49. *We consider the progressive computation function* $\alpha \in \widehat{\Pi}_1, \alpha = 0,1,0,1,0,1,...$ *for which we get* $\forall k \in \mathbf{N}, \varphi^\alpha(k) = 2k+1$.

Theorem 121. *a)* φ^α *is bijective;*

b) φ^α *is strictly increasing and* $(\varphi^\alpha(k)) \in \widehat{Seq'}$;

c) $\forall k \in \mathbf{N}$,

$$\varphi^\alpha(k) \geq k; \tag{4.1}$$

d) the following statements are equivalent:

$$\forall k \in \mathbf{N}, \varphi^\alpha(k) = k, \tag{4.2}$$

$$\forall k \in \mathbf{N}, \alpha^k \neq (0,..,0), \tag{4.3}$$

$$supp\ \alpha = \mathbf{N}. \tag{4.4}$$

Proof. c) We denote like before $supp\ \alpha = \{j_0, j_1, j_2, ...\}$. Then $j_0 \geq 0$ and we suppose that for an arbitrary $k \in \mathbf{N}$ we have $j_k \geq k$. We infer $j_{k+1} \geq j_k + 1 \geq k+1$.

d) (4.2)$\Longrightarrow$(4.3) Let $k \in \mathbf{N}$ arbitrary, for which we have $\varphi^\alpha(k) = k$. We infer that $(0,...,0) \neq \alpha^{\varphi^\alpha(k)} = \alpha^k$.

(4.3)$\Longrightarrow$(4.4) $supp\ \alpha = \{k | k \in \mathbf{N}, \alpha^k \neq (0,...,0)\} = \mathbf{N}$.

(4.4)$\Longrightarrow$(4.2) We can write

$$\mathbf{N} = \{k | k \in \mathbf{N}, \alpha^k \neq (0,...,0)\} = \{\varphi^\alpha(k) | k \in \mathbf{N}\}. \tag{4.5}$$

We suppose against all reason that (4.2) is false, i.e. $\exists k \in \mathbf{N}, \varphi^\alpha(k) \neq k$. As in this case $\varphi^\alpha(k) > k$ from (4.1) and because φ^α is strictly increasing, we infer that $\forall k' \in \mathbf{N}, \varphi^\alpha(k') \neq k$ and this represents a contradiction with (4.5). Statement (4.2) holds. □

Theorem 122. *a) Let* $\alpha \in \widehat{\Pi}_n$ *and we define* $\beta \in \widehat{\Pi}'_n$ *by* $\forall k \in \mathbf{N}$,

$$\beta^k = \alpha^{\varphi^\alpha(k)}. \tag{4.6}$$

Then $\forall k \in \mathbf{N}, \beta^k \neq (0,...,0)$ *and* $\beta \in \widehat{\Pi}_n$.

b) The sequence $(t_k) \in Seq$ *is considered and we define* $\forall k \in \mathbf{N}$,

$$t'_k = t_{\varphi^\alpha(k)}. \tag{4.7}$$

Then $(t'_k) \in Seq$.

c) the computation functions

$$\rho(t) = \alpha^0 \cdot \chi_{\{t_0\}}(t) \oplus \alpha^1 \cdot \chi_{\{t_1\}}(t) \oplus ... \oplus \alpha^k \cdot \chi_{\{t_k\}}(t) \oplus ... \tag{4.8}$$

$$\rho'(t) = \beta^0 \cdot \chi_{\{t'_0\}}(t) \oplus \beta^1 \cdot \chi_{\{t'_1\}}(t) \oplus ... \oplus \beta^k \cdot \chi_{\{t'_k\}}(t) \oplus ... \tag{4.9}$$

are progressive and moreover $\rho'(t) = \rho(t)$.

Proof. a) For an arbitrary $k \in \mathbf{N}$, we can write that $\beta^k = \alpha^{\varphi^{\alpha}(k)} \in \{\alpha^j | j \in supp\ \alpha\}$, thus $\beta^k \neq (0,...,0)$. Let us notice now that

$$\widehat{\omega}(\alpha) = \widehat{\omega}(\beta) \text{ or } \widehat{\omega}(\alpha) = \widehat{\omega}(\beta) \cup \{(0,...,0)\}, \tag{4.10}$$

wherefrom we obtain the implications

$$\alpha \in \widehat{\Pi}_n \Longrightarrow \forall i \in \{1,...,n\}, \exists \mu' \in \widehat{\omega}(\alpha), \mu'_i = 1$$

$$\overset{(4.10)}{\Longrightarrow} \forall i \in \{1,...,n\}, \exists \mu' \in \widehat{\omega}(\beta), \mu'_i = 1 \Longrightarrow \beta \in \widehat{\Pi}_n.$$

b) The statement is true because (t'_k) is a strictly increasing subsequence of (t_k), resulting that it is also unbounded from above.

c) The progressiveness of ρ' is a consequence of a), b). The equality $\rho'(t) = \rho(t)$ holds because β is that subsequence of α that contains exactly the non null terms and (t'_k) represents the appropriate subsequence of (t_k). □

Theorem 123. *Let $\mu \in \mathbf{B}^n, \alpha, \beta \in \widehat{\Pi}_n, \rho, \rho' \in \Pi_n$ and $(t_k), (t'_k) \in Seq$ with $\forall k \in \mathbf{N}$,*

$$\beta^k = \alpha^{\varphi^{\alpha}(k)} \tag{4.11}$$

and

$$\rho(t) = \alpha^0 \cdot \chi_{\{t_0\}}(t) \oplus \alpha^1 \cdot \chi_{\{t_1\}}(t) \oplus ... \oplus \alpha^k \cdot \chi_{\{t_k\}}(t) \oplus ... \tag{4.12}$$

$$\rho'(t) = \beta^0 \cdot \chi_{\{t'_0\}}(t) \oplus \beta^1 \cdot \chi_{\{t'_1\}}(t) \oplus ... \oplus \beta^k \cdot \chi_{\{t'_k\}}(t) \oplus ... \tag{4.13}$$

We define

$$J_{-1} = \{-1,...,\varphi^{\alpha}(0) - 1\},$$
$$\forall k \in \mathbf{N}, J_k = \{\varphi^{\alpha}(k), \varphi^{\alpha}(k)+1,...,\varphi^{\alpha}(k+1) - 1\}.$$

The following statements are true:

a) $\forall k \in \mathbf{N}_{\_}, \forall p \in J_k$,

$$\widehat{\Phi}^{\beta}(\mu, k) = \widehat{\Phi}^{\alpha}(\mu, p), \tag{4.14}$$

in particular $\forall k \in \mathbf{N}$,

$$\widehat{\Phi}^{\beta}(\mu, k) = \widehat{\Phi}^{\alpha}(\mu, \varphi^{\alpha}(k)); \tag{4.15}$$

b) we have

$$\widehat{Or}^{\beta}(\mu) = \widehat{Or}^{\alpha}(\mu), \tag{4.16}$$

$$\widehat{\omega}^{\beta}(\mu) = \widehat{\omega}^{\alpha}(\mu); \tag{4.17}$$

c) we also have:

$$Or^{\rho'}(\mu) = Or^{\rho}(\mu), \tag{4.18}$$

$$\omega^{\rho'}(\mu) = \omega^{\rho}(\mu). \tag{4.19}$$

Proof. a) We notice first of all that $\forall k \in \mathbf{N}_{\_}, J_k \neq \varnothing$ and that the family (J_k) is a partition of $\mathbf{N}_{\_}$. We make the proof by induction on k. We take $k=-1$, when $\widehat{\Phi}^{\beta}(\mu,-1)=\mu$ and we have two possibilities.

Case $\varphi^{\alpha}(0)=0$

Then $J_{-1}=\{-1\}, p=-1$ and $\widehat{\Phi}^{\alpha}(\mu,-1)=\mu.$

Case $\varphi^{\alpha}(0)>0$

We have in addition to the previous situation the possibility that $p\geq 0, \alpha^0=...=\alpha^p=(0,...,0)$, thus $\widehat{\Phi}^{\alpha}(\mu,p)=\Phi^{\alpha^0...\alpha^p}(\mu)=\mu.$

We suppose now that (4.14) is true for $k\geq 0$:

$$\widehat{\Phi}^{\beta}(\mu,k)=\widehat{\Phi}^{\alpha}(\mu,\varphi^{\alpha}(k+1)-1)=\Phi^{\alpha^0...\alpha^{\varphi^{\alpha}(k+1)-1}}(\mu)$$

and we prove it for $k+1$, when

$$\Phi^{\beta}(\mu,k+1)=\Phi^{\beta^{k+1}}(\Phi^{\beta}(\mu,k))=\Phi^{\beta^{k+1}}(\Phi^{\alpha^0...\alpha^{\varphi^{\alpha}(k+1)-1}}(\mu))$$

$$=\Phi^{\alpha^0...\alpha^{\varphi^{\alpha}(k+1)-1}\beta^{k+1}}(\mu)\overset{(4.11)}{=}\Phi^{\alpha^0...\alpha^{\varphi^{\alpha}(k+1)}}(\mu).$$

We have two possibilities.

Case $\varphi^{\alpha}(k+2)=\varphi^{\alpha}(k+1)+1$

In this situation $J_{k+1}=\{\varphi^{\alpha}(k+1)\}, p=\varphi^{\alpha}(k+1)$ and $\widehat{\Phi}^{\alpha}(\mu,\varphi^{\alpha}(k+1))=\Phi^{\alpha^0...\alpha^{\varphi^{\alpha}(k+1)}}(\mu).$

Case $\varphi^{\alpha}(k+2)>\varphi^{\alpha}(k+1)+1$

We have in addition to the previous situation the possibility that $p\geq\varphi^{\alpha}(k+1)+1$ when $\alpha^{\varphi^{\alpha}(k+1)+1}=...=\alpha^p=(0,...,0)$ and we infer

$$\widehat{\Phi}^{\alpha}(\mu,p)=\Phi^{\alpha^{\varphi^{\alpha}(k+1)+1}...\alpha^p}(\Phi^{\alpha^0...\alpha^{\varphi^{\alpha}(k+1)}}(\mu))=\Phi^{\alpha^0...\alpha^{\varphi^{\alpha}(k+1)}}(\mu).$$

b) We prove (4.16):

$$\widehat{Or}^{\beta}(\mu)=\{\widehat{\Phi}^{\beta}(\mu,k)|k\in\mathbf{N}_{\_}\}\overset{(4.14)}{=}\{\widehat{\Phi}^{\alpha}(\mu,p)|k\in\mathbf{N}_{\_},p\in J_k\}$$

$$=\{\widehat{\Phi}^{\alpha}(\mu,p)|p\in\mathbf{N}_{\_}\}=\widehat{Or}^{\alpha}(\mu).$$

In order to prove (4.17), we suppose that $k_1,k_2\in\mathbf{N}_{\_}$ fulfill

$$\widehat{\omega}^{\beta}(\mu)=\{\widehat{\Phi}^{\beta}(\mu,k)|k\geq k_1\},$$

$$\widehat{\omega}^{\alpha}(\mu)=\{\widehat{\Phi}^{\alpha}(\mu,k)|k\geq k_2\}$$

and we take arbitrarily some $k'\geq\max\{k_1,k_2\}$. We infer

$$\widehat{\omega}^{\beta}(\mu)=\{\widehat{\Phi}^{\beta}(\mu,k)|k\geq k'\}\overset{(4.14)}{=}\{\widehat{\Phi}^{\alpha}(\mu,p)|k\geq k',p\in J_k\}$$

$$=\{\widehat{\Phi}^{\alpha}(\mu,p)|p\geq\varphi^{\alpha}(k')\}\overset{(4.1)}{=}\{\widehat{\Phi}^{\alpha}(\mu,p)|p\geq k'\}=\widehat{\omega}^{\alpha}(\mu).$$

c) We apply Theorem 6 a), page 11. We infer:

$$Or^{\rho'}(\mu)=\widehat{Or}^{\beta}(\mu)\overset{(4.16)}{=}\widehat{Or}^{\alpha}(\mu)=Or^{\rho}(\mu),$$

$$\omega^{\rho'}(\mu)=\widehat{\omega}^{\beta}(\mu)\overset{(4.17)}{=}\widehat{\omega}^{\alpha}(\mu)=\omega^{\rho}(\mu).$$

□

Remark 163. *Theorem 123 shows how we can work with computation functions that contain non null terms only.*

Definition 37. *We say that the computation functions* $\alpha,\beta\in\widehat{\Pi}_n$ *are* ***equivalent*** *if* $\forall k\in\mathbf{N}$,

$$\alpha^{\varphi^{\alpha}(k)}=\beta^{\varphi^{\beta}(k)}. \tag{4.20}$$

In this case we denote $\alpha\sim\beta$.

The computation functions $\rho,\rho'\in\Pi_n$ *are called* ***equivalent*** *if* $\alpha,\beta\in\widehat{\Pi}_n$ *and* $(t_k),(t'_k)\in$ *Seq exist such that (4.12), (4.13) hold and* $\alpha\sim\beta$. *We use the notation* $\rho\approx\rho'$.

Corollary 7. *Let* $\mu\in\mathbf{B}^n,\alpha,\beta\in\widehat{\Pi}_n$ *and* $\rho,\rho'\in\Pi_n$ *such that* $\alpha\sim\beta$ *and* $\rho\approx\rho'$. *We have* $\forall k\in\mathbf{N}_{\_},\widehat{\Phi}^{\alpha}(\mu,\varphi^{\alpha}(k))=\widehat{\Phi}^{\beta}(\mu,\varphi^{\beta}(k)),\ \widehat{Or}^{\alpha}(\mu)=\widehat{Or}^{\beta}(\mu),\ \widehat{\omega}^{\alpha}(\mu)=\widehat{\omega}^{\beta}(\mu),\ Or^{\rho}(\mu)=Or^{\rho'}(\mu),\ \omega^{\rho}(\mu)=\omega^{\rho'}(\mu)$.

Proof. This follows from Theorem 123, page 183. □

Remark 164. *Let* $\rho,\rho'\in\Pi_n$ *be given by*

$$\rho(t)=\alpha^0\cdot\chi_{\{t_0\}}(t)\oplus\alpha^1\cdot\chi_{\{t_1\}}(t)\oplus\ldots\oplus\alpha^k\cdot\chi_{\{t_k\}}(t)\oplus\ldots$$

$$\rho'(t)=\beta^0\cdot\chi_{\{t'_0\}}(t)\oplus\beta^1\cdot\chi_{\{t'_1\}}(t)\oplus\ldots\oplus\beta^k\cdot\chi_{\{t'_k\}}(t)\oplus\ldots$$

where $(t_k),(t'_k)\in$ *Seq and* $\alpha,\beta\in\widehat{\Pi}_n$. *We omit the null terms in the previous sums and we get*

$$\rho(t)=\alpha^{\varphi^{\alpha}(0)}\cdot\chi_{\{t_{\varphi^{\alpha}(0)}\}}(t)\oplus\alpha^{\varphi^{\alpha}(1)}\cdot\chi_{\{t_{\varphi^{\alpha}(1)}\}}(t)\oplus\ldots\oplus\alpha^{\varphi^{\alpha}(k)}\cdot\chi_{\{t_{\varphi^{\alpha}(k)}\}}(t)\oplus\ldots$$

$$\rho'(t)=\beta^{\varphi^{\beta}(0)}\cdot\chi_{\{t'_{\varphi^{\beta}(0)}\}}(t)\oplus\beta^{\varphi^{\beta}(1)}\cdot\chi_{\{t'_{\varphi^{\beta}(1)}\}}(t)\oplus\ldots\oplus\beta^{\varphi^{\beta}(k)}\cdot\chi_{\{t'_{\varphi^{\beta}(k)}\}}(t)\oplus\ldots$$

We ask that $\alpha\sim\beta$, *thus* $\rho\approx\rho'$ *and this means that* $\forall k\in\mathbf{N},\alpha^{\varphi^{\alpha}(k)}=\beta^{\varphi^{\beta}(k)}$. *We can define now the function* $h:\mathbf{R}\longrightarrow\mathbf{R}$ *by* $\forall t\in\mathbf{R}$,

$$h(t)=\begin{cases} t'_{\varphi^{\beta}(0)}+\frac{t'_{\varphi^{\beta}(1)}-t'_{\varphi^{\beta}(0)}}{t_{\varphi^{\alpha}(1)}-t_{\varphi^{\alpha}(0)}}\cdot(t-t_{\varphi^{\alpha}(0)}),\ if\, t<t_{\varphi^{\alpha}(1)},\\ t'_{\varphi^{\beta}(1)}+\frac{t'_{\varphi^{\beta}(2)}-t'_{\varphi^{\beta}(1)}}{t_{\varphi^{\alpha}(2)}-t_{\varphi^{\alpha}(1)}}\cdot(t-t_{\varphi^{\alpha}(1)}),\ if\, t\in[t_{\varphi^{\alpha}(1)},t_{\varphi^{\alpha}(2)}),\\ t'_{\varphi^{\beta}(2)}+\frac{t'_{\varphi^{\beta}(3)}-t'_{\varphi^{\beta}(2)}}{t_{\varphi^{\alpha}(3)}-t_{\varphi^{\alpha}(2)}}\cdot(t-t_{\varphi^{\alpha}(2)}),\ if\, t\in[t_{\varphi^{\alpha}(2)},t_{\varphi^{\alpha}(3)}),\\ \ldots \end{cases}$$

We get that h is continuous, strictly increasing and it fulfills $\forall k\in\mathbf{N},t'_{\varphi^{\beta}(k)}=h(t_{\varphi^{\alpha}(k)})$. *Moreover, we get* $\rho'(t)=\rho(h(t))$ *and* $\Phi^{\rho'}(\mu,t)=\Phi^{\rho}(\mu,h(t))$.

Chapter 12

A Wider Point of View: Control and Systems

In Sections 1 and 2 we show how the concept of flow is generalized in the presence of an input and by the concept of system. In fact the flows that are generated by the asynchronous iterations of the function $\Phi : \mathbf{B}^n \to \mathbf{B}^n$ can be considered as deterministic initialized autonomous asynchronous systems.

The chapter ends with a short discussion on flows and systems.

1. Control

Definition 38. *Let the function* $\Psi : \mathbf{B}^n \times \mathbf{B}^m \to \mathbf{B}^n$ *and* $\lambda \in \mathbf{B}^n$. *We define* $\Psi^\lambda : \mathbf{B}^n \times \mathbf{B}^m \to \mathbf{B}^n$ *in the following way:* $\forall \mu \in \mathbf{B}^n, \forall \nu \in \mathbf{B}^m$,

$$\Psi^\lambda(\mu,\nu) = (\overline{\lambda_1} \cdot \mu_1 \oplus \lambda_1 \cdot \Psi_1(\mu,\nu), \ldots, \overline{\lambda_n} \cdot \mu_n \oplus \lambda_n \cdot \Psi_n(\mu,\nu)).$$

Definition 39. *a) The functions* $\mathbf{B}^n \times \widehat{S}^{(m)} \times \mathbf{N}_{\_} \times \widehat{\Pi}_n \ni (\mu, \widehat{u}, k, \alpha) \longmapsto \widehat{\Psi}^{-\alpha}(\mu, \widehat{u}, k) \in \mathbf{B}^n$ *defined by* $\forall k \in \mathbf{N}_{\_}$,

$$\widehat{\Psi}^{-\alpha}(\mu,\widehat{u},k) = \begin{cases} \mu, \; if\; k = -1, \\ \Psi^{\alpha^k}(\widehat{\Psi}^{-\alpha}(\mu,\widehat{u},k-1), \widehat{u}(k-1)), if\, k \geq 0 \end{cases}$$

and $\mathbf{B}^n \times \widehat{S}^{(m)} \times \mathbf{N}_{\_} \times \widehat{\Pi}_n \ni (\mu, \widehat{u}, k, \alpha) \longmapsto \widehat{\Psi}^{\alpha}(\mu, \widehat{u}, k) \in \mathbf{B}^n$ *defined by* $\forall k \in \mathbf{N}_{\_}$,

$$\widehat{\Psi}^{\alpha}(\mu,\widehat{u},k) = \begin{cases} \mu, \; if\; k = -1, \\ \Psi^{\alpha^k}(\widehat{\Psi}^{\alpha}(\mu,\widehat{u},k-1), \widehat{u}(k)), if\, k \geq 0 \end{cases}$$

are called (discrete time) ***evolution functions***. $\mathbf{B}^n$ *is the* ***state space***, μ *is the* ***initial state,*** $\widehat{u}$ *is the* ***input*** *(or* ***control****) and* α *is the* ***computation function***. *The signals*

$$\widehat{x}(k) = \widehat{\Psi}^{-\alpha}(\mu,\widehat{u},k),$$

$$\widehat{y}(k) = \widehat{\Psi}^{\alpha}(\mu,\widehat{u},k)$$

where $k \in \mathbf{N}_{\_}$ *are called* ***states***.

b) We define the functions $\mathbf{B}^n \times S^{(m)} \times \mathbf{R} \times \Pi_n \ni (\mu, u, t, \rho) \longmapsto \Psi^{-\rho}(\mu, u, t) \in \mathbf{B}^n$, $\mathbf{B}^n \times S^{(m)} \times \mathbf{R} \times \Pi_n \ni (\mu, u, t, \rho) \longmapsto \Psi^{\rho}(\mu, u, t) \in \mathbf{B}^n$ *in the following way. Let* $\forall t \in \mathbf{R}$,

$$\rho(t) = \alpha^0 \cdot \chi_{\{t_0\}}(t) \oplus \alpha^1 \cdot \chi_{\{t_1\}}(t) \oplus \ldots \oplus \alpha^k \cdot \chi_{\{t_k\}}(t) \oplus \ldots$$

where $\alpha \in \widehat{\Pi}_n$ *and* $(t_k) \in Seq$, *and if* $\widehat{u} \in \widehat{S}^{(m)}$ *is given by* $\forall k \in \mathbf{N}_{\_}$,

$$\widehat{u}(k) = \begin{cases} u(-\infty + 0), k = -1, \\ u(t_k), k \geq 0, \end{cases}$$

then

$$\begin{aligned} \Psi^{-\rho}(\mu, u, t) = \widehat{\Psi}^{-\alpha}(\mu, \widehat{u}, -1) \cdot \chi_{(-\infty, t_0)}(t) \oplus \widehat{\Psi}^{-\alpha}(\mu, \widehat{u}, 0) \cdot \chi_{[t_0, t_1)}(t) \oplus \ldots \\ \ldots \oplus \widehat{\Psi}^{-\alpha}(\mu, \widehat{u}, k) \cdot \chi_{[t_k, t_{k+1})}(t) \oplus \ldots \end{aligned} \tag{1.1}$$

$$\begin{aligned} \Psi^{\rho}(\mu, u, t) = \widehat{\Psi}^{\alpha}(\mu, \widehat{u}, -1) \cdot \chi_{(-\infty, t_0)}(t) \oplus \widehat{\Psi}^{\alpha}(\mu, \widehat{u}, 0) \cdot \chi_{[t_0, t_1)}(t) \oplus \ldots \\ \ldots \oplus \widehat{\Psi}^{\alpha}(\mu, \widehat{u}, k) \cdot \chi_{[t_k, t_{k+1})}(t) \oplus \ldots \end{aligned} \tag{1.2}$$

These functions are called (real time) ***evolution functions***. $\mathbf{B}^n$ *is the* ***state space***, μ *is the* ***initial state,*** *u is the* ***input*** *(or* ***control****) and* ρ *is the* ***computation function***. *The signals*

$$x(t) = \Psi^{-\rho}(\mu, u, t),$$

$$y(t) = \Psi^{\rho}(\mu, u, t)$$

are called ***states***.

Definition 40. *a) We fix in the argument of the discrete time evolution functions* $\mu \in \mathbf{B}^n$, $\widehat{u} \in \widehat{S}^{(m)}$ *and* $\alpha \in \widehat{\Pi}_n$. *The signals* $\widehat{\Psi}^{-\alpha}(\mu, \widehat{u}, \cdot), \widehat{\Psi}^{\alpha}(\mu, \widehat{u}, \cdot) \in \widehat{S}^{(n)}$ *are called (discrete time)* ***flows***.

b) We fix in the argument of the real time evolution functions $\mu \in \mathbf{B}^n$, $u \in S^{(m)}$ *and* $\rho \in \Pi_n$. *The signals* $\Psi^{-\rho}(\mu, u, \cdot), \Psi^{\rho}(\mu, u, \cdot) \in S^{(n)}$ *are called (real time)* ***flows***.

Remark 165. *When an input exists, there are at least two ways of defining the flows in discrete time and at least two ways of defining the flows in real time. No essential differences exist between* $\widehat{\Psi}^{-\alpha}(\mu, \widehat{u}, \cdot)$ *and* $\widehat{\Psi}^{\alpha}(\mu, \widehat{u}, \cdot)$, *respectively between* $\Psi^{-\rho}(\mu, u, \cdot)$ *and* $\Psi^{\rho}(\mu, u, \cdot)$ *and for this reason we choose to make use in the following, without loosing the generality, of* $\widehat{\Psi}^{\alpha}(\mu, \widehat{u}, \cdot)$ *and* $\Psi^{\rho}(\mu, u, \cdot)$.

2. Systems

Notation 23. *For some set H, we use the notation*

$$P^*(H) = \{A | A \subset H, A \neq \varnothing\}.$$

Definition 41. *a) A multi-valued function* $\widehat{f} : \widehat{U} \to P^*(\widehat{S}^{(n)})$, $\widehat{U} \in P^*(\widehat{S}^{(m)})$ *is called discrete time* ***asynchronous system***. $\widehat{U}$ *is the* ***input set*** *and any* $\widehat{u} \in \widehat{U}$ *is called* ***(admissible) input***. *The functions* $\widehat{x} \in \widehat{f}(\widehat{u})$ *are called* ***(possible) states***.

b) Let $U \in P^*(S^{(m)})$. *A multi-valued function* $f : U \to P^*(S^{(n)})$ *is called real time* ***asynchronous system***. *U is called the* ***input set*** *and its elements* $u \in U$ *are called* ***(admissible) inputs***, *while the functions* $x \in f(u)$ *are called* ***(possible) states***.

Definition 42. *We consider the function* $\Psi : \mathbf{B}^n \times \mathbf{B}^m \to \mathbf{B}^n$.

a) The system $\widehat{\Xi}_\Psi : \widehat{S}^{(m)} \to P^*(\widehat{S}^{(n)})$, *defined by* $\forall \widehat{u} \in \widehat{S}^{(m)}$,

$$\widehat{\Xi}_\Psi(\widehat{u}) = \{\widehat{\Psi}^\alpha(\mu, \widehat{u}, \cdot) | \mu \in \mathbf{B}^n, \alpha \in \widehat{\Pi}_n\}$$

is called the discrete time ***universal regular asynchronous system*** *that is generated by the function* Ψ.

b) The system $\Xi_\Psi : S^{(m)} \to P^*(S^{(n)})$ *defined by* $\forall u \in S^{(m)}$,

$$\Xi_\Phi(u) = \{\Phi^\rho(\mu, u, \cdot) | \mu \in \mathbf{B}^n, \rho \in \Pi_n\}$$

is called the real time ***universal regular asynchronous system*** *that is generated by the function* Ψ.

Definition 43. *a) Let the system* $\widehat{f} : \widehat{U} \to P^*(\widehat{S}^{(n)}), \widehat{U} \in P^*(\widehat{S}^{(m)})$. *The function* $\widehat{i}_0 : \widehat{U} \to P^*(\mathbf{B}^n)$ *defined by* $\forall \widehat{u} \in \widehat{U}$,

$$\widehat{i}_0(\widehat{u}) = \{\widehat{x}(-1) | \widehat{x} \in \widehat{f}(\widehat{u})\}$$

is called the ***initial state function*** *of* $\widehat{f}$.

b) Let the system $f : U \to P^*(S^{(n)}), U \in P^*(S^{(m)})$. *The function* $i_0 : U \to P^*(\mathbf{B}^n)$ *defined by* $\forall u \in U$,

$$i_0(u) = \{x(-\infty + 0) | x \in f(u)\}$$

is called the ***initial state function*** *of* f.

Notation 24. *For* $\widehat{f}, f$ *like previously we use the notations*

$$\widehat{\Delta} = \{(\mu, \widehat{u}) | \widehat{u} \in \widehat{U}, \mu \in \widehat{i}_0(\widehat{u})\}, \tag{2.1}$$

$$\Delta = \{(\mu, u) | u \in U, \mu \in i_0(u)\}. \tag{2.2}$$

Definition 44. *a) A system* $\widehat{f} : \widehat{U} \to P^*(\widehat{S}^{(n)}), \widehat{U} \in P^*(\widehat{S}^{(m)})$ *is called* ***regular*** *if* Ψ *exists with the property that* $\forall \widehat{u} \in \widehat{U}, \widehat{f}(\widehat{u}) \subset \widehat{\Xi}_\Psi(\widehat{u})$. *In this case the function* Ψ *is called the* ***generator function*** *of* $\widehat{f}$ *and the notation is* $\widehat{f} \subset \widehat{\Xi}_\Psi$.

b) A system $f : U \to P^*(S^{(n)}), U \in P^*(S^{(m)})$ *is called* ***regular*** *if* Ψ *exists such that* $\forall u \in U, f(u) \subset \Xi_\Psi(u)$. *The function* Ψ *is called the* ***generator function*** *of* f *and we use to write* $f \subset \Xi_\Psi$.

Theorem 124. *a) If* $\widehat{f} \subset \widehat{\Xi}_\Psi$, *then the function* $\widehat{\pi} : \widehat{\Delta} \to P^*(\widehat{\Pi}_n)$ *exists such that* $\forall \widehat{u} \in \widehat{U}$,

$$\widehat{f}(\widehat{u}) = \{\widehat{\Psi}^\alpha(\mu, \widehat{u}, \cdot) | \mu \in \widehat{i}_0(\widehat{u}), \alpha \in \widehat{\pi}(\mu, \widehat{u})\}. \tag{2.3}$$

b) If $\widehat{f}$ *is defined by the existence of* $\widehat{i}_0 : \widehat{U} \to P^*(\mathbf{B}^n)$ *and* $\widehat{\pi} : \widehat{\Delta} \to P^*(\widehat{\Pi}_n)$ *such that (2.3) holds, then* $\widehat{f} \subset \widehat{\Xi}_\Psi$.

Proof. a) Let $\widehat{u} \in \widehat{U}$ arbitrary. From the fact that $\widehat{f} \subset \widehat{\Xi}_\Psi$ and $\widehat{f}(\widehat{u}) \neq \varnothing$ we infer that $\forall \mu \in \widehat{i}_0(\widehat{u})$, the set $\{\alpha | \alpha \in \widehat{\Pi}_n, \widehat{\Psi}^\alpha(\mu, \widehat{u}, \cdot) \in \widehat{f}(\widehat{u})\}$ is non-empty. We define $\widehat{\pi} : \widehat{\Delta} \to P^*(\widehat{\Pi}_n)$ by $\forall (\mu, \widehat{u}) \in \widehat{\Delta}$,

$$\widehat{\pi}(\mu, \widehat{u}) = \{\alpha | \alpha \in \widehat{\Pi}_n, \widehat{\Psi}^\alpha(\mu, \widehat{u}, \cdot) \in \widehat{f}(\widehat{u})\}.$$

Property (2.3) is fulfilled.

b) Obvious. □

Theorem 125. *a) If $f \subset \Xi_\Psi$, then the function $\pi : \Delta \to P^*(\Pi_n)$ exists such that $\forall u \in U$,*

$$f(u) = \{\Psi^\rho(\mu, u, \cdot) | \mu \in i_0(u), \rho \in \pi(\mu, u)\}. \tag{2.4}$$

b) If f is defined by the existence of $i_0 : U \to P^*(\mathbf{B}^n)$ and $\pi : \Delta \to P^*(\Pi_n)$ such that (2.4) holds, then $f \subset \Xi_\Psi$.

Definition 45. *If the initial state functions $\widehat{i_0} : \widehat{U} \to P^*(\mathbf{B}^n)$, $i_0 : U \to P^*(\mathbf{B}^n)$ of $\widehat{f}, f$ fulfill the property that $\mu \in \mathbf{B}^n$ exists with*

$$\forall \widehat{u} \in \widehat{U}, \widehat{i_0}(\widehat{u}) = \{\mu\},$$

$$\forall u \in U, i_0(u) = \{\mu\},$$

then the systems $\widehat{f}, f$ are called ***initialized****.*

Definition 46. *Let the asynchronous systems $\widehat{f} : \widehat{U} \to P^*(\widehat{S^{(n)}}), f : U \to P^*(S^{(n)})$. If $\forall \widehat{u} \in \widehat{U}, card(\widehat{f}(\widehat{u})) = 1$, then $\widehat{f}$ is called* ***deterministic*** *and it is called* ***nondeterministic*** *otherwise. If $\forall u \in U, card(f(u)) = 1$, then f is called* ***deterministic*** *and it is called* ***nondeterministic*** *otherwise.*

Definition 47. *The asynchronous systems $\widehat{f}, f$ are considered. If the set $\widehat{X} \in P^*(\widehat{S^{(n)}})$ exists such that $\forall \widehat{u} \in \widehat{U}, \widehat{f}(\widehat{u}) = \widehat{X}$, the system $\widehat{f}$ is called* ***autonomous*** *and if the set $X \in P^*(S^{(n)})$ exists such that $\forall u \in U, f(u) = X$, the system f is called* ***autonomous*** *too.*

Example 50. *Let $\Psi : \mathbf{B}^n \times \mathbf{B}^m \longrightarrow \mathbf{B}^n$ and $\mu \in \mathbf{B}^n$, $\alpha \in \widehat{\Pi}_n$, $\rho \in \Pi_n$ fixed. The associations $\widehat{S^{(m)}} \ni \widehat{u} \longrightarrow \widehat{\Psi}^\alpha(\mu, \widehat{u}, \cdot) \in \widehat{S^{(n)}}$, $S^{(m)} \ni u \longrightarrow \Psi^\rho(\mu, u, \cdot) \in S^{(n)}$ define the deterministic initialized systems $\forall \widehat{u} \in \widehat{U}$,*

$$\widehat{f}(\widehat{u}) = \{\widehat{\Psi}^\alpha(\mu, \widehat{u}, \cdot)\}$$

and $\forall u \in U$,

$$f(u) = \{\Psi^\rho(\mu, u, \cdot)\}.$$

Example 51. *Let $\Phi : \mathbf{B}^n \longrightarrow \mathbf{B}^n$ and $\mu \in \mathbf{B}^n$, $\alpha \in \widehat{\Pi}_n$, $\rho \in \Pi_n$ fixed. The flows $\widehat{\Phi}^\alpha(\mu, \cdot), \Phi^\rho(\mu, \cdot)$ define the deterministic initialized autonomous systems $\forall \widehat{u} \in \widehat{U}$,*

$$\widehat{f}(\widehat{u}) = \{\widehat{\Phi}^\alpha(\mu, \cdot)\}$$

and $\forall u \in U$,

$$f(u) = \{\Phi^\rho(\mu, \cdot)\}.$$

3. Discussion

Remark 166. *We consider the concept of nondeterministic asynchronous system as representing a multi-valued function that associates to an admissible cause $\widehat{u}, u$, several possible effects $\widehat{x} \in \widehat{f}(\widehat{u}), x \in f(u)$. The multi-valued character of the cause-effect relation represented by $\widehat{f}, f$ is given by the insufficient knowledge of the modeled object (of its delays), as well as by the existence of some variable parameters (the tension of the power supply source, the temperature of the environment) that are ignored in modeling. Modeling must take into account all the possibilities.*

Remark 167. *The regular asynchronous systems are those whose states are flows* $\widehat{x}(k) = \widehat{\Psi}^{\alpha}(\mu,\widehat{u},k), x(t) = \Psi^{\rho}(\mu,u,t)$, *where* $\mu,\widehat{u},u,\alpha,\rho$ *run through some sets of values.*

Remark 168. *The attribute 'universal' from Definition 42, page 189 means maximal in the sense of the inclusion.*

Remark 169. *A system is deterministic if the cause-effect relation is uni-valued, i.e. if* $\widehat{f}, f$ *are uni-valued functions. In general, the asynchronous systems (considered as models) are nondeterministic.*

Remark 170. *An autonomous system can be defined, in principle, in three different, non-equivalent ways (in Definition 47 our option was for the last of these possibilities):*
- *a system without input;*
- *a system with exactly one input:* $card(\widehat{U}) = card(U) = 1$;
- *a system with input, but that does not depend on the input.*

Remark 171. *The Examples 50 and 51 make us consider abusively that the functions* $\Psi : \mathbf{B}^n \times \mathbf{B}^m \longrightarrow \mathbf{B}^n$ *are deterministic initialized asynchronous systems and respectively the functions* $\Phi : \mathbf{B}^n \longrightarrow \mathbf{B}^n$ *are deterministic initialized autonomous asynchronous systems.*

Remark 172. *We discuss in further details the possibilities of defining the real time flows with control. Let* $\mu \in \mathbf{B}^n, \alpha \in \widehat{\Pi}_n, \rho \in \Pi_n, (t_k) \in Seq$ *and we suppose that*

$$\rho(t) = \alpha^0 \cdot \chi_{\{t_0\}}(t) \oplus \alpha^1 \cdot \chi_{\{t_1\}}(t) \oplus \alpha^2 \cdot \chi_{\{t_2\}}(t) \oplus ... \tag{3.1}$$

holds. The input is $u \in S^{(m)}$. *If we define* $\widehat{u} \in \widehat{S}^{(n)}$ *by* $\forall k \in \mathbf{N}_{\_}$,

$$\widehat{u}(k) = \begin{cases} u(-\infty+0),\ if\ k=-1, \\ u(t_k),\ if\ k \geq 0 \end{cases}$$

then $(1.1)_{page\ 188}$, $(1.2)_{page\ 188}$ *give*

$$\begin{aligned}\Psi^{-\rho}(\mu,u,t) = \mu \cdot \chi_{(-\infty,t_0)}(t) \oplus \Psi^{\alpha^0}(\mu,u(-\infty+0)) \cdot \chi_{[t_0,t_1)}(t) \\ \oplus \Psi^{\alpha^1}(\Psi^{\alpha^0}(\mu,u(-\infty+0)),u(t_0)) \cdot \chi_{[t_1,t_2)}(t) \oplus ...\end{aligned} \tag{3.2}$$

$$\begin{aligned}\Psi^{\rho}(\mu,u,t) = \mu \cdot \chi_{(-\infty,t_0)}(t) \oplus \Psi^{\alpha^0}(\mu,u(t_0)) \cdot \chi_{[t_0,t_1)}(t) \\ \oplus \Psi^{\alpha^1}(\Psi^{\alpha^0}(\mu,u(t_0)),u(t_1)) \cdot \chi_{[t_1,t_2)}(t) \oplus ...\end{aligned} \tag{3.3}$$

A direct definition of these flows, that does not require $\widehat{\Psi}^{-\alpha}, \widehat{\Psi}^{\alpha}$ *like Definition 39 is:*

$$\begin{aligned}\Psi^{-\rho}(\mu,u,t) = \mu \cdot \chi_{(-\infty,t_0)}(t) \oplus \Psi^{\alpha^0}(\mu,u(t_0-0)) \cdot \chi_{[t_0,t_1)}(t) \\ \oplus \Psi^{\alpha^1}(\Psi^{\alpha^0}(\mu,u(t_0-0)),u(t_1-0)) \cdot \chi_{[t_1,t_2)}(t) \oplus ...\end{aligned} \tag{3.4}$$

$$\begin{aligned}\Psi^{\rho}(\mu,u,t) = \mu \cdot \chi_{(-\infty,t_0)}(t) \oplus \Psi^{\alpha^0}(\mu,u(t_0)) \cdot \chi_{[t_0,t_1)}(t) \\ \oplus \Psi^{\alpha^1}(\Psi^{\alpha^0}(\mu,u(t_0)),u(t_1)) \cdot \chi_{[t_1,t_2)}(t) \oplus ...\end{aligned} \tag{3.5}$$

Definitions (3.4), (3.5) take in consideration on one hand that ρ *(not u!) shows how* Ψ *is computed, and on the other hand that two possibilities of computing* Ψ *exist, making use either of the values* $(u(t_k-0))_{k\in\mathbf{N}}$, *or of the values* $(u(t_k))_{k\in\mathbf{N}}$. *We have reproduced twice*

$\Psi^{\rho}(\mu,u,t)$ in (3.3) and (3.5) for the sake of the symmetry of the exposure only, there is no difference. The only difference comes by comparing $\Psi^{-\rho}(\mu,u,t)$ in (3.2) and (3.4). Indeed, $u(-\infty+0)=u(t_0-0)$, $u(t_0)=u(t_1-0),...$ are guaranteed to be true only if

$$u(t)=u(-\infty+0)\cdot\chi_{(-\infty,t_0)}(t)\oplus u(t_0)\cdot\chi_{[t_0,t_1)}(t)\oplus u(t_1)\cdot\chi_{[t_1,t_2)}(t)\oplus... \quad (3.6)$$

((t_k) is compatible with both ρ and u) otherwise (3.2) and (3.4) give different flows. Either we accept the situation, or, given the sequences $(t'_k),(t''_k)\in Seq$ with

$$\rho(t)=\alpha^0\cdot\chi_{\{t'_0\}}(t)\oplus\alpha^1\cdot\chi_{\{t'_1\}}(t)\oplus\alpha^2\cdot\chi_{\{t'_2\}}(t)\oplus... \quad (3.7)$$

$$u(t)=u(-\infty+0)\cdot\chi_{(-\infty,t''_0)}(t)\oplus u(t''_0)\cdot\chi_{[t''_0,t''_1)}(t)\oplus u(t''_1)\cdot\chi_{[t''_1,t''_2)}(t)\oplus... \quad (3.8)$$

we define $(t_k)\in Seq$ making (3.1), (3.6) true in the following way: we index the elements of the set $\{t'_k|k\in\mathbf{N}\}\cup\{t''_k|k\in\mathbf{N}\}$ in the increasing order and denote the new sequence with (t_k). Now (3.2) and (3.4) coincide.

Chapter 13

Eventually Constant Flows

In the first section we mention some generic eventual constancy properties, showing what basically results when we replace the eventual constancy of $\widehat{x}, x$ with the eventual constancy of $\widehat{\Phi}^{\alpha}(\mu,\cdot), \Phi^{\rho}(\mu,\cdot)$.

The specific eventual constancy properties of the flows $\widehat{\Phi}^{\alpha}(\mu,\cdot), \Phi^{\rho}(\mu,\cdot)$ are presented in Section 2, and they refer to the fact that the orbit contains a fixed point of Φ. The existence of this fixed point makes us call the initial values of the state eventually fixed points of Φ and this definition, together with an example, occur in Section 3.

The equivalence of the discrete time and of the real time eventual constancy is presented in Section 4.

Section 5 contains a property of eventual constancy.

1. Generic Eventual Constancy Properties

Remark 173. *Let $\Phi : \mathbf{B}^n \to \mathbf{B}^n, \mu \in \mathbf{B}^n, \alpha \in \widehat{\Pi}_n$ and $\rho \in \Pi_n$. The properties of the flows $\widehat{\Phi}^{\alpha}(\mu,\cdot), \Phi^{\rho}(\mu,\cdot)$ are two kinds: generic and specific. By generic properties we mean these properties that coincide with the (general) properties of the signals $\widehat{x}, x$ rewritten for $\widehat{x}(k) = \widehat{\Phi}^{\alpha}(\mu,k), x(t) = \Phi^{\rho}(\mu,t)$. The specific properties are the properties which are not generic. A consequence of this remark is that we can write corollaries for all of Theorem 15, Theorem 17, Theorem 18, Theorem 19 and Theorem 20 from Chapter 3, wherefrom generic properties of eventual constancy of the flows result. We give an example.*

Corollary 8. *a) The properties*

$$\exists \mu' \in \mathbf{B}^n, \exists k' \in \mathbf{N}_{\_}, \forall k \geq k', \widehat{\Phi}^{\alpha}(\mu,k) = \mu', \tag{1.1}$$

$$\exists \mu' \in \mathbf{B}^n, \exists k' \in \mathbf{N}_{\_}, \{k', k'+1, k'+2, ...\} \subset \widehat{\mathbf{T}}^{\alpha}_{\mu,\mu'}, \tag{1.2}$$

$$\exists \mu' \in \mathbf{B}^n, \widehat{\omega}^{\alpha}(\mu) = \{\mu'\}, \tag{1.3}$$

stating the eventual constancy of $\widehat{\Phi}^{\alpha}(\mu,\cdot)$ are equivalent.

b) The properties

$$\exists \mu' \in \mathbf{B}^n, \exists t' \in \mathbf{R}, \forall t \geq t', \Phi^{\rho}(\mu,t) = \mu', \tag{1.4}$$

$$\exists \mu' \in \mathbf{B}^n, \exists t' \in \mathbf{R}, [t', \infty) \subset \mathbf{T}^{\rho}_{\mu,\mu'}, \tag{1.5}$$

$$\exists \mu' \in \mathbf{B}^n, \omega^{\rho}(\mu) = \{\mu'\}, \tag{1.6}$$

stating the eventual constancy of $\Phi^{\rho}(\mu, \cdot)$ *are also equivalent.*

Proof. This is a restatement of Theorem 15, page 25. □

2. Specific Eventual Constancy Properties

Remark 174. *The properties of eventual constancy to follow are specific to the flows and they are not related with periodicity. On the other hand we could have written in (2.1),...,(2.4)* $\mu' \in \widehat{\omega}^{\alpha}(\mu), \mu' \in \omega^{\rho}(\mu)$ *instead of* $\mu' \in \widehat{Or}^{\alpha}(\mu), \mu' \in Or^{\rho}(\mu)$ *since* μ', *the final value of* $\widehat{\Phi}^{\alpha}(\mu, \cdot), \Phi^{\rho}(\mu, \cdot)$, *belongs to the* $\omega-$*limit set* $\widehat{\omega}^{\alpha}(\mu), \omega^{\rho}(\mu)$.

Theorem 126. *Let* $\mu \in \mathbf{B}^n, \alpha \in \widehat{\Pi}_n$ *and* $\rho \in \Pi_n$.

a) The following statements are equivalent with the fact that $\widehat{\Phi}^{\alpha}(\mu, \cdot)$ *is eventually constant*:

$$\exists \mu' \in \widehat{Or}^{\alpha}(\mu), \Phi(\mu') = \mu', \tag{2.1}$$

$$\exists \mu' \in \widehat{Or}^{\alpha}(\mu), \forall \lambda \in \mathbf{B}^n, \Phi^{\lambda}(\mu') = \mu'. \tag{2.2}$$

b) The following statements are equivalent with the eventual constancy of the flow $\Phi^{\rho}(\mu, \cdot)$:

$$\exists \mu' \in Or^{\rho}(\mu), \Phi(\mu') = \mu', \tag{2.3}$$

$$\exists \mu' \in Or^{\rho}(\mu), \forall \lambda \in \mathbf{B}^n, \Phi^{\lambda}(\mu') = \mu'. \tag{2.4}$$

Proof. a) (1.1)$\Longrightarrow$(2.1) $\mu' \in \mathbf{B}^n$ and $k' \in \mathbf{N}\_$ exist such that

$$\forall k \geq k', \widehat{\Phi}^{\alpha}(\mu, k) = \mu'. \tag{2.5}$$

Obviously $\mu' \in \widehat{Or}^{\alpha}(\mu)$ and we prove that $\Phi(\mu') = \mu'$. We denote $I_k = \{i | i \in \{1, ..., n\}, \alpha_i^k = 1\}, k \in \mathbf{N}$ and we know (property (6.3)$_{page\ 167}$) that $N \in \mathbf{N}$ exists with $I_{k'+1} \cup ... \cup I_{k'+N} = \{1, ..., n\}$. We have

$$\begin{aligned} \widehat{\Phi}^{\alpha}(\mu, k'+1) &= \Phi^{\alpha^{k'+1}}(\widehat{\Phi}^{\alpha}(\mu, k')) \overset{(2.5)}{=} \Phi^{\alpha^{k'+1}}(\mu') \\ &= \begin{cases} \Phi_i(\mu'), i \in I_{k'+1} \\ \mu'_i, i \in \{1, ..., n\} \setminus I_{k'+1} \end{cases} \overset{(2.5)}{=} \mu', \end{aligned} \tag{2.6}$$

$$\forall i \in I_{k'+1}, \Phi_i(\mu') \overset{(2.6)}{=} \mu'_i, \tag{2.7}$$

$$\ldots$$

$$\begin{aligned} \widehat{\Phi}^{\alpha}(\mu, k'+N) &= \Phi^{\alpha^{k'+N}}(\widehat{\Phi}^{\alpha}(\mu, k'+N-1)) \overset{(2.5)}{=} \Phi^{\alpha^{k'+N}}(\mu') \\ &= \begin{cases} \Phi_i(\mu'), i \in I_{k'+N} \\ \mu'_i, i \in \{1, ..., n\} \setminus I_{k'+N} \end{cases} \overset{(2.5)}{=} \mu', \end{aligned} \tag{2.8}$$

$$\forall i \in I_{k'+N}, \Phi_i(\mu') \overset{(2.8)}{=} \mu'_i, \tag{2.9}$$

$$\forall i \in \{1,...,n\}, \Phi_i(\mu') \overset{(2.7),...,(2.9)}{=} \mu'_i. \tag{2.10}$$

(2.1)$\Longrightarrow$(2.2) $\mu' \in \widehat{Or}^{\alpha}(\mu)$ exists such that $\Phi(\mu') = \mu'$ and let $\lambda \in \mathbf{B}^n$. We infer

$$\Phi^{\lambda}(\mu') = \begin{cases} \Phi_i(\mu'), \lambda_i = 1 \\ \mu'_i, \lambda_i = 0 \end{cases} = \mu'.$$

(2.2)$\Longrightarrow$(1.1) $\mu' \in \widehat{Or}^{\alpha}(\mu)$ exists such that

$$\forall \lambda \in \mathbf{B}^n, \Phi^{\lambda}(\mu') = \mu'. \tag{2.11}$$

Let $k' \in \mathbf{N}\_$ with the property that $\mu' = \widehat{\Phi}^{\alpha}(\mu, k')$. Then

$$\widehat{\Phi}^{\alpha}(\mu, k'+1) = \Phi^{\alpha^{k'+1}}(\widehat{\Phi}^{\alpha}(\mu, k')) = \Phi^{\alpha^{k'+1}}(\mu') \overset{(2.11)}{=} \mu', \tag{2.12}$$

$$\widehat{\Phi}^{\alpha}(\mu, k'+2) = \Phi^{\alpha^{k'+2}}(\widehat{\Phi}^{\alpha}(\mu, k'+1)) \overset{(2.12)}{=} \Phi^{\alpha^{k'+2}}(\mu') \overset{(2.11)}{=} \mu', \tag{2.13}$$

$$\dots$$

□

3. Eventually Fixed Points

Definition 48. *If $\widehat{\Phi}^{\alpha}(\mu, \cdot)$ is eventually constant, then μ is called* ***eventually*** α***–fixed point*** *of Φ (or* ***eventually*** α***– equilibrium point, eventually*** α***–rest point*** *of $\widehat{\Phi}^{\alpha}(\mu, \cdot)$).*

If $\Phi^{\rho}(\mu, \cdot)$ is eventually constant, then μ is called ***eventually*** ρ***– fixed point*** *of Φ (or* ***eventually*** ρ***–equilibrium point, eventually*** ρ***–rest point*** *of $\Phi^{\rho}(\mu, \cdot)$).*

Example 52. *For $\Phi : \mathbf{B}^2 \to \mathbf{B}^2, \forall \mu \in \mathbf{B}^2$,*

$$\Phi(\mu_1, \mu_2) = (\mu_1\overline{\mu_2} \cup \mu_2, \overline{\mu_1})$$

and

$$\rho(t) = (1,1) \cdot \chi_{\{0\}}(t) \oplus (1,1) \cdot \chi_{\{1\}}(t) \oplus (1,1) \cdot \chi_{\{2\}}(t) \oplus \dots$$

we have

$$\Phi^{\rho}((0,0),t) = (0,1) \cdot \chi_{[0,1)}(t) \oplus (1,1) \cdot \chi_{[1,2)}(t) \oplus (1,0) \cdot \chi_{[2,\infty)}(t).$$

This happens because $\Phi(1,0) = (1,0)$, i.e. $(1,0)$ is a fixed point of Φ and $(0,0)$ is an eventually ρ–fixed point of Φ. In fact, a brief study shows that $(0,0)$ is eventually ρ–fixed point of Φ for any $\rho \in \Pi_2$.

4. Discrete Time vs Real Time

Theorem 127. *$\mu \in \mathbf{B}^n, \alpha \in \widehat{\Pi}_n$ and $\rho \in \Pi_n$ are given. If*

$$\rho(t) = \alpha^0 \cdot \chi_{\{t_0\}}(t) \oplus \alpha^1 \cdot \chi_{\{t_1\}}(t) \oplus ... \oplus \alpha^k \cdot \chi_{\{t_k\}}(t) \oplus ... \tag{4.1}$$

holds for some $(t_k) \in Seq$, then the eventual constancies of $\widehat{\Phi}^{\alpha}(\mu, \cdot)$ and $\Phi^{\rho}(\mu, \cdot)$ are equivalent.

Proof. The eventual constancies of $\widehat{\Phi}^{\alpha}(\mu, \cdot)$ and $\Phi^{\rho}(\mu, \cdot)$ are equivalent from Theorem 126, page 194, both of them, with the existence of $\mu' \in \widehat{Or}^{\alpha}(\mu) = Or^{\rho}(\mu)$ that fulfills $\Phi(\mu') = \mu'$. □

Corollary 9. *Let μ and α, ρ, (t_k) making (4.1) true. If $\widehat{\Phi}^{\alpha}(\mu, \cdot)$ and $\Phi^{\rho}(\mu, \cdot)$ are eventually constant, then $\mu' \in \mathbf{B}^n$ exists such that*

$$\widehat{\omega}^{\alpha}(\mu) = \omega^{\rho}(\mu) = \{\mu'\}, \tag{4.2}$$

$$\lim_{k \to \infty} \widehat{\Phi}^{\alpha}(\mu, k) = \lim_{t \to \infty} \Phi^{\rho}(\mu, t) = \mu'. \tag{4.3}$$

Moreover, we have that $\Phi(\mu') = \mu'$.

Proof. The fact that $\widehat{\omega}^{\alpha}(\mu) = \omega^{\rho}(\mu)$ is known from Theorem 6, page 11. Corollary 8, page 193 implies the existence of $\mu' \in \mathbf{B}^n$ such that (4.2), (4.3) hold and $\Phi(\mu') = \mu'$ results from Theorem 126, page 194. □

5. A Property of Eventual Constancy

Theorem 128. *Let $\mu \in \mathbf{B}^n, \alpha \in \widehat{\Pi}_n$ and $\rho \in \Pi_n$ be given.*

a) If μ is an eventually $\alpha-$fixed point of Φ and $\widetilde{\mu} = \widehat{\Phi}^{\alpha}(\mu, \widetilde{k}), \widetilde{k} \geq -1$ then $\widetilde{\mu}$ is eventually $\widehat{\sigma}^{\widetilde{k}+1}(\alpha)-$fixed point of Φ; in this case, if k' is a limit of equilibrium of $\widehat{\Phi}^{\alpha}(\mu, \cdot)$, then $\max\{-1, k' - \widetilde{k} - 1\}$ is a limit of equilibrium of $\widehat{\Phi}^{\widehat{\sigma}^{\widetilde{k}+1}(\alpha)}(\widetilde{\mu}, \cdot)$.

b) We presume that μ is an eventually $\rho-$fixed point of Φ and $\widetilde{\mu} = \Phi^{\rho}(\mu, \widetilde{t} - 0), \widetilde{t} \in \mathbf{R}$; then $\widetilde{\mu}$ is an eventually $\sigma^{\widetilde{t}}(\rho)-$fixed point of Φ and if t' is a limit of equilibrium of $\Phi^{\rho}(\mu, \cdot)$, we have that $\max\{t', \widetilde{t}\}$ is a limit of equilibrium of $\Phi^{\sigma^{\widetilde{t}}(\rho)}(\widetilde{\mu}, \cdot)$.

Proof. a) If $\exists \lim_{k \to \infty} \widehat{\Phi}^{\alpha}(\mu, k)$ and

$$\widetilde{\mu} = \widehat{\Phi}^{\alpha}(\mu, \widetilde{k}), \tag{5.1}$$

then we must prove that $\exists \lim_{k \to \infty} \widehat{\Phi}^{\widehat{\sigma}^{\widetilde{k}+1}(\alpha)}(\widetilde{\mu}, k)$. More exactly, for $\widetilde{\mu} \in \widehat{Or}^{\alpha}(\mu)$, $k', \widetilde{k} \in \mathbf{N}_{\_}$ like in the hypothesis, we have the existence of $\mu' \in \widehat{Or}^{\alpha}(\mu)$ such that

$$\forall k \geq k', \widehat{\Phi}^{\alpha}(\mu, k) = \mu', \tag{5.2}$$

and we must prove that $\forall k \geq \max\{-1, k' - \widetilde{k} - 1\}$,

$$\widehat{\Phi}^{\widehat{\sigma}^{\widetilde{k}+1}(\alpha)}(\widetilde{\mu}, k) = \mu' \tag{5.3}$$

is true. Let for this $k \geq \max\{-1, k' - \widetilde{k} - 1\}$ be arbitrary, fixed. As far as $k + \widetilde{k} + 1 \geq k'$, we can make use of (5.2) and we have

$$\widehat{\Phi}^{\widehat{\sigma}^{\widetilde{k}+1}(\alpha)}(\widetilde{\mu}, k) \overset{(5.1)}{=} \widehat{\Phi}^{\widehat{\sigma}^{\widetilde{k}+1}(\alpha)}(\widehat{\Phi}^{\alpha}(\mu, \widetilde{k}), k)$$

$$\overset{(3.4)_{page\ 178}}{=} \widehat{\sigma}^{\widetilde{k}+1}(\widehat{\Phi}^{\alpha}(\mu, \cdot))(k) = \widehat{\Phi}^{\alpha}(\mu, k + \widetilde{k} + 1) \overset{(5.2)}{=} \mu'.$$

b) We take $\widetilde{\mu} \in Or^{\rho}(\mu)$, $t', \widetilde{t} \in \mathbf{R}$ having the property that $\mu' \in Or^{\rho}(\mu)$ exists with

$$\forall t \geq t', \Phi^{\rho}(\mu, t) = \mu', \tag{5.4}$$

$$\widetilde{\mu} = \Phi^{\rho}(\mu, \widetilde{t} - 0) \tag{5.5}$$

and the truth of $\forall t \geq \max\{t', \widetilde{t}\}$,

$$\Phi^{\sigma^{\widetilde{t}}(\rho)}(\widetilde{\mu}, t) = \mu' \tag{5.6}$$

is to be proved. Let $t \geq \max\{t', \widetilde{t}\}$ be arbitrary. We infer

$$\Phi^{\sigma^{\widetilde{t}}(\rho)}(\widetilde{\mu}, t) \overset{(5.5)}{=} \Phi^{\sigma^{\widetilde{t}}(\rho)}(\Phi^{\rho}(\mu, \widetilde{t} - 0), t)$$

$$\overset{(3.5)_{page\ 178}}{=} \sigma^{\widetilde{t}}(\Phi^{\rho}(\mu, \cdot))(t) = \Phi^{\rho}(\mu, t) \overset{(5.4)}{=} \mu'.$$

□

Chapter 14

Constant Flows

In the first section we show how the generic properties of constancy of the flows can be constructed.

The second section presents some specific constancy properties. The point is here that the constant flows $\widehat{\Phi}^{\alpha}(\mu,\cdot),\Phi^{\rho}(\mu,\cdot)$ are exactly those for which $\Phi(\mu)=\mu$. This implies the conclusion represented by the fourth section, that the discrete time constancy of $\widehat{\Phi}^{\alpha}(\mu,\cdot)$ and the real time constancy of $\Phi^{\rho}(\mu,\cdot)$ are equivalent.

In the last section we give two further properties of constancy.

1. Generic Constancy Properties

Remark 175. *Let $\Phi:\mathbf{B}^n\to\mathbf{B}^n,\mu\in\mathbf{B}^n,\alpha\in\widehat{\Pi}_n$ and $\rho\in\Pi_n$. The generic properties of constancy of the flows result by the replacement in Theorem 22, Theorem 23, Theorem 24 and Theorem 25 from Chapter 4 of the signals $\widehat{x},x$ with $\widehat{\Phi}^{\alpha}(\mu,\cdot),\Phi^{\rho}(\mu,\cdot)$. Here is an example.*

Corollary 10. *a) The following requirements stating the constancy of $\widehat{\Phi}^{\alpha}(\mu,\cdot)$ are equivalent:*

$$\forall k\in\mathbf{N}_{\_},\widehat{\Phi}^{\alpha}(\mu,k)=\mu, \tag{1.1}$$

$$\widehat{\mathbf{T}}^{\alpha}_{\mu,\mu}=\mathbf{N}_{\_}, \tag{1.2}$$

$$\widehat{Or}^{\alpha}(\mu)=\{\mu\}. \tag{1.3}$$

b) The following requirements stating the constancy of $\Phi^{\rho}(\mu,\cdot)$ are also equivalent:

$$\forall t\in\mathbf{R},\Phi^{\rho}(\mu,t)=\mu, \tag{1.4}$$

$$\mathbf{T}^{\rho}_{\mu,\mu}=\mathbf{R}, \tag{1.5}$$

$$Or^{\rho}(\mu)=\{\mu\}. \tag{1.6}$$

Proof. This is a restatement of Theorem 22, page 45 with $\widehat{x}(k)=\widehat{\Phi}^{\alpha}(\mu,k),k\in\mathbf{N}_{\_}$ and $x(t)=\Phi^{\rho}(\mu,t),t\in\mathbf{R}$. □

2. Specific Constancy Properties

Remark 176. *Theorem 129 contains specific constancy properties only, (2.1) and (2.2) cannot be even stated in the general case of the signals. All these properties are timeless.*

Theorem 129. *Let $\Phi : \mathbf{B}^n \to \mathbf{B}^n, \mu \in \mathbf{B}^n, \alpha \in \widehat{\Pi}_n$ and $\rho \in \Pi_n$. Any of the following statements is equivalent with the constancy of both $\widehat{\Phi}^{\alpha}(\mu, \cdot)$ and $\Phi^{\rho}(\mu, \cdot)$:*

$$\Phi(\mu) = \mu, \tag{2.1}$$

$$\forall \lambda \in \mathbf{B}^n, \Phi^{\lambda}(\mu) = \mu, \tag{2.2}$$

$$\widehat{\omega}^{\alpha}(\mu) = \{\mu\}, \tag{2.3}$$

$$\omega^{\rho}(\mu) = \{\mu\}. \tag{2.4}$$

Proof. (1.1)$\Longrightarrow$(2.1) We denote $I_k = \{i | i \in \{1, ..., n\}, \alpha_i^k = 1\}, k \in \mathbf{N}$ and we notice, from the fact that α is progressive, that $N \in \mathbf{N}$ exists with $I_0 \cup I_1 \cup ... \cup I_N = \{1, ..., n\}$. We have:

$$\widehat{\Phi}^{\alpha}(\mu, 0) = \Phi^{\alpha^0}(\mu) = \begin{cases} \Phi_i(\mu), i \in I_0 \\ \mu_i, i \in \{1, ..., n\} \setminus I_0 \end{cases} \overset{(1.1)}{=} \mu, \tag{2.5}$$

$$\forall i \in I_0, \Phi_i(\mu) \overset{(2.5)}{=} \mu_i, \tag{2.6}$$

$$...$$

$$\widehat{\Phi}^{\alpha}(\mu, N) = \Phi^{\alpha^N}(\widehat{\Phi}^{\alpha}(\mu, N-1)) = \Phi^{\alpha^N}(\mu)$$
$$= \begin{cases} \Phi_i(\mu), i \in I_N \\ \mu_i, i \in \{1, ..., n\} \setminus I_N \end{cases} \overset{(1.1)}{=} \mu, \tag{2.7}$$

$$\forall i \in I_N, \Phi_i(\mu) \overset{(2.7)}{=} \mu_i, \tag{2.8}$$

$$\forall i \in \{1, ..., n\}, \Phi_i(\mu) \overset{(2.6), ..., (2.8)}{=} \mu_i. \tag{2.9}$$

(2.1)$\Longrightarrow$(2.2) Let $\lambda \in \mathbf{B}^n$ arbitrary. We see that

$$\Phi^{\lambda}(\mu) = \begin{cases} \Phi_i(\mu), i \in \{1, ..., n\}, \lambda_i = 1 \\ \mu_i, i \in \{1, ..., n\}, \lambda_i = 0 \end{cases} \overset{(2.1)}{=} \mu.$$

(2.2)$\Longrightarrow$(2.3) We compute in succession:

$$\widehat{\Phi}^{\alpha}(\mu, 0) = \Phi^{\alpha^0}(\widehat{\Phi}^{\alpha}(\mu, -1)) = \Phi^{\alpha^0}(\mu) \overset{(2.2)}{=} \mu, \tag{2.10}$$

$$\widehat{\Phi}^{\alpha}(\mu, 1) = \Phi^{\alpha^1}(\widehat{\Phi}^{\alpha}(\mu, 0)) \overset{(2.10)}{=} \Phi^{\alpha^1}(\mu) \overset{(2.2)}{=} \mu, \tag{2.11}$$

$$\widehat{\Phi}^{\alpha}(\mu, 2) = \Phi^{\alpha^2}(\widehat{\Phi}^{\alpha}(\mu, 1)) \overset{(2.11)}{=} \Phi^{\alpha^2}(\mu) \overset{(2.2)}{=} \mu, \tag{2.12}$$

$$...$$

thus $\widehat{Or}^{\alpha}(\mu) = \{\mu\}$. As $\emptyset \neq \widehat{\omega}^{\alpha}(\mu) \subset \widehat{Or}^{\alpha}(\mu)$, we infer that $\widehat{\omega}^{\alpha}(\mu) = \{\mu\}$.

(2.3)$\Longrightarrow$(2.4) Some $k' \in \mathbf{N}\_$ exists such that $\widehat{\omega}^{\alpha}(\mu) = \{\widehat{\Phi}^{\alpha}(\mu,k)|k \geq k'\}$ thus, from the hypothesis,

$$\forall k \geq k', \widehat{\Phi}^{\alpha}(\mu,k) = \mu. \tag{2.13}$$

We infer similarly with the proof (1.1)$\Longrightarrow$(2.1) the fact that (2.13) implies $\Phi(\mu) = \mu$, i.e. $N \geq 1$ exists having the property that $I_{k'+1} \cup I_{k'+2} \cup ... \cup I_{k'+N} = \{1,...,n\}$ and we have

$$\widehat{\Phi}^{\alpha}(\mu,k'+1) = \Phi^{\alpha^{k'+1}}(\widehat{\Phi}^{\alpha}(\mu,k')) \overset{(2.13)}{=} \Phi^{\alpha^{k'+1}}(\mu)$$

$$= \begin{cases} \Phi_i^{\alpha^{k'+1}}(\mu), i \in I_{k'+1} \\ \mu_i, i \in \{1,...,n\} \setminus I_{k'+1} \end{cases} \overset{(2.13)}{=} \mu,$$

wherefrom $\forall i \in I_{k'+1}, \Phi_i(\mu) = \mu_i$ etc.

We show that $\Phi(\mu) = \mu$ implies $\omega^{\rho}(\mu) = \{\mu\}$ and we suppose for this that

$$\rho(t) = \beta^0 \cdot \chi_{\{t_0\}}(t) \oplus \beta^1 \cdot \chi_{\{t_1\}}(t) \oplus ... \oplus \beta^k \cdot \chi_{\{t_k\}}(t) \oplus ..., \tag{2.14}$$

where $\beta \in \widehat{\Pi}_n, (t_k) \in Seq$. We can see that

$$\widehat{\Phi}^{\beta}(\mu,0) = \Phi^{\beta^0}(\mu) = \begin{cases} \Phi_i(\mu), i \in \{1,...,n\}, \beta_i^0 = 1 \\ \mu_i, i \in \{1,...,n\}, \beta_i^0 = 0 \end{cases} = \mu, \tag{2.15}$$

$$\begin{aligned} \widehat{\Phi}^{\beta}(\mu,1) &= \Phi^{\beta^1}(\widehat{\Phi}^{\beta}(\mu,0)) \overset{(2.15)}{=} \Phi^{\beta^1}(\mu) \\ &= \begin{cases} \Phi_i(\mu), i \in \{1,...,n\}, \beta_i^1 = 1 \\ \mu_i, i \in \{1,...,n\}, \beta_i^1 = 0 \end{cases} = \mu, \end{aligned} \tag{2.16}$$

$$\begin{aligned} \widehat{\Phi}^{\beta}(\mu,2) &= \Phi^{\beta^2}(\widehat{\Phi}^{\beta}(\mu,1)) \overset{(2.16)}{=} \Phi^{\beta^2}(\mu) \\ &= \begin{cases} \Phi_i(\mu), i \in \{1,...,n\}, \beta_i^2 = 1 \\ \mu_i, i \in \{1,...,n\}, \beta_i^2 = 0 \end{cases} = \mu, \end{aligned} \tag{2.17}$$

$$...$$

thus

$$\begin{aligned} \Phi^{\rho}(\mu,t) = \widehat{\Phi}^{\beta}(\mu,-1) \cdot \chi_{(-\infty,t_0)}(t) \oplus \widehat{\Phi}^{\beta}(\mu,0) \cdot \chi_{[t_0,t_1)}(t) \oplus ... \\ \oplus \widehat{\Phi}^{\beta}(\mu,k) \cdot \chi_{[t_k,t_{k+1})}(t) \oplus ... = \mu. \end{aligned}$$

We have obtained that $Or^{\rho}(\mu) = \{\mu\}$, wherefrom $\varnothing \neq \omega^{\rho}(\mu) \subset Or^{\rho}(\mu)$ and $\omega^{\rho}(\mu) = \{\mu\}$.

(2.4)$\Longrightarrow$(1.1) We suppose that ρ is of the form (2.14), where $\beta \in \widehat{\Pi}_n, (t_k) \in Seq$. Some $t' \in \mathbf{R}$ exists with $\omega^{\rho}(\mu) = \{\Phi^{\rho}(\mu,t)|t \geq t'\}$ and we can presume without loss that t' is a term of the sequence (t_k), thus we can suppose that $\omega^{\rho}(\mu) = \{\Phi^{\rho}(\mu,t)|t \geq t_{k'}\}$. From

$$\Phi^{\rho}(\mu,t) = \widehat{\Phi}^{\beta}(\mu,-1) \cdot \chi_{(-\infty,t_0)}(t) \oplus \widehat{\Phi}^{\beta}(\mu,0) \cdot \chi_{[t_0,t_1)}(t) \oplus ... \oplus \widehat{\Phi}^{\beta}(\mu,k) \cdot \chi_{[t_k,t_{k+1})}(t) \oplus ...$$

we obtain, taking into account the hypothesis (2.4), that

$$\{\mu\} = \omega^{\rho}(\mu) = \{\Phi^{\rho}(\mu,t)|t \geq t_{k'}\} = \{\widehat{\Phi}^{\beta}(\mu,k)|k \geq k'\},$$

i.e.

$$\forall k \geq k', \widehat{\Phi}^{\beta}(\mu,k) = \mu. \tag{2.18}$$

Similarly with the proof of (2.3)$\Longrightarrow$(2.4) (and (1.1)$\Longrightarrow$(2.1)) it is shown that (2.18) implies $\Phi(\mu) = \mu$. Furthermore, this easily implies that (1.1) is fulfilled. □

3. Fixed Points

Definition 49. *If the flow $\widehat{\Phi}^{\alpha}(\mu,\cdot)$ is constant, then μ is called a* ***fixed point*** *of Φ (or an* ***equilibrium point****, or a* ***rest point of*** $\widehat{\Phi}^{\alpha}(\mu,\cdot)$*).*

If $\Phi^{\rho}(\mu,\cdot)$ is a constant flow, then μ is called a ***fixed point*** *of Φ (or an* ***equilibrium point****, or a* ***rest point of*** $\Phi^{\rho}(\mu,\cdot)$*).*

Example 53. *The identity function $1_{\mathbf{B}^n}$ has the property that any $\mu \in \mathbf{B}^n$ is a fixed point. The flows $(1_{\mathbf{B}^n})^{\alpha}(\mu,\cdot),(1_{\mathbf{B}^n})^{\rho}(\mu,\cdot)$ are all constant, when $\mu \in \mathbf{B}^n,\alpha \in \widehat{\Pi}_n$ and $\rho \in \Pi_n$.*

Example 54. *The constant function*

$$\exists \mu' \in \mathbf{B}^n, \forall \mu \in \mathbf{B}^n, \Phi(\mu) = \mu'$$

has exactly one fixed point, μ'. For any choice of $\alpha \in \widehat{\Pi}_n, \rho \in \Pi_n$, the flows $\widehat{\Phi}^{\alpha}(\mu,\cdot),\Phi^{\rho}(\mu,\cdot)$ are variable (eventually constant) when $\mu \neq \mu'$ and constant otherwise.

Example 55. *We consider the signal $x \in S^{(1)}$,*

$$x(t) = \chi_{[0,1)}(t).$$

We have $x(-\infty+0) = 0$ and $\omega(x) = \{0\}$, i.e.(2.4) is fulfilled. As x is not constant, functions $\Phi : \mathbf{B} \to \mathbf{B}, \rho \in \Pi_1$ with the property $x(t) = \Phi^{\rho}(0,t)$ do not exist.

4. Discrete Time vs Real Time

Remark 177. *The relation*

$$\rho(t) = \alpha^0 \cdot \chi_{\{t_0\}}(t) \oplus \alpha^1 \cdot \chi_{\{t_1\}}(t) \oplus ... \oplus \alpha^k \cdot \chi_{\{t_k\}}(t) \oplus ... \tag{4.1}$$

between $\alpha \in \widehat{\Pi}_n$ and $\rho \in \Pi_n$, where $(t_k) \in Seq$, is superfluous in order that the constancy of $\widehat{\Phi}^{\alpha}(\mu,\cdot)$ is equivalent with the constancy of $\Phi^{\rho}(\mu,\cdot)$: (2.1)$_{page\,200}$ of Theorem 129, where α and ρ do not occur shows that the discrete time and the real time constancies of $\widehat{\Phi}^{\alpha}(\mu,\cdot)$ and $\Phi^{\rho}(\mu,\cdot)$ are always equivalent.

5. Properties of Constancy

Theorem 130. *Let $\mu \in \mathbf{B}^n, \alpha \in \widehat{\Pi}_n, \rho \in \Pi_n$.*

a) $\widehat{\Phi}^{\alpha}(\mu,\cdot)$ is constant if and only if it is eventually constant with the initial value=the final value.

b) $\Phi^{\rho}(\mu,\cdot)$ is constant if and only if it is eventually constant with the initial value=the final value.

Proof. The real time equivalence is proved like this. We have $\Phi^{\rho}(\mu,\cdot)$ is constant $\Longrightarrow$ $Or^{\rho}(\mu) = \{\mu\} \Longrightarrow \omega^{\rho}(\mu) = \{\mu\}$, stating the eventual constancy of $\Phi^{\rho}(\mu,\cdot)$ with the initial state equal to the final state.

Conversely, $\Phi^{\rho}(\mu,\cdot)$ is eventually constant with the initial state equal to the final state $\overset{Corollary\ 8,\ page\ 193}{\Longrightarrow} \omega^{\rho}(\mu) = \{\mu\} \overset{Theorem\ 129,\ page\ 200}{\Longrightarrow} \Phi^{\rho}(\mu,\cdot)$ is constant. □

Theorem 131. *Let $\mu \in \mathbf{B}^n, \alpha \in \widehat{\Pi}_n, \rho \in \Pi_n$.*

a) The following statements are equivalent:

a.1) μ is a fixed point of Φ;

a.2) μ is an eventually $\alpha-$fixed point of Φ with the property that any $k' \in \mathbf{N}\_$ is a limit of equilibrium;

a.3) μ is an eventually $\rho-$fixed point of Φ having the property that any $t' \in \mathbf{R}$ is a limit of equilibrium.

b) Any of a.1), a.2), a.3) implies any of:

b.1) μ is an eventually $\beta-$fixed point of Φ, $\forall \beta \in \widehat{\Pi}_n$;

b.2) μ is an eventually $\rho'-$fixed point of Φ, $\forall \rho' \in \Pi_n$.

Proof. a) Let us notice first of all the equivalence of the statements:

$$\exists \mu' \in \widehat{Or}^{\alpha}(\mu), \forall k' \in \mathbf{N}_{\_}, \forall k \geq k', \widehat{\Phi}^{\alpha}(\mu, k) = \mu', \tag{5.1}$$

$$\forall k \in \mathbf{N}_{\_}, \widehat{\Phi}^{\alpha}(\mu, k) = \mu, \tag{5.2}$$

$$\Phi(\mu) = \mu, \tag{5.3}$$

$$\forall t \in \mathbf{R}, \Phi^{\rho}(\mu, t) = \mu, \tag{5.4}$$

$$\exists \mu' \in Or^{\rho}(\mu), \forall t' \in \mathbf{R}, \forall t \geq t', \Phi^{\rho}(\mu, t) = \mu'. \tag{5.5}$$

Indeed, if we take in (5.1) $k' = -1$, then $\mu' = \mu$ results and (5.2) is true. The implication (5.2)$\Longrightarrow$(5.3) coincides with $(1.1)_{page\ 199} \overset{Theorem\ 129}{\Longrightarrow} (2.1)_{page\ 200}$. The implication (5.3)$\Longrightarrow$(5.1) is obvious if we take $\mu' = \mu$ and then, for an arbitrary $k' \in \mathbf{N}_{\_}$, we prove by induction on k that $\forall k \geq k', \widehat{\Phi}^{\alpha}(\mu, k) = \mu$ holds. The implications (5.5)$\Longrightarrow$(5.4)$\Longrightarrow$(5.3)$\Longrightarrow$(5.5) are proved similarly.

But (5.1), (5.3), (5.5) coincide with the statements a.2), a.1), a.3).

b) is obvious. □

Chapter 15

The Periodicity of the Flows

The Sections 1, 2, 4, 5 are dedicated to the eventually periodic values of the flows, the eventually periodic flows, the periodic values of the flows and the periodic flows. They result by a brief run through the issues of Chapter 5, Chapter 6, Chapter 7, Chapter 8, by replacing $\widehat{x}, x$ with $\widehat{\Phi}^{\alpha}(\mu,\cdot), \Phi^{\rho}(\mu,\cdot)$. Sections 3 and 6 introduce double eventual periodicity and respectively double periodicity, i.e. the situation when the computation functions and the flows are together eventually periodic, respectively they are together periodic. The fundamental result is represented by Theorem 135, stating that double eventual periodicity is implied by the eventual periodicity of the computation functions.

1. Eventually Periodic Values of the Flows

Remark 178. *Let $\mu \in \mathbf{B}^n, \alpha \in \widehat{\Pi}_n, \rho \in \Pi_n$. The defining equivalent properties of eventual periodicity of $\mu' \in \widehat{\omega}^{\alpha}(\mu), \mu' \in \omega^{\rho}(\mu)$ with the period $p \geq 1, T > 0$ are inferred from Theorem 27, page 66, for $\widehat{x}(k) = \widehat{\Phi}^{\alpha}(\mu,k)$ and $x(t) = \Phi^{\rho}(\mu,t)$. We give the example of $(1.1)_{page\,66}$, $(1.6)_{page\,66}$ that become:*

$$\begin{cases} \exists k' \in \mathbf{N}_{\_}, \forall k \in \widehat{\mathbf{T}}^{\alpha}_{\mu,\mu'} \cap \{k', k'+1, k'+2, ...\}, \\ \{k+zp | z \in \mathbf{Z}\} \cap \{k', k'+1, k'+2, ...\} \subset \widehat{\mathbf{T}}^{\alpha}_{\mu,\mu'}, \end{cases} \tag{1.1}$$

$$\exists t'_1 \in \mathbf{R}, \forall t \in \mathbf{T}^{\rho}_{\mu,\mu'} \cap [t'_1, \infty), \{t+zT | z \in \mathbf{Z}\} \cap [t'_1, \infty) \subset \mathbf{T}^{\rho}_{\mu,\mu'}. \tag{1.2}$$

Remark 179. *When rewriting Theorem 27 we use the fact that for any $k'' \in \mathbf{N}$ and any $t'' \in \mathbf{R}$ we infer*

$$\widehat{\mathbf{T}}^{\widehat{\sigma}^{k''}(\widehat{x})}_{\mu'} = \widehat{\mathbf{T}}^{\widehat{\sigma}^{k''}(\widehat{\Phi}^{\alpha}(\mu,\cdot))}_{\mu'} = \widehat{\mathbf{T}}^{\widehat{\Phi}^{\widehat{\sigma}^{k''}(\alpha)}(\widehat{\Phi}^{\alpha}(\mu,k''-1),\cdot)}_{\mu'} = \widehat{\mathbf{T}}^{\widehat{\sigma}^{k''}(\alpha)}_{\widehat{\Phi}^{\alpha}(\mu,k''-1),\mu'},$$

$$I^{\sigma^{t''}(x)} = I^{\sigma^{t''}(\Phi^{\rho}(\mu,\cdot))} = I^{\Phi^{\sigma^{t''}(\rho)}(\Phi^{\rho}(\mu,t''-0),\cdot)} = I^{\sigma^{t''}(\rho)}_{\Phi^{\rho}(\mu,t''-0)},$$

$$\mathbf{T}^{\sigma^{t''}(x)}_{\mu'} = \mathbf{T}^{\sigma^{t''}(\Phi^{\rho}(\mu,\cdot))}_{\mu'} = \mathbf{T}^{\Phi^{\sigma^{t''}(\rho)}(\Phi^{\rho}(\mu,t''-0),\cdot)}_{\mu'} = \mathbf{T}^{\sigma^{t''}(\rho)}_{\Phi^{\rho}(\mu,t''-0),\mu'}.$$

Remark 180. *We suppose that $\mu' \in \widehat{\omega}^{\alpha}(\mu), \mu' \in \omega^{\rho}(\mu)$ are eventually periodic. Then Theorem 30, page 70 shows that the least $k' \in \mathbf{N}\_$ such that*

$$\begin{cases} \forall k \in \widehat{\mathbf{T}}^{\alpha}_{\mu,\mu'} \cap \{k', k'+1, k'+2, ...\}, \\ \{k+zp | z \in \mathbf{Z}\} \cap \{k', k'+1, k'+2, ...\} \subset \widehat{\mathbf{T}}^{\alpha}_{\mu,\mu'} \end{cases}$$

holds -exists and, if $\Phi^{\rho}(\mu, \cdot)$ is not constant, then the least $t'_1 \in \mathbf{R}$ such that

$$\forall t \in \mathbf{T}^{\rho}_{\mu,\mu'} \cap [t'_1, \infty), \{t + zT | z \in \mathbf{Z}\} \cap [t'_1, \infty) \subset \mathbf{T}^{\rho}_{\mu,\mu'}$$

holds -exists.

Remark 181. *We presume -see Theorem 32, page 76- that*

$$\rho(t) = \alpha^0 \cdot \chi_{\{t_0\}}(t) \oplus \alpha^1 \cdot \chi_{\{t_0+h\}}(t) \oplus ... \oplus \alpha^k \cdot \chi_{\{t_0+kh\}}(t) \oplus ... \tag{1.3}$$

is true, where $t_0 \in \mathbf{R}$, $h > 0$ and we take some $\mu' \in \widehat{\omega}^{\alpha}(\mu) = \omega^{\rho}(\mu)$, where $\Phi(\mu') \neq \mu'$ (this avoids the eventual constancy of the flows). If $p \geq 1, k' \in \mathbf{N}\_$ exist such that

$$\forall k \in \widehat{\mathbf{T}}^{\alpha}_{\mu,\mu'} \cap \{k', k'+1, k'+2, ...\}, \{k+zp | z \in \mathbf{Z}\} \cap \{k', k'+1, k'+2, ...\} \subset \widehat{\mathbf{T}}^{\alpha}_{\mu,\mu'} \tag{1.4}$$

then $t' \in \mathbf{R}$ exists making

$$\forall t \in \mathbf{T}^{\rho}_{\mu,\mu'} \cap [t', \infty), \{t + zT | z \in \mathbf{Z}\} \cap [t', \infty) \subset \mathbf{T}^{\rho}_{\mu,\mu'} \tag{1.5}$$

true for $T = ph$. And if $T > 0$ and $t' \in \mathbf{R}$ exist for which (1.5) is true, then $\frac{T}{h} \in \{1, 2, 3, ...\}$ and $k' \in \mathbf{N}\_$ exists such that (1.4) is true for $p = \frac{T}{h}$.

Remark 182. *Let $\mu' \in \widehat{\omega}^{\alpha}(\mu), \mu' \in \omega^{\rho}(\mu)$ be eventually periodic. The fact that the sum, the difference and the multiples of periods are periods is inferred from Theorem 33, page 78 and Theorem 34, page 79.*

Notation 25. *We denote with $\widehat{P}^{\alpha}_{\mu,\mu'}$, the set of the periods of $\mu' \in \widehat{\omega}^{\alpha}(\mu)$:*

$$\widehat{P}^{\alpha}_{\mu,\mu'} = \{p | p \geq 1, \text{(1.1) holds}\},$$

and with $P^{\rho}_{\mu,\mu'}$ the set of the periods of $\mu' \in \omega^{\rho}(\mu)$:

$$P^{\rho}_{\mu,\mu'} = \{T | T > 0, \text{(1.2) holds}\}.$$

Remark 183. *The form of the sets of periods of $\mu' \in \widehat{\omega}^{\alpha}(\mu), \mu' \in \omega^{\rho}(\mu)$, in particular the existence of the prime period $\widetilde{p}, \widetilde{T}$ are inferred from Theorem 35, page 79. We obviously have*

$$\widehat{P}^{\alpha}_{\mu,\mu'} = \{\widetilde{p}, 2\widetilde{p}, 3\widetilde{p}, ...\},$$

$$P^{\rho}_{\mu,\mu'} = \{\widetilde{T}, 2\widetilde{T}, 3\widetilde{T}, ...\}$$

when $\Phi^{\rho}(\mu, \cdot)$ is not eventually constant.

Remark 184. *The necessity and the sufficiency conditions of eventual periodicity of* $\mu' \in \widehat{\omega}^{\alpha}(\mu), \mu' \in \omega^{\rho}(\mu)$ *are obtained by applying Theorem 37, page 81,..., Theorem 40, page 83. In discrete time, the fact that* $p \in \widehat{P}^{\alpha}_{\mu,\mu'}$ *means the existence of* $k' \in \mathbf{N}_{\_}, k_1 \geq 1$ *and* $n_1, n_2, ..., n_{k_1} \in \{k', k'+1, ..., k'+p-1\}$ *with*

$$\widehat{\mathbf{T}}^{\alpha}_{\mu,\mu'} \cap \{k', k'+1, k'+2, ...\} = \bigcup_{k \in \mathbf{N}} \{n_1 + kp, n_2 + kp, ..., n_{k_1} + kp\}$$

and in real time, the fact that $T \in P^{\rho}_{\mu,\mu'}$ *means the existence of* $t' \in \mathbf{R}$, $k_1 \geq 1$ *and* $a_1, b_1, a_2, b_2, ..., a_{k_1}, b_{k_1} \in \mathbf{R}$, *such that*

$$t' \leq a_1 < b_1 < a_2 < b_2 < ... < a_{k_1} < b_{k_1} \leq t' + T,$$

$$\mathbf{T}^{\rho}_{\mu,\mu'} \cap [t', \infty) =$$
$$\bigcup_{k \in \mathbf{N}} ([a_1 + kT, b_1 + kT) \cup [a_2 + kT, b_2 + kT) \cup ... \cup [a_{k_1} + kT, b_{k_1} + kT))$$

are true. In the special case when previously $k_1 = 1$ *-see Theorem 41, page 84 and Theorem 42, page 84- p and T are prime periods of* μ'.

Remark 185. *If* $\mu' \in \widehat{\omega}^{\alpha}(\mu), \mu' \in \omega^{\rho}(\mu)$ *is eventually periodic and the flow* $\widehat{\Phi}^{\alpha}(\mu, \cdot), \Phi^{\rho}(\mu, \cdot)$ *is constant in an interval of length* $\geq$ *the period, then the flow is eventually constant. This relation between eventual periodicity and eventual constancy is a consequence of Theorem 43, page 85.*

Remark 186. *We suppose that* $\widehat{P}^{\alpha}_{\mu,\mu'} \neq \varnothing, P^{\rho}_{\mu,\mu'} \neq \varnothing$ *-see Corollary 4, page 130- and let* $k', k'' \in \mathbf{N}$, $t', t'' \in \mathbf{R}$ *arbitrary. We infer* $\widehat{P}^{\widehat{\sigma}^{k'}(\alpha)}_{\widehat{\Phi}^{\alpha}(\mu,k'-1),\mu'} = \widehat{P}^{\widehat{\sigma}^{k''}(\alpha)}_{\widehat{\Phi}^{\alpha}(\mu,k''-1),\mu'}$ *and* $P^{\sigma^{t'}(x)}_{\Phi^{\rho}(\mu,t'-0),\mu'} = P^{\sigma^{t''}(x)}_{\Phi^{\rho}(\mu,t''-0),\mu'}$.

2. Eventually Periodic Flows

Remark 187. *We consider* $\mu \in \mathbf{B}^n, \alpha \in \widehat{\Pi}_n, \rho \in \Pi_n$. *The equivalent properties of eventual periodicity of the flows are obtained from Theorem 44, page 88 and Theorem 45, page 90 by the replacement of* $\widehat{x}, x$ *with* $\widehat{\Phi}^{\alpha}(\mu, \cdot), \Phi^{\rho}(\mu, \cdot)$. *We give the examples of (1.1)*$_{page\,88}$, *(1.5)*$_{page\,88}$, *(2.1)*$_{page\,90}$ *and (2.4)*$_{page\,90}$ *which become, for* $p \geq 1, T > 0$:

$$\begin{cases} \forall \mu' \in \widehat{\omega}^{\alpha}(\mu), \exists k' \in \mathbf{N}_{\_}, \forall k \in \widehat{\mathbf{T}}^{\alpha}_{\mu,\mu'} \cap \{k', k'+1, k'+2, ...\}, \\ \{k + zp | z \in \mathbf{Z}\} \cap \{k', k'+1, k'+2, ...\} \subset \widehat{\mathbf{T}}^{\alpha}_{\mu,\mu'}, \end{cases} \tag{2.1}$$

$$\begin{cases} \forall \mu' \in \omega^{\rho}(\mu), \exists t' \in I^{\rho}_{\mu}, \\ \exists t'_1 \geq t', \forall t \in \mathbf{T}^{\rho}_{\mu,\mu'} \cap [t'_1, \infty), \{t + zT | z \in \mathbf{Z}\} \cap [t'_1, \infty) \subset \mathbf{T}^{\rho}_{\mu,\mu'}, \end{cases} \tag{2.2}$$

$$\exists k' \in \mathbf{N}_{\_}, \forall k \geq k', \widehat{\Phi}^{\alpha}(\mu, k) = \widehat{\Phi}^{\alpha}(\mu, k+p), \tag{2.3}$$

$$\exists t'_1 \in \mathbf{R}, \forall t \geq t'_1, \Phi^{\rho}(\mu, t) = \Phi^{\rho}(\mu, t+T). \tag{2.4}$$

Notation 26. *We use the notation* $\widehat{P}_\mu^\alpha$ *for the set of the periods of* $\widehat{\Phi}^\alpha(\mu,\cdot)$:

$$\widehat{P}_\mu^\alpha = \{p | p \geq 1, (2.1) \text{ holds}\},$$

and the notation P_μ^ρ *for the set of the periods of* $\Phi^\rho(\mu,\cdot)$:

$$P_\mu^\rho = \{T | T > 0, (2.2) \text{ holds}\}.$$

Remark 188. *The eventually periodic flows reach all the points of the omega limit set in a time interval with the length of a period. This result is obtained from Theorem 47, page 93.*

Remark 189. *If* $p \in \widehat{P}_\mu^\alpha, T \in P_\mu^\rho$ *then the bound of the limit of periodicity is represented by the least* $k' \in \mathbf{N}_\_, t_1' \in \mathbf{R}$ *having the property that*

$$\forall k \geq k', \widehat{\Phi}^\alpha(\mu,k) = \widehat{\Phi}^\alpha(\mu,k+p),$$

$$\forall t \geq t_1', \Phi^\rho(\mu,t) = \Phi^\rho(\mu,t+T).$$

The existence of these bounds results from Theorem 50, page 95.

Remark 190. *We suppose that the relation between* α *and* ρ *is given by*

$$\rho(t) = \alpha^0 \cdot \chi_{\{t_0\}}(t) \oplus \alpha^1 \cdot \chi_{\{t_0+h\}}(t) \oplus ... \oplus \alpha^k \cdot \chi_{\{t_0+kh\}}(t) \oplus ... \tag{2.5}$$

where $t_0 \in \mathbf{R}$ *and* $h > 0$. *If, see Theorem 52, page 98,* $p \geq 1, k' \in \mathbf{N}_\_$ *exist such that*

$$\forall k \geq k', \widehat{\Phi}^\alpha(\mu,k) = \widehat{\Phi}^\alpha(\mu,k+p), \tag{2.6}$$

holds, then $t' \in \mathbf{R}$ *exists making*

$$\forall t \geq t', \Phi^\rho(\mu,t) = \Phi^\rho(\mu,t+T) \tag{2.7}$$

true for $T = ph$. *From Theorem 53, page 99, we have also the inverse implication under the form: if* $\widehat{\Phi}^\alpha(\mu,\cdot), \Phi^\rho(\mu,\cdot)$ *are not eventually constant and* $T > 0, t' \in \mathbf{R}$ *exist such that (2.7) is fulfilled, then* $\frac{T}{h} \in \{1,2,3,...\}$ *and* $k' \in \mathbf{N}_\_$ *exists such that (2.6) is true for* $p = \frac{T}{h}$.

Remark 191. *The eventually periodic flows have the property that the sum, the difference and the multiples of periods are periods. These are consequences of Theorem 54, page 100 and Theorem 55, page 100.*

Remark 192. *We can apply Theorem 56, page 101 to get the form of* $\widehat{P}_\mu^\alpha, P_\mu^\rho$: *if* $\widehat{\Phi}^\alpha(\mu,\cdot)$ *is eventually periodic then* $\widetilde{p} \geq 1$ *exists with* $\widehat{P}_\mu^\alpha = \{\widetilde{p}, 2\widetilde{p}, 3\widetilde{p}, ...\}$, *while if* $\Phi^\rho(\mu,\cdot)$ *is eventually periodic and not eventually constant then* $\widetilde{T} > 0$ *exists such that* $P_\mu^\rho = \{\widetilde{T}, 2\widetilde{T}, 3\widetilde{T}, ...\}$.

Remark 193. *The necessary and the sufficient conditions of eventual periodicity of* $\widehat{\Phi}^\alpha(\mu,\cdot), \Phi^\rho(\mu,\cdot)$ *result from Theorem 59, page 102,..., Theorem 62, page 104. Let* $\widehat{\omega}^\alpha(\mu) = \{\mu^1, ..., \mu^s\}, \omega^\rho = \{\mu^1, ..., \mu^s\}$. *Then* $p \in \widehat{P}_\mu^\alpha$ *if and only if* $k' \in \mathbf{N}_\_, n_1^i, n_2^i, ..., n_{k_i}^i \in \{k', k'+1, ..., k'+p-1\}$ *exist,* $k_i \geq 1$, *such that*

$$\widehat{\mathbf{T}}^\alpha_{\mu,\mu^i} \cap \{k', k'+1, k'+2, ...\} = \bigcup_{k \in \mathbf{N}} \{n_1^i + kp, n_2^i + zp, ..., n_{k_i}^i + kp\}$$

holds for $i \in \{1,...,s\}$ *and* $T \in P^{\rho}_{\mu}$ *if and only if* $t', a^i_1, b^i_1, a^i_2, b^i_2, ..., a^i_{k_i}, b^i_{k_i} \in \mathbf{R}$ *exist,* $k_i \geq 1$, *such that*

$$t' \leq a^i_1 < b^i_1 < a^i_2 < b^i_2 < ... < a^i_{k_i} < b^i_{k_i} \leq t' + T,$$

$$\mathbf{T}^{\rho}_{\mu,\mu^i} \cap [t', \infty) =$$
$$\bigcup_{k \in \mathbf{N}} ([a^i_1 + kT, b^i_1 + kT) \cup [a^i_2 + kT, b^i_2 + kT) \cup ... \cup [a^i_{k_i} + kT, b^i_{k_i} + kT))$$

hold for $i \in \{1,...,s\}$. *If previously* $k_i = 1, i \in \{1,...,s\}$, *see Theorem 63, page 104 and Theorem 64, page 104, then* p, T *are prime periods.*

Remark 194. *In the discrete time eventual periodicity property*

$$\begin{cases} \exists p \geq 1, \exists k' \in \mathbf{N}_{\_}, \forall \mu' \in \widehat{\omega}^{\alpha}(\mu), \forall k \in \widehat{\mathbf{T}}^{\alpha}_{\mu,\mu'} \cap \{k', k'+1, k'+2, ...\}, \\ \{k + zp | z \in \mathbf{Z}\} \cap \{k', k'+1, k'+2, ...\} \subset \widehat{\mathbf{T}}^{\alpha}_{\mu,\mu'}, \end{cases}$$

see Theorem 65, page 105, the order of the quantifiers $\exists p \geq 1, \exists k' \in \mathbf{N}_{\_}, \forall \mu' \in \widehat{\omega}^{\alpha}(\mu)$ *is not important; however, in the real time case*

$$\begin{cases} \exists T > 0, \exists t' \in \mathbf{R}, \forall \mu' \in \omega^{\rho}(\mu), \\ \forall t \in \mathbf{T}^{\rho}_{\mu,\mu'} \cap [t', \infty), \{t + zT | z \in \mathbf{Z}\} \cap [t', \infty) \subset \mathbf{T}^{\rho}_{\mu,\mu'}, \end{cases}$$

the independence of the eventual periodicity property on the order of the quantifiers $\exists T > 0, \exists t' \in \mathbf{R}, \forall \mu' \in \omega^{\rho}(\mu)$ *is not proved so far.*

Remark 195. *The relation between the prime periods* $\widetilde{p}_i = \min \widehat{P}^{\alpha}_{\mu,\mu^i}$, $\widetilde{T}_i = \min P^{\rho}_{\mu,\mu^i}$ *of the eventually periodic points* $\mu^i \in \widehat{Or}^{\alpha}(\mu), \mu^i \in Or^{\rho}(\mu), i \in \{1,...,s\}$ *and the prime period* $\widetilde{p} = \min \widehat{P}^{\alpha}_{\mu}, \widetilde{T} = \min P^{\rho}_{\mu}$ *of the eventually periodic flows* $\widehat{\Phi}^{\alpha}(\mu,\cdot), \Phi^{\rho}(\mu,\cdot)$ *is inferred from Theorem 66, page 107:*

$$\widetilde{p} = n_1 \widetilde{p}_1 = ... = n_s \widetilde{p}_s,$$
$$\widetilde{T} = n_1 \widetilde{T}_1 = ... = n_s \widetilde{T}_s,$$

where $n_1 \geq 1, ..., n_s \geq 1$ *are relatively prime.*

3. Double Eventual Periodicity

Remark 196. *Let* $\mu \in \mathbf{B}^n, \alpha \in \widehat{\Pi}_n$ *and* $\rho \in \Pi_n$. *We wish to study the situation when* α, ρ *are eventually periodic together with the flows* $\widehat{\Phi}^{\alpha}(\mu,\cdot), \Phi^{\rho}(\mu,\cdot)$:

$$\begin{aligned} &\exists p \geq 1, \exists p' \geq 1, \exists k' \in \mathbf{N}, \\ &\forall k \geq k', \alpha^k = \alpha^{k+p} \text{ and } \forall k \geq k', \widehat{\Phi}^{\alpha}(\mu,k) = \widehat{\Phi}^{\alpha}(\mu,k+p'), \end{aligned} \tag{3.1}$$

$$\begin{aligned} &\exists T > 0, \exists T' > 0, \exists t' \in \mathbf{R}, \\ &\forall t \geq t', \rho(t) = \rho(t+T) \text{ and } \forall t \geq t', \Phi^{\rho}(\mu,t) = \Phi^{\rho}(\mu,t+T'). \end{aligned} \tag{3.2}$$

Note that previously the limits of periodicity k' *of* $\alpha, \widehat{\Phi}^{\alpha}(\mu,\cdot)$ *and* t' *of* $\rho, \Phi^{\rho}(\mu,\cdot)$ *coincide, but the periods* p, p' *and* T, T' *are different.*

Theorem 132. *We ask that $t_0 \in \mathbf{R}$ and $h > 0$ exist such that*

$$\rho(t) = \alpha^0 \cdot \chi_{\{t_0\}}(t) \oplus \alpha^1 \cdot \chi_{\{t_0+h\}}(t) \oplus \ldots \oplus \alpha^k \cdot \chi_{\{t_0+kh\}}(t) \oplus \ldots \tag{3.3}$$

Then (3.1), (3.2) are equivalent.

Proof. We infer from (3.3) that

$$\Phi^{\rho}(\mu,t) = \widehat{\Phi}^{\alpha}(\mu,-1) \cdot \chi_{(-\infty,t_0)}(t) \oplus \widehat{\Phi}^{\alpha}(\mu,0) \cdot \chi_{[t_0,t_0+h)}(t) \oplus$$

$$\ldots \oplus \widehat{\Phi}^{\alpha}(\mu,k) \cdot \chi_{[t_0+kh,t_0+(k+1)h)}(t) \oplus \ldots$$

$$= \mu \cdot \chi_{(-\infty,t_0)}(t) \oplus \Phi^{\alpha^0}(\mu) \cdot \chi_{[t_0,t_0+h)}(t) \oplus \ldots \oplus \Phi^{\alpha^0 \ldots \alpha^k}(\mu) \cdot \chi_{[t_0+kh,t_0+(k+1)h)}(t) \oplus \ldots$$

(3.1)$\Longrightarrow$(3.2) The hypothesis states the existence of $p \geq 1, p' \geq 1$ and $k' \in \mathbf{N}$ such that

$$\forall k \geq k', \alpha^k = \alpha^{k+p}, \tag{3.4}$$

$$\forall k \geq k', \widehat{\Phi}^{\alpha}(\mu,k) = \widehat{\Phi}^{\alpha}(\mu,k+p') \tag{3.5}$$

are true. We use the notations $t' = t_0 + k'h, T = ph$ and $T' = p'h$. Let $t \geq t'$ arbitrary and we have two possibilities.

Case $\exists k \geq k', t = t_0 + kh$

In this case $t + T = t + ph = t_0 + (k+p)h$ and we can write

$$\rho(t) = \rho(t_0 + kh) = \alpha^k \overset{(3.4)}{=} \alpha^{k+p} = \rho(t_0 + (k+p)h) = \rho(t+T).$$

Case $\exists k \geq k', t \in (t_0 + kh, t_0 + (k+1)h)$

We have $t + T = t + ph \in (t_0 + (k+p)h, t_0 + (k+p+1)h)$ and

$$\rho(t) = (0,\ldots,0) = \rho(t+T).$$

On the other hand, let us suppose that $t \geq t'$ is arbitrary and $k \geq k'$ exists such that $t \in [t_0 + kh, t_0 + (k+1)h)$. We infer that $t + T' = t + p'h \in [t_0 + (k+p')h, t_0 + (k+p'+1)h)$ and

$$\Phi^{\rho}(\mu,t) = \widehat{\Phi}^{\alpha}(\mu,k) \overset{(3.5)}{=} \widehat{\Phi}^{\alpha}(\mu,k+p') = \Phi^{\rho}(\mu,t+T').$$

(3.2)$\Longrightarrow$(3.1) We have from the hypothesis the existence of $T > 0, T' > 0$ and $t' \in R$ such that

$$\forall t \geq t', \rho(t) = \rho(t+T), \tag{3.6}$$

$$\forall t \geq t', \Phi^{\rho}(\mu,t) = \Phi^{\rho}(\mu,t+T') \tag{3.7}$$

are satisfied. Without loosing the generality we suppose that $k' \in \mathbf{N}$ exists with $t' = t_{k'} = t_0 + k'h$. We denote $p = \frac{T}{h}, p' = \frac{T'}{h}$. We show that $\frac{T}{h} \in \{1,2,3,\ldots\}$. Let us suppose against all reason that this is not true, meaning that some $k_1 \in \mathbf{N}$ exists with $k_1 < \frac{T}{h} < k_1 + 1$, i.e. $k_1 h < T < (k_1+1)h$. This fact implies that $\forall k \geq k'$, we have $t_0 + kh \geq t_0 + k'h = t'$ and

$$t_0 + (k+k_1)h < t_0 + kh + T < t_0 + (k+k_1+1)h,$$

$$\alpha^k = \rho(t_0 + kh) \overset{(3.6)}{=} \rho(t_0 + kh + T) = (0,\ldots,0),$$

contradiction with the supposition that $\alpha \in \widehat{\Pi}_n$. We have shown that $p = \frac{T}{h} \in \{1,2,3,...\}$ and we infer $\forall k \geq k'$,

$$\alpha^k = \rho(t_0 + kh) \overset{(3.6)}{=} \rho(t_0 + kh + T) = \rho(t_0 + (k+p)h) = \alpha^{k+p}.$$

Furthermore, there are two possibilities.

Case $p' = \frac{T'}{h} \in \{1,2,3,...\}$

In this situation we have $\forall k \geq k', t_0 + kh \geq t'$ again and

$$\widehat{\Phi}^\alpha(\mu,k) = \Phi^\rho(\mu,t_0+kh) \overset{(3.7)}{=} \Phi^\rho(\mu,t_0+kh+T')$$

$$= \Phi^\rho(\mu,t_0+(k+p')h) = \widehat{\Phi}^\alpha(\mu,k+p').$$

Case $p' = \frac{T'}{h} \notin \{1,2,3,...\}$

when some $k_2 \in \mathbf{N}$ exists with $k_2 < \frac{T'}{h} < k_2+1$, i.e. $k_2h < T' < (k_2+1)h$. We claim that in this case we have $\forall t \geq t'$,

$$\Phi^\rho(\mu,t) = \widehat{\Phi}^\alpha(\mu,k'). \tag{3.8}$$

Let us suppose against all reason that this is not true, namely that $k \geq k'$ exists with

$$\Phi^\rho(\mu,t_0+(k+1)h-0) = \Phi^\rho(\mu,t_0+kh) \neq \Phi^\rho(\mu,t_0+(k+1)h). \tag{3.9}$$

As far as $t_0 + (k+1)h + T' \in (t_0 + (k+k_2+1)h, t_0 + (k+k_2+2)h)$, we get

$$\Phi^\rho(\mu,t_0+(k+1)h+T'-0) = \Phi^\rho(\mu,t_0+(k+1)h+T') \tag{3.10}$$

thus we can write

$$\Phi^\rho(\mu,t_0+kh) = \Phi^\rho(\mu,t_0+(k+1)h-0) \overset{(3.7)}{=} \Phi^\rho(\mu,t_0+(k+1)h+T'-0)$$

$$\overset{(3.10)}{=} \Phi^\rho(\mu,t_0+(k+1)h+T') \overset{(3.7)}{=} \Phi^\rho(\mu,t_0+(k+1)h)$$

representing a contradiction with (3.9). We have inferred the truth of $\forall t \geq t'$,(3.8), wherefrom $\forall k \geq k'$,

$$\widehat{\Phi}^\alpha(\mu,k) = \widehat{\Phi}^\alpha(\mu,k')$$

and $\forall k \geq k'$,

$$\widehat{\Phi}^\alpha(\mu,k) = \widehat{\Phi}^\alpha(\mu,k+1)$$

holds. □

Definition 50. *We presume that (3.1) is true; then $\widehat{\Phi}^\alpha(\mu,\cdot)$ is called* ***double eventually periodic****. And if (3.2) is true, then $\Phi^\rho(\mu,\cdot)$ is called* ***double eventually periodic****.*

4. Periodic Values of the Flows

Remark 197. *Let* $\mu \in \mathbf{B}^n, \alpha \in \widehat{\Pi}_n, \rho \in \Pi_n$ *and the function* $\Phi : \mathbf{B}^n \longrightarrow \mathbf{B}^n$. *The periodic values of the flows are the periodic points* $\mu' \in \widehat{Or}^{\alpha}(\mu), \mu' \in Or^{\rho}(\mu)$ *as they are defined by the equivalent properties from Theorem 67, page 110, where we put* $\widehat{x}(k) = \widehat{\Phi}^{\alpha}(\mu,k), x(t) = \Phi^{\rho}(\mu,t)$. *For example the properties (1.1)*$_{page\ 110}$, *(1.7)*$_{page\ 110}$ *stating the periodicity of* μ' *become:*

$$\forall k \in \widehat{\mathbf{T}}^{\alpha}_{\mu,\mu'}, \{k+zp | z \in \mathbf{Z}\} \cap \mathbf{N}_{\_} \subset \widehat{\mathbf{T}}^{\alpha}_{\mu,\mu'},$$

$$\exists t' \in I^{\rho}_{\mu}, \forall t \in \mathbf{T}^{\rho}_{\mu,\mu'} \cap [t',\infty), \{t+zT | z \in \mathbf{Z}\} \cap [t',\infty) \subset \mathbf{T}^{\rho}_{\mu,\mu'},$$

where $p \geq 1$ *and* $T > 0$.

Remark 198. *We can use Theorem 68, page 113 and Corollary 3, page 116 to indicate, in case of periodicity of* $\mu' \in Or^{\rho}(\mu)$, *which are the bounds* $t'_0 < t_0$ *with the property that* $\forall t' \in [t'_0, t_0)$, *we have*

$$t' \in I^{\rho}_{\mu} \text{ and } \forall t \in \mathbf{T}^{\rho}_{\mu,\mu'} \cap [t',\infty), \{t+zT | z \in \mathbf{Z}\} \cap [t',\infty) \subset \mathbf{T}^{\rho}_{\mu,\mu'},$$

and for any $t' \notin [t'_0, t_0)$, *the previous statement is false.*

Remark 199. *We suppose that* $t_0 \in \mathbf{R}$, $h > 0$ *exist such that* α *and* ρ *are related by*

$$\rho(t) = \alpha^0 \cdot \chi_{\{t_0\}}(t) \oplus \alpha^1 \cdot \chi_{\{t_0+h\}}(t) \oplus \ldots \oplus \alpha^k \cdot \chi_{\{t_0+kh\}}(t) \oplus \ldots$$

thus we have

$$\Phi^{\rho}(\mu,t) = \widehat{\Phi}^{\alpha}(\mu,-1) \cdot \chi_{(-\infty,t_0)}(t) \oplus \widehat{\Phi}^{\alpha}(\mu,0) \cdot \chi_{[t_0,t_0+h)}(t) \oplus \ldots$$
$$\ldots \oplus \widehat{\Phi}^{\alpha}(\mu,k) \cdot \chi_{[t_0+kh,t_0+(k+1)h)}(t) \oplus \ldots$$

If $\widehat{\Phi}^{\alpha}(\mu,\cdot), \Phi^{\rho}(\mu,\cdot)$ *are not constant, we can apply Theorem 70, page 118 to characterize the connection between the periodic values of the discrete time and the real time flows. We get that the periodicity of* $\mu' \in \widehat{Or}^{\alpha}(\mu)$ *with the period* $p \geq 1$ *implies the periodicity of* $\mu' \in Or^{\rho}(\mu)$ *with the period* $T = ph$ *and the periodicity of* $\mu' \in Or^{\rho}(\mu)$ *with the period* T *implies the eventual periodicity of* $\mu' \in \widehat{Or}^{\alpha}(\mu)$ *with the period* $p = \frac{T}{h}$.

Remark 200. *From Theorem 71, page 119 we get that* $\forall p \in \widehat{P}^{\alpha}_{\mu,\mu'}, \forall p' \in \widehat{P}^{\alpha}_{\mu,\mu'}$,

$$(p+p' \in \widehat{P}^{\alpha}_{\mu,\mu'} \text{ and } p > p') \Longrightarrow p-p' \in \widehat{P}^{\alpha}_{\mu,\mu'}$$

and $\forall T \in P^{\rho}_{\mu,\mu'}, \forall T' \in P^{\rho}_{\mu,\mu'}$,

$$(T+T' \in P^{\rho}_{\mu,\mu'} \text{ and } T > T') \Longrightarrow T-T' \in P^{\rho}_{\mu,\mu'}$$

hold. Similarly, we can use Theorem 72, page 120 stating that the multiples of periods are periods.

Remark 201. *Theorem 73, page 121 gives the form of the sets of the periods:*

$$\widehat{P}^{\alpha}_{\mu,\mu'} = \{\widetilde{p}, 2\widetilde{p}, 3\widetilde{p}, ...\},$$

$$P^{\rho}_{\mu,\mu'} = \{\widetilde{T}, 2\widetilde{T}, 3\widetilde{T}, ...\}$$

whenever $\Phi^{\rho}(\mu,\cdot)$ *is not constant. In particular, this shows the existence of the prime periods* $\widetilde{p} \geq 1, \widetilde{T} > 0$.

Remark 202. *If* $\widehat{\Phi}^{\alpha}(\mu,\cdot)$ *is not constant and* $p \geq 1$ *is arbitrary, then* $p \in \widehat{P}^{\alpha}_{\mu,\mu'}$ *implies from Theorem 75, page 121 the existence of* $n_1, n_2, ..., n_{k_1} \in \{-1, 0, ..., p-2\}, k_1 \geq 1$ *that fulfill*

$$\widehat{\mathbf{T}}^{\alpha}_{\mu,\mu'} = \bigcup_{k\in\mathbf{N}} \{n_1 + kp, n_2 + kp, ..., n_{k_1} + kp\}. \quad (4.1)$$

The inverse statement is represented by Theorem 78, page 124: if $n_1, n_2, ..., n_{k_1} \in \{-1, 0, ..., p-2\}$, $k_1 \geq 1$ *fulfill (4.1), then* $p \in \widehat{P}^{\alpha}_{\mu,\mu'}$. *In the special case when in (4.1) we have* $k_1 = 1$, *Theorem 81, page 126 shows that* p *is the prime period of* μ'. *Similar statements hold for the real time case.*

Remark 203. *Theorem 84, page 128 applied here gives the relation between periodicity and eventual periodicity, which is: if* $\mu' \in \widehat{Or}^{\alpha}(\mu)$ *is a periodic point then, for any* $\widetilde{k} \in \mathbf{N}$, *we have* $\widehat{P}^{\alpha}_{\mu,\mu'} = \widehat{P}^{\sigma^{\widetilde{k}}(\alpha)}_{\mu,\mu'}$ *and respectively if* $\mu' \in Or^{\rho}(\mu)$ *is a periodic point then, for any* $\widetilde{t} \in \mathbf{R}$, *we have* $P^{\rho}_{\mu,\mu'} = P^{\sigma^{\widetilde{t}}(\rho)}_{\mu,\mu'}$.

5. Periodic Flows

Remark 204. *Let the function* $\Phi : \mathbf{B}^n \to \mathbf{B}^n$ *and* $\mu \in \mathbf{B}^n$, $\alpha \in \widehat{\Pi}_n$, $\rho \in \Pi_n$. *Here are two properties that are equivalent with the periodicity of* $\widehat{\Phi}^{\alpha}(\mu,\cdot), \Phi^{\rho}(\mu,\cdot)$:

$$\forall \mu' \in \widehat{Or}^{\alpha}(\mu), \forall k \in \widehat{\mathbf{T}}^{\alpha}_{\mu,\mu'}, \{k + zp | z \in \mathbf{Z}\} \cap \mathbf{N}_{\_} \subset \widehat{\mathbf{T}}^{\alpha}_{\mu,\mu'},$$

$$\forall \mu' \in Or^{\rho}(\mu), \exists t' \in I^{\rho}_{\mu}, \forall t \in \mathbf{T}^{\rho}_{\mu,\mu'} \cap [t', \infty), \{t + zT | z \in \mathbf{Z}\} \cap [t', \infty) \subset \mathbf{T}^{\rho}_{\mu,\mu'},$$

as inferred from Theorem 85, statements $(1.1)_{page\ 134}$, $(1.7)_{page\ 134}$. *They refer to the periodicity of all the points* $\mu' \in \widehat{Or}^{\alpha}(\mu), \mu' \in Or^{\rho}(\mu)$. *Perhaps a more intuitive perspective on periodicity is given by the second group of periodicity properties from Theorem 86, where* $(2.1)_{page\ 136}$, $(2.4)_{page\ 136}$ *become*

$$\forall k \in \mathbf{N}_{\_}, \widehat{\Phi}^{\alpha}(\mu, k) = \widehat{\Phi}^{\alpha}(\mu, k + p),$$

$$\exists t' \in I^{\rho}_{\mu}, \forall t \geq t', \Phi^{\rho}(\mu, t) = \Phi^{\rho}(\mu, t + T).$$

Remark 205. *We know from Theorem 87, page 137 that if* $\widehat{\Phi}^{\alpha}(\mu,\cdot), \Phi^{\rho}(\mu,\cdot)$ *are periodic, then* $\widehat{\omega}^{\alpha}(\mu) = \widehat{Or}^{\alpha}(\mu), \omega^{\rho}(\mu) = Or^{\rho}(\mu)$. *The inverse implications are not true.*

Remark 206. *If $\widehat{\Phi}^{\alpha}(\mu,\cdot),\Phi^{\rho}(\mu,\cdot)$ are periodic with the period p,T, Theorem 89, page 138 shows that all the points of the orbit are reached during a period:*

$$\forall k\in\mathbf{N}_{\_},\widehat{Or}^{\alpha}(\mu)=\{\widehat{\Phi}^{\alpha}(\mu,i)|i\in\{k,k+1,...,k+p-1\}\},$$

respectively $t'\in I^{\rho}_{\mu}$ exists with

$$\forall t\geq t',Or^{\rho}(\mu)=\{\Phi^{\rho}(\mu,\xi)|\xi\in[t,t+T)\}.$$

Remark 207. *In case that $T\in P^{\rho}_{\mu}$, Theorem 91, page 139 and Corollary 5, page 141 give the bounds $t'_0,t_0\in\mathbf{R}$ such that $\forall t'\in[t'_0,t_0)$, we have*

$$t'\in I^{\rho}_{\mu} \text{ and } \forall t\geq t',\Phi^{\rho}(\mu,t)=\Phi^{\rho}(\mu,t+T) \tag{5.1}$$

and if $t'\notin[t'_0,t_0)$, then (5.1) is false.

Remark 208. *We presume that $t_0\in\mathbf{R}$, $h>0$ exist such that α and ρ are related by*

$$\rho(t)=\alpha^0\cdot\chi_{\{t_0\}}(t)\oplus\alpha^1\cdot\chi_{\{t_0+h\}}(t)\oplus...\oplus\alpha^k\cdot\chi_{\{t_0+kh\}}(t)\oplus...$$

In this case Theorem 93, page 144 shows that $p\in\widehat{P}^{\alpha}_{\mu}\Longrightarrow T=ph\in P^{\rho}_{\mu}$. The inverse implication results from Theorem 94, page 144: if $\widehat{\Phi}^{\alpha}(\mu,\cdot),\Phi^{\rho}(\mu,\cdot)$ are not constant and $T\in P^{\rho}_{\mu}$, then $\frac{T}{h}\in\{1,2,3,...\}$ and $k'\in\mathbf{N}_{\_}$ exists making

$$\forall k\geq k',\widehat{\Phi}^{\alpha}(\mu,k)=\widehat{\Phi}^{\alpha}(\mu,k+p),$$

$$\forall k\in\{-1,0,...,k'\},\widehat{\Phi}^{\alpha}(\mu,k)=\mu$$

true for $p=\frac{T}{h}$, i.e. a special case of eventual periodicity.

Remark 209. *From Theorem 95, page 146 and Theorem 96, page 146 we get that the sum of periods, the difference of periods and the multiples of periods of $\widehat{\Phi}^{\alpha}(\mu,\cdot),\Phi^{\rho}(\mu,\cdot)$ are periods.*

Remark 210. *The form of the sets of periods $\widehat{P}^{\alpha}_{\mu}=\{\widetilde{p},2\widetilde{p},3\widetilde{p},...\},P^{\rho}_{\mu}=\{\widetilde{T},2\widetilde{T},3\widetilde{T},\ ...\}$ results from Theorem 97, page 147.*

Remark 211. *Theorem 99, page 147,..., Theorem 102, page 149 give necessary conditions and sufficient conditions of periodicity of $\widehat{\Phi}^{\alpha}(\mu,\cdot),\Phi^{\rho}(\mu,\cdot)$. For example, if $\widehat{Or}^{\alpha}(\mu)=\{\mu^1,...,\mu^s\},s\geq 2$, the fact that $p\in\widehat{P}^{\alpha}_{\mu}$ implies the existence for any $i\in\{1,...,s\}$, of $n^i_1,n^i_2,...,n^i_{k_i}\in\{-1,0,...,p-2\}$, $k_i\geq 1$, such that*

$$\widehat{\mathbf{T}}^{\alpha}_{\mu,\mu^i}=\bigcup_{k\in\mathbf{N}}\{n^i_1+kp,n^i_2+kp,...,n^i_{k_i}+kp\}$$

and similar properties result for the real time case. In the special case when previously $\forall i\in\{1,...,s\}$, $k_i=1$, we get the interesting property that p is the prime period, see also Theorems 103 and 104, page 150 (the last theorem refers to the real time).

Remark 212. *We infer from Theorem 105, page 151 that if* $\widehat{\Phi}^{\alpha}(\mu,\cdot)$ *is periodic then* $\forall\widetilde{k}\in\mathbf{N}$, $\widehat{\sigma}^{\widetilde{k}}(\widehat{\Phi}^{\alpha}(\mu,\cdot))$ *is periodic and* $\widehat{P}^{\alpha}_{\mu}=\widehat{P}^{\widehat{\sigma}^{\widetilde{k}}(\alpha)}_{\widehat{\Phi}^{\alpha}(\mu,\widetilde{k}-1)}$ *and if* $\Phi^{\rho}(\mu,\cdot)$ *is periodic, then* $\forall\widetilde{t}\in\mathbf{R}$, *we have that* $\sigma^{\widetilde{t}}(\Phi^{\rho}(\mu,\cdot))$ *is periodic and* $P^{\rho}_{\mu}=P^{\sigma^{\widetilde{t}}(\rho)}_{\Phi^{\rho}(\mu,\widetilde{t}-0)}$.

Remark 213. *The equivalence of the statements*

$$\begin{cases} \exists T>0,\exists t'\in I^{\rho}_{\mu},\forall\mu'\in Or^{\rho}(\mu),\\ \forall t\in\mathbf{T}^{\rho}_{\mu,\mu'}\cap[t',\infty),\{t+zT|z\in\mathbf{Z}\}\cap[t',\infty)\subset\mathbf{T}^{\rho}_{\mu,\mu'},\end{cases}$$

$$\begin{cases} \exists T>0,\forall\mu'\in Or^{\rho}(\mu),\exists t'\in I^{\rho}_{\mu},\\ \forall t\in\mathbf{T}^{\rho}_{\mu,\mu'}\cap[t',\infty),\{t+zT|z\in\mathbf{Z}\}\cap[t',\infty)\subset\mathbf{T}^{\rho}_{\mu,\mu'},\end{cases}$$

$$\begin{cases} \exists t'\in I^{\rho}_{\mu},\forall\mu'\in Or^{\rho}(\mu),\exists T>0,\\ \forall t\in\mathbf{T}^{\rho}_{\mu,\mu'}\cap[t',\infty),\{t+zT|z\in\mathbf{Z}\}\cap[t',\infty)\subset\mathbf{T}^{\rho}_{\mu,\mu'},\end{cases}$$

$$\begin{cases} \forall\mu'\in Or^{\rho}(\mu),\exists T>0,\exists t'\in I^{\rho}_{\mu},\\ \forall t\in\mathbf{T}^{\rho}_{\mu,\mu'}\cap[t',\infty),\{t+zT|z\in\mathbf{Z}\}\cap[t',\infty)\subset\mathbf{T}^{\rho}_{\mu,\mu'}\end{cases}$$

is not proved, see Theorem 106, page 152.

6. Double Periodicity

Definition 51. *Let* $\mu\in\mathbf{B}^n,\alpha\in\widehat{\Pi}_n$ *and* $\rho\in\Pi_n$. *If*

$$\exists p\geq1,\exists p'\geq1,\forall k\in\mathbf{N},\alpha^k=\alpha^{k+p}\text{ and }\forall k\in\mathbf{N}_{\_},\widehat{\Phi}^{\alpha}(\mu,k)=\widehat{\Phi}^{\alpha}(\mu,k+p') \tag{6.1}$$

is true, $\widehat{\Phi}^{\alpha}(\mu,\cdot)$ *is called* ***double periodic*** *and if*

$$\begin{array}{c}\exists T>0,\exists T'>0,\exists t'\in I^{\rho}\cap I^{\rho}_{\mu},\forall t\geq t',\rho(t)=\rho(t+T)\text{ and}\\ \text{and }\forall t\geq t',\Phi^{\rho}(\mu,t)=\Phi^{\rho}(\mu,t+T')\end{array} \tag{6.2}$$

is true, $\Phi^{\rho}(\mu,\cdot)$ *is called* ***double periodic*** *too.*

Remark 214. *Double periodicity means the existence of two periodicities, of the computation function and of the flow. We notice that, like at Definition 50, page 211, the periods are different; in addition, (6.2) indicates the existence of a common initial time for the computation function and the flow.*

Theorem 133. *We ask that* $t_0\in\mathbf{R}$ *and* $h>0$ *exist such that*

$$\rho(t)=\alpha^0\cdot\chi_{\{t_0\}}(t)\oplus\alpha^1\cdot\chi_{\{t_0+h\}}(t)\oplus\ldots\oplus\alpha^k\cdot\chi_{\{t_0+kh\}}(t)\oplus\ldots \tag{6.3}$$

In this situation (6.1) implies (6.2).

Proof. We infer from (6.3) that

$$\Phi^{\rho}(\mu,t)=\widehat{\Phi}^{\alpha}(\mu,-1)\cdot\chi_{(-\infty,t_0)}(t)\oplus\widehat{\Phi}^{\alpha}(\mu,0)\cdot\chi_{[t_0,t_0+h)}(t)\oplus\ldots$$

$$\ldots\oplus\widehat{\Phi}^{\alpha}(\mu,k)\cdot\chi_{[t_0+kh,t_0+(k+1)h)}(t)\oplus\ldots$$

The hypothesis states the existence of $p \geq 1$ and $p' \geq 1$ such that

$$\forall k \in \mathbf{N}, \alpha^k = \alpha^{k+p}, \tag{6.4}$$

$$\forall k \in \mathbf{N}_{\_}, \widehat{\Phi}^{\alpha}(\mu, k) = \widehat{\Phi}^{\alpha}(\mu, k + p') \tag{6.5}$$

and we show that for $T = ph, T' = p'h$ and an arbitrary $t' \in (t_0 - h, t_0)$, the properties

$$\forall t \geq t', \rho(t) = \rho(t + T), \tag{6.6}$$

$$\forall t \geq t', \Phi^{\rho}(\mu, t) = \Phi^{\rho}(\mu, t + T') \tag{6.7}$$

hold. Let $t \geq t'$ arbitrary also and we have two possibilities.

Case $t \in [t', t_0)$

We get $t + T \in [t' + ph, t_0 + ph) \subset (t_0 + (p-1)h, t_0 + ph)$ and $t + T' \in [t' + p'h, t_0 + p'h) \subset [t_0 + (p'-1)h, t_0 + p'h)$, thus

$$\rho(t) = \rho(t + T) = (0, ..., 0),$$

$$\mu = \Phi^{\rho}(\mu, t) = \widehat{\Phi}^{\alpha}(\mu, -1) \overset{(6.5)}{=} \widehat{\Phi}^{\alpha}(\mu, p' - 1) = \Phi^{\rho}(\mu, t + T').$$

Case $t \geq t_0$

This is a special case of Theorem 132, page 210 with $k' = 0$ and $t' = t_0$. □

Remark 215. *The inverse of Theorem 133 is true under a weaker form. Let us suppose that real time double periodicity holds, i.e. (6.2) is true, and the relation between the discrete time and the real time computation functions is expressed by (6.3). Then $\widehat{\Phi}^{\alpha}(\mu, \cdot)$ is double eventually periodic only, due to reasons that have been addressed at Theorem 94, page 144. The proof of this statement makes use of Theorem 132.*

Theorem 134. *a) If $\alpha \in \widehat{\Pi}_n$ is periodic $\exists p \geq 1$,*

$$\forall k \in \mathbf{N}, \alpha^k = \alpha^{k+p} \tag{6.8}$$

then, $\forall \mu \in \mathbf{B}^n$, the flow $\widehat{\Phi}^{\alpha}(\mu, \cdot)$ is double eventually periodic: $\exists p' \geq 1, \exists k' \in \mathbf{N}_{\_}$,

$$\forall k \geq k', \widehat{\Phi}^{\alpha}(\mu, k) = \widehat{\Phi}^{\alpha}(\mu, k + p'). \tag{6.9}$$

b) If $\rho \in \Pi_n$ is periodic $\exists T > 0, \exists t' \in I^{\rho}$,

$$\forall t \geq t', \rho(t) = \rho(t + T) \tag{6.10}$$

then, $\forall \mu \in \mathbf{B}^n$, the flow $\Phi^{\rho}(\mu, \cdot)$ is double eventually periodic: $\exists T' > 0, \exists t'' \in \mathbf{R}$,

$$\forall t \geq t'', \Phi^{\rho}(\mu, t) = \Phi^{\rho}(\mu, t + T'). \tag{6.11}$$

Proof. a) We denote with $\mu^1, \mu^2, \mu^3, ...$ the following values

$$\mu^1 = \widehat{\Phi}^{\alpha}(\mu, p - 1) = \Phi^{\alpha^0 ... \alpha^{p-1}}(\mu),$$

$$\mu^2 = \widehat{\Phi}^{\alpha}(\mu, 2p - 1) = \Phi^{\alpha^0 ... \alpha^{p-1} \alpha^p ... \alpha^{2p-1}}(\mu) = \Phi^{\alpha^0 ... \alpha^{p-1} \alpha^0 ... \alpha^{p-1}}(\mu)$$

$$= \Phi^{\alpha^0...\alpha^{p-1}}(\Phi^{\alpha^0...\alpha^{p-1}}(\mu)),$$

$$\mu^3 = \widehat{\Phi}^{\alpha}(\mu, 3p-1) = \Phi^{\alpha^0...\alpha^{p-1}\alpha^p...\alpha^{2p-1}\alpha^{2p}...\alpha^{3p-1}}(\mu)$$

$$= \Phi^{\alpha^0...\alpha^{p-1}\alpha^0...\alpha^{p-1}\alpha^0...\alpha^{p-1}}(\mu) = \Phi^{\alpha^0...\alpha^{p-1}}(\Phi^{\alpha^0...\alpha^{p-1}}(\Phi^{\alpha^0...\alpha^{p-1}}(\mu))),$$

$$...$$

In the sequence $\mu^1, \mu^2, \mu^3, ...$ let $k_1 \geq 1$ be a rank with the property that $k_2 > k_1$ exists such that $\mu^{k_1} = \mu^{k_2}$. We infer:

$$\mu^{k_1} = \widehat{\Phi}^{\alpha}(\mu, k_1 p - 1) = \Phi^{\alpha^0...\alpha^{k_1 p-1}}(\mu),$$

$$\mu^{k_2} = \widehat{\Phi}^{\alpha}(\mu, k_2 p - 1) = \Phi^{\alpha^0...\alpha^{k_2 p-1}}(\mu).$$

Furthermore:

$$\widehat{\Phi}^{\alpha}(\mu, k_1 p) = \Phi^{\alpha^{k_1 p}}(\widehat{\Phi}^{\alpha}(\mu, k_1 p - 1)) = \Phi^{\alpha^0}(\mu^{k_1}) = \Phi^{\alpha^0}(\mu^{k_2})$$

$$= \Phi^{\alpha^{k_2 p}}(\mu^{k_2}) = \Phi^{\alpha^{k_2 p}}(\widehat{\Phi}^{\alpha}(\mu, k_2 p - 1)) = \widehat{\Phi}^{\alpha}(\mu, k_2 p),$$

$$\widehat{\Phi}^{\alpha}(\mu, k_1 p + 1) = \Phi^{\alpha^{k_1 p+1}}(\widehat{\Phi}^{\alpha}(\mu, k_1 p)) = \Phi^{\alpha^1}(\widehat{\Phi}^{\alpha}(\mu, k_1 p))$$

$$= \Phi^{\alpha^1}(\widehat{\Phi}^{\alpha}(\mu, k_2 p)) = \Phi^{\alpha^{k_2 p+1}}(\widehat{\Phi}^{\alpha}(\mu, k_2 p)) = \widehat{\Phi}^{\alpha}(\mu, k_2 p + 1),$$

$$...$$

in other words

$$\forall k \geq k_1 p - 1, \widehat{\Phi}^{\alpha}(\mu, k) = \widehat{\Phi}^{\alpha}(\mu, k + (k_2 - k_1)p).$$

b) We have the existence of $(t_k) \in Seq, \alpha \in \widehat{\Pi}_n$ with the property that

$$\rho(t) = \alpha^0 \cdot \chi_{\{t_0\}}(t) \oplus \alpha^1 \cdot \chi_{\{t_1\}}(t) \oplus ... \oplus \alpha^k \cdot \chi_{\{t_k\}}(t) \oplus ... \tag{6.12}$$

Without loss we consider that $\alpha^0 \neq (0, ..., 0)$ and $I^{\rho} = (-\infty, t_0)$, thus in (6.10) we have $t' < t_0$. We obtain in succession from (6.10), (6.12):

$t = t_0 : p \geq 1$ exists such that $t_0 + T = t_p$ and

$$\alpha^0 = \rho(t_0) = \rho(t_0 + T) = \rho(t_p) = \alpha^p;$$

$t \in (t_0, t_1)$:

$$\rho(t) = \rho(t + T) = (0, ..., 0); \tag{6.13}$$

$t = t_1 : t_1 + T = t_{p+1}$ and

$$\alpha^1 = \rho(t_1) = \rho(t_1 + T) = \rho(t_{p+1}) = \alpha^{p+1};$$

$t \in (t_1, t_2)$: (6.13) holds;

$$...$$

$t = t_p : t_p + T = t_{2p}$ and

$$\alpha^p = \rho(t_p) = \rho(t_p + T) = \rho(t_{2p}) = \alpha^{2p};$$

$t \in (t_p, t_{p+1})$: (6.13) holds;

...

thus

$$\rho(t) = \alpha^0 \cdot \chi_{\{t_0\}}(t) \oplus \alpha^1 \cdot \chi_{\{t_1\}}(t) \oplus ... \tag{6.14}$$
$$... \oplus \alpha^{p-1} \cdot \chi_{\{t_{p-1}\}}(t) \oplus \alpha^0 \cdot \chi_{\{t_0+T\}}(t) \oplus \alpha^1 \cdot \chi_{\{t_1+T\}}(t) \oplus ...$$
$$... \oplus \alpha^{p-1} \cdot \chi_{\{t_{p-1}+T\}}(t) \oplus \alpha^0 \cdot \chi_{\{t_0+2T\}}(t) \oplus \alpha^1 \cdot \chi_{\{t_1+2T\}}(t) \oplus ...$$

and two periodicities exist here, the periodicity of the computation function $\alpha \in \widehat{\Pi}_n$ in the sense that (6.8) holds and the periodicity of the location of the values α^k on the real time axis, at time instants that repeat periodically.

The first periodicity shows, from a), the existence of $k_1 < k_2$ such that (6.9) holds with $p' = (k_2 - k_1)p$ and $k' = k_1 p - 1$. We analyze the second periodicity by putting the numbers $k \in \mathbf{N}$ under the form $k = mp + q$, where $m \in \mathbf{N}$ and $q \in \{0, 1, ..., p-1\}$. We infer from (6.14) that

$$t_k = t_{mp+q} = t_q + mT \tag{6.15}$$

and, more general, for any $m' \in \mathbf{N}$ we have that

$$t_{k+m'p} = t_{mp+q+m'p} = t_q + mT + m'T \overset{(6.15)}{=} t_k + m'T. \tag{6.16}$$

Let now $t \geq t_{k'}$ arbitrary, fixed, for which $k \geq k'$ exists with the property that $t \in [t_k, t_{k+1})$; we get that

$$\begin{aligned} t_{k+p'} = t_{k+(k_2-k_1)p} &\overset{(6.16)}{=} t_k + (k_2 - k_1)T \leq t + (k_2 - k_1)T \\ < t_{k+1} + (k_2 - k_1)T &\overset{(6.16)}{=} t_{k+1+(k_2-k_1)p} = t_{k+1+p'}, \end{aligned} \tag{6.17}$$

$$\Phi^\rho(\mu, t) = \Phi^\rho(\mu, t_k) = \widehat{\Phi}^\alpha(\mu, k) \overset{(6.9)}{=} \widehat{\Phi}^\alpha(\mu, k + p') = \Phi^\rho(\mu, t_{k+p'})$$

$$= \Phi^\rho(\mu, t_{k+(k_2-k_1)p}) \overset{(6.16)}{=} \Phi^\rho(\mu, t_k + (k_2 - k_1)T) \overset{(6.17)}{=} \Phi^\rho(\mu, t + (k_2 - k_1)T).$$

We infer that $\Phi^\rho(\mu, \cdot)$ is eventually periodic with the limit of periodicity $t'' = t_{k'}$ and the period $T' = (k_2 - k_1)T$. □

Remark 216. *Note that in both cases a) and b) in the previous theorem the compatibility between p, p' and T, T' is expressed by the existence of $n' \geq 1$ (namely $n' = k_2 - k_1$) such that $p' = n'p, T' = n'T$.*

Theorem 135. *Let $\mu \in \mathbf{B}^n$. a) If $\alpha \in \widehat{\Pi}_n$ is eventually periodic*

$$\exists p \geq 1, \exists k' \in \mathbf{N}, \forall k \geq k', \alpha^k = \alpha^{k+p} \tag{6.18}$$

then $\widehat{\Phi}^\alpha(\mu, \cdot)$ is double eventually periodic.

b) If $\rho \in \Pi_n$ is eventually periodic

$$\exists T > 0, \exists t' \in \mathbf{R}, \forall t \geq t', \rho(t) = \rho(t + T) \tag{6.19}$$

then $\Phi^\rho(\mu, \cdot)$ is double eventually periodic.

Proof. a) The hypothesis states the existence of $p \geq 1$ and $k' \in \mathbf{N}$ having the property that

$$\forall k \geq k', \alpha^k = \alpha^{k+p}. \tag{6.20}$$

We denote

$$\mu' = \widehat{\Phi}^{\alpha}(\mu, k'-1), \tag{6.21}$$

$$\beta = \widehat{\sigma}^{k'}(\alpha), \tag{6.22}$$

thus the relation between $\widehat{\Phi}^{\alpha}(\mu,\cdot)$ and $\widehat{\Phi}^{\beta}(\mu',\cdot)$ is given by

$$\forall k \geq k'-1, \widehat{\Phi}^{\alpha}(\mu,k) \overset{(3.7)_{page\ 180}}{=} \widehat{\Phi}^{\beta}(\mu', k-k'). \tag{6.23}$$

The sequence $\beta \in \widehat{\Pi}_n$ is periodic:

$$\forall k \in \mathbf{N}, \beta^k \overset{(6.22)}{=} \alpha^{k+k'} \overset{(6.20)}{=} \alpha^{k+k'+p} \overset{(6.22)}{=} \beta^{k+p},$$

and in this situation we can apply the reasoning from Theorem 134 in order to infer that $\widehat{\Phi}^{\beta}(\mu',\cdot)$ is double eventually periodic. Hence $1 \leq k_1 < k_2$ exist, defined by

$$\widehat{\Phi}^{\beta}(\mu', k_1 p - 1) = \widehat{\Phi}^{\beta}(\mu', k_2 p - 1) \tag{6.24}$$

such that the eventual periodicity property

$$\forall k \geq k_1 p - 1, \widehat{\Phi}^{\beta}(\mu', k) = \widehat{\Phi}^{\beta}(\mu', k + (k_2 - k_1)p) \tag{6.25}$$

holds. We infer that $\forall k \geq k' + k_1 p - 1$,

$$\widehat{\Phi}^{\alpha}(\mu,k) \overset{(6.23)}{=} \widehat{\Phi}^{\beta}(\mu', k-k') \overset{(6.25)}{=} \widehat{\Phi}^{\beta}(\mu', k-k'+(k_2-k_1)p) \overset{(6.23)}{=} \widehat{\Phi}^{\alpha}(\mu, k+(k_2-k_1)p).$$

b) We put $\rho \in \Pi_n$ under the form

$$\rho(t) = \alpha^0 \cdot \chi_{\{t_0\}}(t) \oplus \alpha^1 \cdot \chi_{\{t_1\}}(t) \oplus \ldots \oplus \alpha^k \cdot \chi_{\{t_k\}}(t) \oplus \ldots \tag{6.26}$$

where $\alpha \in \widehat{\Pi}_n$ and $(t_k) \in Seq$. The hypothesis states the existence of $T > 0$ and $t' \in \mathbf{R}$ such that

$$\forall t \geq t', \rho(t) = \rho(t+T) \tag{6.27}$$

and we suppose, without loosing the generality, that in the previous equation $\rho(t') = (0,\ldots,0)$. Let us define

$$k' = \min\{k | k \in \mathbf{N}, t_k > t'\}.$$

We define

$$\mu' = \widehat{\Phi}^{\alpha}(\mu, k'-1)$$

and $\rho' \in \Pi_n$,

$$\rho'(t) = \beta^0 \cdot \chi_{\{t'_0\}}(t) \oplus \beta^1 \cdot \chi_{\{t'_1\}}(t) \oplus \ldots \oplus \beta^k \cdot \chi_{\{t'_k\}}(t) \oplus \ldots \tag{6.28}$$

where $\beta \in \widehat{\Pi}_n$ and $(t'_k) \in Seq$ are given by

$$\beta = \widehat{\sigma}^{k'}(\alpha),$$

$$\forall k \in \mathbf{N}, t'_k = t_{k+k'}.$$

This means that the relation between $\Phi^{\rho}(\mu,\cdot)$ and $\Phi^{\rho'}(\mu',\cdot)$ is given by

$$\forall t \geq t', \Phi^{\rho}(\mu,t) \overset{(3.8)_{page\ 180}}{=} \Phi^{\rho'}(\mu',t).$$

The equation

$$\forall t \geq t', \rho'(t) = \rho(t) = \rho(t+T) = \rho'(t+T) \tag{6.29}$$

where $t' \in I^{\rho'}$ (because $(-\infty, t_{k'}) \subset I^{\rho'}$ and $t' < t_{k'}$) indicates the periodicity of ρ'.

We can follow now the reasoning from Theorem 134, showing that $p \geq 1$ exists with

$$\rho'(t) = \beta^0 \cdot \chi_{\{t'_0\}}(t) \oplus \beta^1 \cdot \chi_{\{t'_1\}}(t) \oplus \ldots \tag{6.30}$$

$$\ldots \oplus \beta^{p-1} \cdot \chi_{\{t'_{p-1}\}}(t) \oplus \beta^0 \cdot \chi_{\{t'_0+T\}}(t) \oplus \beta^1 \cdot \chi_{\{t'_1+T\}}(t) \oplus \ldots$$

$$\ldots \oplus \beta^{p-1} \cdot \chi_{\{t'_{p-1}+T\}}(t) \oplus \beta^0 \cdot \chi_{\{t'_0+2T\}}(t) \oplus \beta^1 \cdot \chi_{\{t'_1+2T\}}(t) \oplus \ldots$$

This means that β is periodic with the period p and the location of the values of β^k on the real time axis is periodic with the period T. The existence of $k_1 < k_2$ results such that (6.24), (6.25) are fulfilled.

We denote $t'' = t'_{k_1 p-1}$ and let $t \geq t''$ arbitrary. Some $k \geq k_1 p - 1$ exists with the property that $t \in [t'_k, t'_{k+1})$. Because $t'' \geq t'_0 = t_{k'} > t'$ and

$$t'_{k+(k_2-k_1)p} = t'_k + (k_2-k_1)T \leq t + (k_2-k_1)T < t'_{k+1} + (k_2-k_1)T = t'_{k+1+(k_2-k_1)p}$$

we can write that

$$\Phi^{\rho}(\mu,t) = \Phi^{\rho'}(\mu',t) = \Phi^{\rho'}(\mu',t'_k) = \widehat{\Phi}^{\beta}(\mu',k) \overset{(6.25)}{=} \widehat{\Phi}^{\beta}(\mu',k+(k_2-k_1)p)$$

$$= \Phi^{\rho'}(\mu',t'_{k+(k_2-k_1)p}) = \Phi^{\rho'}(\mu',t'_k+(k_2-k_1)T) = \Phi^{\rho'}(\mu',t+(k_2-k_1)T)$$

$$= \Phi^{\rho}(\mu,t+(k_2-k_1)T).$$

This proves the eventual periodicity of $\Phi^{\rho}(\mu,\cdot)$ with the period $(k_2-k_1)T$ and the limit of periodicity t''. □

Notations

Lemmas

The purpose of this appendix is that of presenting results that are necessary in the proofs of some theorems. Several lemmas are interesting by themselves too.

Lemma 1. *Let the signals* $\widehat{x} \in \widehat{S}^{(n)}, x \in S^{(n)}$ *and we consider the following statements:*

$$\mu \in \widehat{Or}(\widehat{x}), \tag{0.1}$$

$$\mu \in \widehat{\omega}(\widehat{x}), \tag{0.2}$$

$$\widehat{\mathbf{T}}^{\widehat{x}}_{\mu} \cap \{k', k'+1, k'+2, ...\} \neq \varnothing, \tag{0.3}$$

$$\forall k \in \widehat{\mathbf{T}}^{\widehat{x}}_{\mu} \cap \{k', k'+1, k'+2, ...\}, \{k + zp | z \in \mathbf{Z}\} \cap \{k', k'+1, k'+2, ...\} \subset \widehat{\mathbf{T}}^{\widehat{x}}_{\mu}, \tag{0.4}$$

$$\mu \in Or(x), \tag{0.5}$$

$$\mu \in \omega(x), \tag{0.6}$$

$$\mathbf{T}^{x}_{\mu} \cap [t', \infty) \neq \varnothing, \tag{0.7}$$

$$\forall t \in \mathbf{T}^{x}_{\mu} \cap [t', \infty), \{t + zT | z \in \mathbf{Z}\} \cap [t', \infty) \subset \mathbf{T}^{x}_{\mu}, \tag{0.8}$$

where $k' \in \mathbf{N}\_$ *and* $t' \in \mathbf{R}$. *We have the equivalencies:*

$$((0.1)\ and\ (0.3)\ and\ (0.4)) \Longleftrightarrow ((0.2)\ and\ (0.4)), \tag{0.9}$$

$$((0.5)\ and\ (0.7)\ and\ (0.8)) \Longleftrightarrow ((0.6)\ and\ (0.8)). \tag{0.10}$$

Proof. If (0.1) and (0.3) and (0.4) hold, we can take some $k \in \widehat{\mathbf{T}}^{\widehat{x}}_{\mu} \cap \{k', k'+1, k'+2, ...\}$ arbitrarily. From (0.4) we get $k, k+p, k+2p, ... \in \widehat{\mathbf{T}}^{\widehat{x}}_{\mu}$, thus $\widehat{\mathbf{T}}^{\widehat{x}}_{\mu}$ is infinite and $\mu \in \widehat{\omega}(\widehat{x})$.

Conversely, we suppose that (0.2) and (0.4) hold. Statement (0.1) is trivially fulfilled and as far as $\widehat{\mathbf{T}}^{\widehat{x}}_{\mu}$ is infinite, (0.3) is true also.

Property (0.9) is proved and the proof of (0.10) is similar. □

Remark 217. *Equivalent forms of Lemma 3, Lemma 4, Lemma 5, Lemma 8, and Lemma 10 exist, due to Lemma 1; we can replace in their hypothesis* $\mu \in \widehat{\omega}(\widehat{x})$ *with* $\mu \in \widehat{Or}(\widehat{x})$, $\widehat{\mathbf{T}}^{\widehat{x}}_{\mu} \cap \{k', k'+1, k'+2, ...\} \neq \varnothing$ *and we can also replace* $\mu \in \omega(x)$ *with* $\mu \in Or(x)$, $\mathbf{T}^{x}_{\mu} \cap [t', \infty) \neq \varnothing$.

Lemma 2. *Let* $\mu \in Or(x)$ *and* $t' \in I^{x}$. *Then* $\mathbf{T}^{x}_{\mu} \cap [t', \infty) \neq \varnothing$.

Proof. The hypothesis states that $(-\infty, t'] \subset \mathbf{T}^{x}_{x(-\infty+0)}$ is true. If $\mu = x(-\infty+0)$, when $t' \in \mathbf{T}^{x}_{\mu}$, we have $\mathbf{T}^{x}_{\mu} \cap [t', \infty) \neq \varnothing$ true. And if $\mu \neq x(-\infty+0)$, when $\mathbf{T}^{x}_{\mu} \cap (-\infty, t'] = \varnothing, \mathbf{T}^{x}_{\mu} \neq \varnothing$, we get $\mathbf{T}^{x}_{\mu} \subset (t', \infty)$, thus $\mathbf{T}^{x}_{\mu} \cap [t', \infty) \neq \varnothing$. □

Lemma 3. *a) $\widehat{x} \in \widehat{S}^{(n)}$ is given and we suppose that $\mu \in \widehat{\omega}(\widehat{x})$ is eventually periodic with the period $p \geq 1$ and with the limit of periodicity $k' \in \mathbf{N}_{\_}$. If $k'' \geq k'$, then μ is eventually periodic with the period p and with the limit of periodicity k''.*

b) Let $x \in S^{(n)}$ and we suppose that $\mu \in \omega(x)$ is eventually periodic with the period $T > 0$ and with the limit of periodicity $t' \in \mathbf{R}$. If $t'' \geq t'$, then μ is eventually periodic with the period T and with the limit of periodicity t''.

Proof. b) The hypothesis states that

$$\forall t \in \mathbf{T}_{\mu}^{x} \cap [t', \infty), \{t + zT | z \in \mathbf{Z}\} \cap [t', \infty) \subset \mathbf{T}_{\mu}^{x} \quad (0.11)$$

is true and we must prove

$$\forall t \in \mathbf{T}_{\mu}^{x} \cap [t'', \infty), \{t + zT | z \in \mathbf{Z}\} \cap [t'', \infty) \subset \mathbf{T}_{\mu}^{x} \quad (0.12)$$

for an arbitrary $t'' \geq t'$. Indeed, $\mathbf{T}_{\mu}^{x} \cap [t', \infty), \mathbf{T}_{\mu}^{x} \cap [t'', \infty)$ are both non-empty since $\mathbf{T}_{\mu}^{x}$ is unbounded from above. We take some arbitrary $t \in \mathbf{T}_{\mu}^{x} \cap [t'', \infty)$ and $z \in \mathbf{Z}$ such that $t + zT \geq t''$ holds. Then $t \in \mathbf{T}_{\mu}^{x} \cap [t', \infty)$ and $t + zT \geq t'$ are true, thus we can apply (0.11). We have obtained that $t + zT \in \mathbf{T}_{\mu}^{x}$, i.e. (0.12) is fulfilled. □

Lemma 4. *a) Let $\widehat{x}, \mu \in \widehat{\omega}(\widehat{x})$ that is an eventually periodic point of $\widehat{x}$ with the period $p \geq 1$ and the limit of periodicity $k' \in \mathbf{N}_{\_}$ and let also $k \in \widehat{\mathbf{T}}_{\mu}^{\widehat{x}} \cap \{k', k'+1, k'+2, ...\}$. Then $\{k, k+p, k+2p, ...\} \subset \widehat{\mathbf{T}}_{\mu}^{\widehat{x}}$.*

b) Let $x \in S^{(n)}, \mu \in \omega(x)$ that is eventually periodic with the period $T > 0$ and the limit of periodicity $t' \in \mathbf{R}$ and we suppose that $t_1 < t_2$ fulfill $[t_1, t_2) \subset \mathbf{T}_{\mu}^{x} \cap [t', \infty)$. Then

$$[t_1, t_2) \cup [t_1 + T, t_2 + T) \cup [t_1 + 2T, t_2 + 2T) \cup ... \subset \mathbf{T}_{\mu}^{x}. \quad (0.13)$$

Proof. b) We have the truth of

$$\forall t \in \mathbf{T}_{\mu}^{x} \cap [t', \infty), \{t + zT | z \in \mathbf{Z}\} \cap [t', \infty) \subset \mathbf{T}_{\mu}^{x}. \quad (0.14)$$

Let $k \in \mathbf{N}$ and $t \in [t_1 + kT, t_2 + kT)$ arbitrary. We infer $t - kT \in [t_1, t_2)$, thus

$$\mu = x(t - kT) \overset{(0.14)}{=} x(t),$$

wherefrom $t \in \mathbf{T}_{\mu}^{x}$. Statement (0.13) is proved. □

Lemma 5. *a) $\widehat{x} \in \widehat{S}^{(n)}, \mu \in \widehat{\omega}(\widehat{x})$ are given with the property that μ is eventually periodic with the period $p \geq 1$ and the limit of periodicity $k' \in \mathbf{N}_{\_}$. If $k_1 \geq k'$ and*

$$\widehat{x}(k_1) \neq \mu, \quad (0.15)$$

then $\forall k \in \mathbf{N}$,

$$\widehat{x}(k_1 + kp) \neq \mu. \quad (0.16)$$

b) We suppose that $x \in S^{(n)}$, $\mu \in \omega(x)$ are given and μ is eventually periodic with the period $T > 0$ and the limit of periodicity $t' \in \mathbf{R}$. If $t_1 > t'$ and

$$x(t_1 - 0) \neq \mu, \quad (0.17)$$

then $\forall k \in \mathbf{N}$,

$$x(t_1 + kT - 0) \neq \mu; \tag{0.18}$$

if $t_2 \geq t'$ *and*

$$x(t_2) \neq \mu, \tag{0.19}$$

then $\forall k \in \mathbf{N}$,

$$x(t_2 + kT) \neq \mu. \tag{0.20}$$

Proof. a) The hypothesis states the truth of

$$\forall k \in \widehat{\mathbf{T}}_{\mu}^{\widehat{x}} \cap \{k', k'+1, k'+2, ...\}, \{k + zp | z \in \mathbf{Z}\} \cap \{k', k'+1, k'+2, ...\} \subset \widehat{\mathbf{T}}_{\mu}^{\widehat{x}}. \tag{0.21}$$

Let $k \in \mathbf{N}$ arbitrary and we suppose against all reason that (0.16) is false. We obtain the contradiction:

$$\mu = \widehat{x}(k_1 + kp) \overset{(0.21)}{=} \widehat{x}(k_1) \overset{(0.15)}{\neq} \mu.$$

b) We have from the hypothesis that

$$\forall t \in \mathbf{T}_{\mu}^{x} \cap [t', \infty), \{t + zT | z \in \mathbf{Z}\} \cap [t', \infty) \subset \mathbf{T}_{\mu}^{x} \tag{0.22}$$

holds. Let $k \in \mathbf{N}$ arbitrary. We get the existence of $\varepsilon_1 > 0$ such that $t_1 - \varepsilon_1 \geq t'$ and

$$\forall t \in (t_1 - \varepsilon_1, t_1), x(t) = x(t_1 - 0) \tag{0.23}$$

and respectively the existence of $\varepsilon_2 > 0$ such that $t_1 + kT - \varepsilon_2 \geq t'$ and

$$\forall t \in (t_1 + kT - \varepsilon_2, t_1 + kT), x(t) = x(t_1 + kT - 0). \tag{0.24}$$

We denote $\varepsilon = \min\{\varepsilon_1, \varepsilon_2\}$ and we suppose against all reason that $t'' \in (t_1 + kT - \varepsilon, t_1 + kT)$ exists with $x(t'') = \mu$. We infer

$$t' \leq t_1 + kT - \varepsilon_2 \leq t_1 + kT - \varepsilon < t'' < t_1 + kT, \tag{0.25}$$

$$\mu = x(t'') \overset{(0.24),(0.25)}{=} x(t_1 + kT - 0), \tag{0.26}$$

$$t' \leq t_1 - \varepsilon_1 \leq t_1 - \varepsilon < t'' - kT < t_1, \tag{0.27}$$

$$\mu = x(t'') \overset{(0.22),(0.25),(0.27)}{=} x(t'' - kT) \overset{(0.23),(0.27)}{=} x(t_1 - 0) \overset{(0.17)}{\neq} \mu, \tag{0.28}$$

contradiction. We have obtained that $\forall t \in (t_1 + kT - \varepsilon, t_1 + kT), x(t) \neq \mu$, i.e. (0.18) holds.

Furthermore, if (0.20) is false, against all reason, then we get

$$\mu = x(t_2 + kT) \overset{(0.22)}{=} x(t_2) \overset{(0.19)}{\neq} \mu,$$

contradiction. Statement (0.20) holds. □

Remark 218. *Lemma 5 refers to eventually periodic points* μ *and makes a weaker statement than the appropriate one of the eventually periodic signals. We cannot draw the conclusion here, like at the eventually periodic signals, that* $\widehat{x}(k_1) = \widehat{x}(k_1 + kp), x(t_1 - 0) = x(t_1 + kT - 0), x(t_2) = x(t_2 + kT)$, *but we can state that* $\widehat{x}(k_1) \neq \mu, x(t_1 - 0) \neq \mu, x(t_2) \neq \mu$ *imply* $\widehat{x}(k_1 + kp) \neq \mu, x(t_1 + kT - 0) \neq \mu, x(t_2 + kT) \neq \mu$.

Lemma 6. *Let* $t_1 < t_2, t_1' < t_2', T > 0$ *and* $T' \in (0, t_2 - t_1)$. *Then*

$$([t_1,t_2) \cup [t_1+T, t_2+T) \cup [t_1+2T, t_2+2T) \cup ...) \cap$$
$$\cap([t_1',t_2') \cup [t_1'+T', t_2'+T') \cup [t_1'+2T', t_2'+2T') \cup ...) \neq \varnothing.$$

Proof. Let $k_1 \in \mathbf{N}$ such that $t_1 + k_1 T > t_1'$. In the sequence $t_1', t_1'+T', t_1'+2T', ...$ some $k_2 \in \mathbf{N}$ exists with

$$t_1' + k_2 T' < t_1 + k_1 T, \tag{0.29}$$
$$t_1' + (k_2+1)T' \geq t_1 + k_1 T. \tag{0.30}$$

We get from here that

$$t_1' + (k_2+1)T' \overset{(0.29)}{<} t_1 + k_1 T + T' < t_1 + k_1 T + t_2 - t_1 = t_2 + k_1 T. \tag{0.31}$$

From (0.30) and (0.31) we infer that $t_1' + (k_2+1)T' \in [t_1 + k_1 T, t_2 + k_1 T)$. □

Lemma 7. *a) Let* $\widehat{x}$ *that is not eventually constant and* $\mu \in \widehat{\omega}(\widehat{x})$. *Then*

$$\forall k \in \mathbf{N}, \exists k' > k, \widehat{x}(k'-1) \neq \widehat{x}(k') = \mu.$$

b) We suppose that x *is not eventually constant and we take* $\mu \in \omega(x)$. *We have*

$$\forall t \in \mathbf{R}, \exists t' > t, x(t'-0) \neq x(t') = \mu.$$

Proof. a) We suppose that $\widehat{\omega}(\widehat{x}) = \{\mu^1, ..., \mu^s\}, s \geq 2$ (since $\widehat{x}$ is not eventually constant, see $(1.3)_{page\,25}$), that $\mu = \mu^1$ without loosing the generality and let $k \in \mathbf{N}\_$ arbitrary. $\widehat{\mathbf{T}}^{\widehat{x}}_{\mu^1}, ..., \widehat{\mathbf{T}}^{\widehat{x}}_{\mu^s}$ are all infinite and we define

$$k_1 = \min(\widehat{\mathbf{T}}^{\widehat{x}}_{\mu^2} \cup ... \cup \widehat{\mathbf{T}}^{\widehat{x}}_{\mu^s}) \cap \{k, k+1, k+2, ...\},$$
$$k' = \min \widehat{\mathbf{T}}^{\widehat{x}}_{\mu^1} \cap \{k_1, k_1+1, k_1+2, ...\}.$$

We have $k' > k_1 \geq k$ and

$$\{\mu^2, ..., \mu^s\} \ni \widehat{x}(k'-1) \neq \widehat{x}(k') = \mu^1.$$

b) We put $\omega(x)$ under the form $\omega(x) = \{\mu^1, ..., \mu^2\}$, where $s \geq 2$ (see $(1.6)_{page\,25}$) and $\mu = \mu^1$. Let $t \in \mathbf{R}$ arbitrary. The support sets $\mathbf{T}^x_{\mu^1}, ..., \mathbf{T}^x_{\mu^s}$ are all superiorly unbounded and we define:

$$t_1 = \min(\mathbf{T}^x_{\mu^2} \cup ... \cup \mathbf{T}^x_{\mu^s}) \cap [t, \infty),$$
$$t' = \min \mathbf{T}^x_{\mu^1} \cap [t_1, \infty).$$

The sets $(\mathbf{T}^x_{\mu^2} \cup ... \cup \mathbf{T}^x_{\mu^s}) \cap [t, \infty), \mathbf{T}^x_{\mu^1} \cap [t_1, \infty)$ are of the form $[a,b) \cup [c,d) \cup ...$, thus their minimum exists. We have $t' > t_1 \geq t$ and

$$\{\mu^2, ..., \mu^s\} \ni x(t'-0) \neq x(t') = \mu^1.$$

□

Lemma 8. *Let $x \in S^{(n)}, \mu \in \omega(x)$ be an eventually periodic point with the period $T > 0$ and the limit of periodicity $t' \in \mathbf{R}$. For any $t_1 > t'$,*

$$x(t_1 - 0) \neq x(t_1) = \mu \tag{0.32}$$

implies

$$x(t_1 + T - 0) \neq x(t_1 + T) = \mu. \tag{0.33}$$

Proof. This is a special case of Lemma 5 b) when $x(t_1) = \mu$ and $k = 1$. □

Lemma 9. *Let $x \in S^{(n)}$ and the sequence $T_k \in \mathbf{R}, k \in \mathbf{N}$ that is strictly decreasingly convergent to $T \in \mathbf{R}$. Then $\exists N \in \mathbf{N}, \forall k \geq N$,*

$$x(T_k - 0) = x(T_k) = x(T). \tag{0.34}$$

Proof. Some $\delta > 0$ exists (see Remark 7, page 4) with the property that

$$\forall \xi \in [T, T + \delta), x(\xi) = x(T). \tag{0.35}$$

As $T_k \to T$ strictly decreasingly, $N_\delta \in \mathbf{N}$ exists such that

$$\forall k \geq N_\delta, T < T_k < T + \delta. \tag{0.36}$$

We fix an arbitrary $k \geq N_\delta$. If we take $\varepsilon \in (0, T_k - T)$ arbitrary also, we have

$$T - T_k < -\varepsilon < 0. \tag{0.37}$$

We add T_k to the terms of (0.37) and we obtain, taking into account (0.36) too:

$$T < T_k - \varepsilon < T_k < T + \delta. \tag{0.38}$$

We conclude on one hand that

$$\forall \xi \in (T_k - \varepsilon, T_k), x(\xi) \overset{(0.35),(0.38)}{=} x(T),$$

thus

$$x(T_k - 0) = x(T) \tag{0.39}$$

and on the other hand that

$$x(T_k) \overset{(0.35),(0.38)}{=} x(T). \tag{0.40}$$

By comparing (0.39) with (0.40) we infer (0.34). □

Remark 219. *In Lemma 9 $T_k, k \in \mathbf{N}$ and T are not necessarily related with any property of periodicity of x.*

Lemma 10. *a) We consider* $\widehat{x}$, $p \geq 1$, $k' \in \mathbf{N}_{\_}$ *and* $\mu \in \widehat{\omega}(\widehat{x})$ *such that*

$$\forall k \in \widehat{\mathbf{T}}^{\widehat{x}}_{\mu} \cap \{k', k'+1, k'+2, ...\}, \{k+zp | z \in \mathbf{Z}\} \cap \{k', k'+1, k'+2, ...\} \subset \widehat{\mathbf{T}}^{\widehat{x}}_{\mu} \tag{0.41}$$

holds. We define $n_1, n_2, ..., n_{k_1} \in \mathbf{N}_{\_}, k_1 \geq 1$ *by*

$$\{n_1, n_2, ..., n_{k_1}\} = \widehat{\mathbf{T}}^{\widehat{x}}_{\mu} \cap \{k', k'+1, ..., k'+p-1\}. \tag{0.42}$$

For any $k'' \geq k'$, *with* $n'_1, n'_2, ..., n'_{p_1} \in \mathbf{N}_{\_}, p_1 \geq 1$ *defined by*

$$\{n'_1, n'_2, ..., n'_{p_1}\} = \widehat{\mathbf{T}}^{\widehat{x}}_{\mu} \cap \{k'', k''+1, ..., k''+p-1\}, \tag{0.43}$$

we have $k_1 = p_1$ *and*

$$\bigcup_{k \in \mathbf{N}} \{n_1 + kp, n_2 + kp, ..., n_{k_1} + kp\} =$$
$$= \bigcup_{z \in \mathbf{Z}} \{n'_1 + zp, n'_2 + zp, ..., n'_{k_1} + zp\} \cap \{k', k'+1, k'+2, ...\}.$$

b) Let $x, T > 0, t' \in \mathbf{R}$ *and* $\mu \in \omega(x)$ *with*

$$\forall t \in \mathbf{T}^{x}_{\mu} \cap [t', \infty), \{t + zT | z \in \mathbf{Z}\} \cap [t', \infty) \subset \mathbf{T}^{x}_{\mu}. \tag{0.44}$$

The disjoint intervals $[a_1, b_1), [a_2, b_2), ..., [a_{k_1}, b_{k_1})$, $k_1 \geq 1$ *are defined by*

$$[a_1, b_1) \cup [a_2, b_2) \cup ... \cup [a_{k_1}, b_{k_1}) = \mathbf{T}^{x}_{\mu} \cap [t', t'+T). \tag{0.45}$$

For any $t'' \geq t'$, *we define the disjoint intervals* $[a'_1, b'_1), [a'_2, b'_2), ..., [a'_{p_1}, b'_{p_1})$, $p_1 \geq 1$, *by*

$$[a'_1, b'_1) \cup [a'_2, b'_2) \cup ... \cup [a'_{p_1}, b'_{p_1}) = \mathbf{T}^{x}_{\mu} \cap [t'', t''+T). \tag{0.46}$$

Then we have

$$\bigcup_{k \in \mathbf{N}} ([a_1 + kT, b_1 + kT) \cup [a_2 + kT, b_2 + kT) \cup ... \cup [a_{k_1} + kT, b_{k_1} + kT)) =$$
$$= \bigcup_{z \in \mathbf{Z}} ([a'_1 + zT, b'_1 + zT) \cup [a'_2 + zT, b'_2 + zT) \cup ... \cup [a'_{p_1} + zT, b'_{p_1} + zT)) \cap [t', \infty).$$

Proof. a) Let $k'' \geq k'$ arbitrary. As $\mu \in \widehat{\omega}(\widehat{x})$, we get that $\widehat{\mathbf{T}}^{\widehat{x}}_{\mu}$ is infinite, thus $\widehat{\mathbf{T}}^{\widehat{x}}_{\mu} \cap \{k', k'+1, k'+2, ...\} \neq \emptyset$ and we can apply Theorem 28, page 69, wherefrom $\widehat{\mathbf{T}}^{\widehat{x}}_{\mu} \cap \{k', k'+1, ..., k'+p-1\} \neq \emptyset$, $\widehat{\mathbf{T}}^{\widehat{x}}_{\mu} \cap \{k'', k''+1, ..., k''+p-1\} \neq \emptyset$ hence the definitions (0.42), (0.43) of $n_1, n_2, ..., n_{k_1}$ and $n'_1, n'_2, ..., n'_{p_1}$ make sense.

Let $j \in \{1, ..., k_1\}$ arbitrary. We claim that exactly one term of the sequence $n_j, n_j + p, n_j + 2p, ...$ belongs to $\widehat{\mathbf{T}}^{\widehat{x}}_{\mu} \cap \{k'', k''+1, ..., k''+p-1\}$. Indeed, let us suppose against all reason that no term belongs to $\widehat{\mathbf{T}}^{\widehat{x}}_{\mu} \cap \{k'', k''+1, ..., k''+p-1\}$. As (0.41) implies $\forall k \in \mathbf{N}, n_j + kp \in \widehat{\mathbf{T}}^{\widehat{x}}_{\mu}$, we infer the existence of k having the property that

$$n_j + kp \leq k'' - 1,$$

$$n_j + (k+1)p \geq k'' + p.$$

We infer the contradiction

$$k''+p\leq n_j+(k+1)p\leq k''-1+p.$$

We suppose against all reason that several terms of the sequence $n_j, n_j+p, n_j+2p, ...$ belong to $\widehat{\mathbf{T}}_{\mu}^{\widehat{x}}\cap\{k'',k''+1,...,k''+p-1\}$. This fact implies the existence of $k\in\mathbf{N}$ with

$$n_j+kp\geq k'',$$

$$n_j+(k+1)p\leq k''+p-1,$$

wherefrom we get the contradiction

$$k''+p\leq n_j+(k+1)p\leq k''+p-1.$$

We have shown the existence of the function $\widehat{\Lambda}:\widehat{\mathbf{T}}_{\mu}^{\widehat{x}}\cap\{k',k'+1,...,k'+p-1\}\longrightarrow\widehat{\mathbf{T}}_{\mu}^{\widehat{x}}\cap\{k'',k''+1,...,k''+p-1\}$, $\widehat{\mathbf{T}}_{\mu}^{\widehat{x}}\cap\{k',k'+1,...,k'+p-1\}\ni n_j\longrightarrow n_j+kp\in\widehat{\mathbf{T}}_{\mu}^{\widehat{x}}\cap\{k'',k''+1,...,k''+p-1\}$, where k is chosen conveniently.

We show that $\widehat{\Lambda}$ is injective and we suppose for this, against all reason, that $j_1,j_2\in\{1,...,k_1\}, j_1\neq j_2$, $k_1,k_2\in\mathbf{N}$ exist such that $n_{j_1}+k_1p=n_{j_2}+k_2p$ where, without loss of generality, we have

$$k'\leq n_{j_1}<n_{j_2}\leq k'+p-1.$$

On one hand we have $n_{j_2}-n_{j_1}=(k_1-k_2)p\in\{p,2p,3p,...\}$, thus $n_{j_2}-n_{j_1}\geq p$ and on the other hand we obtain $n_{j_2}-n_{j_1}\leq k'+p-1-k'$, thus $n_{j_2}-n_{j_1}\leq p-1$. The contradiction that we have obtained completes the proof that $\widehat{\Lambda}$ is injective.

We show that $\widehat{\Lambda}$ is surjective and let $j'\in\{1,...,p_1\}$ arbitrary. The fact that exactly one term of the sequence $n'_{j'},n'_{j'}-p,n'_{j'}-2p,...$ belongs to $\widehat{\mathbf{T}}_{\mu}^{\widehat{x}}\cap\{k',k'+1,...,k'+p-1\}$ is proved similarly with the proof of existence of $\widehat{\Lambda}$ and let this term be $n'_{j'}-\overline{k}p=n_j$. Obviously $\widehat{\Lambda}(n_j)=n'_{j'}$.

It has resulted that $\widehat{\Lambda}$ is bijective and $k_1=p_1$.

We prove $\bigcup\limits_{k\in\mathbf{N}}\{n_1+kp,n_2+kp,...,n_{k_1}+kp\}\subset\bigcup\limits_{z\in\mathbf{Z}}\{n'_1+zp,n'_2+zp,...,n'_{k_1}+zp\}\cap\{k',k'+1,k'+2,...\}$ and let $\widetilde{k}\in\bigcup\limits_{k\in\mathbf{N}}\{n_1+kp,n_2+kp,...,n_{k_1}+kp\}$ arbitrary. Some $k\in\mathbf{N}$ and some $j\in\{1,...,k_1\}$ exist with $\widetilde{k}=n_j+kp$. But $kp\geq0$ and $\widetilde{k}\geq n_j\geq k'$. Some $j'\in\{1,...,k_1\}$ and some $\overline{k}\in\mathbf{N}$ exist such that $\widehat{\Lambda}(n_j)=n_j+\overline{k}p=n'_{j'}$, in other words $\widetilde{k}=n'_{j'}-\overline{k}p+kp$. We have proved that $\widetilde{k}\in\bigcup\limits_{z\in\mathbf{Z}}\{n'_1+zp,n'_2+zp,...,n'_{k_1}+zp\}\cap\{k',k'+1,k'+2,...\}$.

We prove that $\bigcup\limits_{z\in\mathbf{Z}}\{n'_1+zp,n'_2+zp,...,n'_{k_1}+zp\}\cap\{k',k'+1,k'+2,...\}\subset\bigcup\limits_{k\in\mathbf{N}}\{n_1+kp,n_2+kp,...,n_{k_1}+kp\}$ and let for this $\widetilde{k}\in\bigcup\limits_{z\in\mathbf{Z}}\{n'_1+zp,n'_2+zp,...,n'_{k_1}+zp\}\cap\{k',k'+1,k'+2,...\}$ arbitrary. Some $j'\in\{1,...,k_1\}$ and some $z\in\mathbf{Z}$ exist with $\widetilde{k}=n'_{j'}+zp\geq k'$. We have the existence of $j\in\{1,...,k_1\}$ and $\overline{k}\in\mathbf{N}$ with $n'_{j'}-\overline{k}p=\widehat{\Lambda}^{-1}(n'_{j'})=n_j$, thus

$\widetilde{k} = n_j + (\overline{k}+z)p$. As $n_j - p \leq k' - 1$, the condition $n_j + (\overline{k}+z)p \geq k'$ implies $\overline{k}+z \in \mathbf{N}$, in other words $\widetilde{k} \in \bigcup_{k\in\mathbf{N}} \{n_1 + kp, n_2 + kp, ..., n_{k_1} + kp\}$.

b) We take an arbitrary $t'' \geq t'$. As far as $\mu \in \omega(x)$, we have that $\mathbf{T}^x_\mu$ is superiorly unbounded and consequently $\mathbf{T}^x_\mu \cap [t',\infty) \neq \varnothing$. We can apply Theorem 28, page 69 and we get $\mathbf{T}^x_\mu \cap [t',t'+T) \neq \varnothing, \mathbf{T}^x_\mu \cap [t'',t''+T) \neq \varnothing$, hence the definitions (0.45) of the disjoint intervals $[a_1,b_1),[a_2,b_2),...,[a_{k_1},b_{k_1})$ and (0.46) of the disjoint intervals $[a'_1,b'_1),[a'_2,b'_2),...,[a'_{p_1},b'_{p_1})$ make sense.

Let $t \in \mathbf{T}^x_\mu \cap [t',t'+T)$ arbitrary. We have from (0.44) that $\forall k \in \mathbf{N}, t+kT \in \mathbf{T}^x_\mu$ and we claim that exactly one term of the sequence $t, t+T, t+2T, ...$ belongs to $\mathbf{T}^x_\mu \cap [t'',t''+T)$. This is proved similarly with a), the supposition that no term of the sequence belongs to $\mathbf{T}^x_\mu \cap [t'',t''+T)$ and the supposition that several terms of the sequence belong to $\mathbf{T}^x_\mu \cap [t'',t''+T)$ give contradictions. The reasoning shows the existence of a function $\Lambda : \mathbf{T}^x_\mu \cap [t',t'+T) \longrightarrow \mathbf{T}^x_\mu \cap [t'',t''+T), \mathbf{T}^x_\mu \cap [t',t'+T) \ni t \longrightarrow t+kT \in \mathbf{T}^x_\mu \cap [t'',t''+T)$, where k is chosen conveniently.

We prove that Λ is injective and let us suppose against all reason that $t_1, t_2 \in \mathbf{T}^x_\mu \cap [t',t'+T), k_1, k_2 \in \mathbf{N}$ exist with the property that $t_1 + k_1T = t_2 + k_2T$. We can suppose without loosing the generality that

$$t' \leq t_1 < t_2 < t'+T.$$

On one hand $t_2 - t_1 = (k_1 - k_2)T \in \{T, 2T, 3T, ...\}$, thus $t_2 - t_1 \geq T$ and on the other hand $t_2 - t_1 < t' + T - t' = T$, contradiction.

The proof that Λ is surjective is made by taking $\widetilde{t} \in \mathbf{T}^x_\mu \cap [t'',t''+T)$ arbitrarily and showing, by making use of (0.44), that exactly one term of the sequence $\widetilde{t}, \widetilde{t}-T, \widetilde{t}-2T, ...$ belongs to $\mathbf{T}^x_\mu \cap [t',t'+T)$.

The conclusion is that Λ is bijective.

We prove $\bigcup_{k\in\mathbf{N}}([a_1+kT,b_1+kT) \cup [a_2+kT,b_2+kT) \cup ... \cup [a_{k_1}+kT,b_{k_1}+kT)) \subset \bigcup_{z\in\mathbf{Z}}([a'_1+zT,b'_1+zT) \cup [a'_2+zT,b'_2+zT) \cup ... \cup [a'_{p_1}+zT,b'_{p_1}+zT)) \cap [t',\infty)$ and let $\widetilde{t} \in \bigcup_{k\in\mathbf{N}}([a_1+kT,b_1+kT) \cup [a_2+kT,b_2+kT) \cup ... \cup [a_{k_1}+kT,b_{k_1}+kT))$ arbitrary. We get the existence of $k \in \mathbf{N}$ and $j \in \{1,...,k_1\}$ such that $\widetilde{t} \in [a_j+kT,b_j+kT)$, thus $\widetilde{t}-kT \in [a_j,b_j)$ and $\widetilde{t}-kT \geq t'$. Furthermore, a unique $\overline{k} \in \mathbf{N}$ exists with the property $\Lambda(\widetilde{t}-kT) = \widetilde{t}-kT+\overline{k}T \in \mathbf{T}^x_\mu \cap [t'',t''+T)$ and a unique $j' \in \{1,...,p_1\}$ exists also with $\widetilde{t}-kT+\overline{k}T \in [a'_{j'},b'_{j'})$, i.e. $\widetilde{t} \in [a'_{j'}+(k-\overline{k})T, b'_{j'}+(k-\overline{k})T)$. As $k-\overline{k} \in \mathbf{Z}$ and $\widetilde{t} \geq a_j+kT \geq a_j \geq t'$, it has resulted that $\widetilde{t} \in \bigcup_{z\in\mathbf{Z}}([a'_1+zT,b'_1+zT) \cup [a'_2+zT,b'_2+zT) \cup ... \cup [a'_{p_1}+zT,b'_{p_1}+zT)) \cap [t',\infty)$.

We prove the inclusion $\bigcup_{z\in\mathbf{Z}}([a'_1+zT,b'_1+zT) \cup [a'_2+zT,b'_2+zT) \cup ... \cup [a'_{p_1}+zT,b'_{p_1}+zT)) \cap [t',\infty) \subset \bigcup_{k\in\mathbf{N}}([a_1+kT,b_1+kT) \cup [a_2+kT,b_2+kT) \cup ... \cup [a_{k_1}+kT,b_{k_1}+kT))$ and let $\widetilde{t} \in \bigcup_{z\in\mathbf{Z}}([a'_1+zT,b'_1+zT) \cup [a'_2+zT,b'_2+zT) \cup ... \cup [a'_{p_1}+zT,b'_{p_1}+zT)) \cap [t',\infty)$ arbitrary. We have the existence of $z \in \mathbf{Z}$ and $j' \in \{1,...,p_1\}$ such that $\widetilde{t} \in [a'_{j'}+zT,b'_{j'}+zT), \widetilde{t} \geq t'$. Thus $\widetilde{t}-zT \in [a'_{j'},b'_{j'})$. Some $\overline{k} \in \mathbf{N}$ exists with $\Lambda^{-1}(\widetilde{t}-zT) = \widetilde{t}-zT-\overline{k}T \in \mathbf{T}^x_\mu \cap [t',t'+T)$, thus $j \in \{1,...,k_1\}$ exists with $\widetilde{t} - zT - \overline{k}T \in [a_j, b_j)$ and finally $\widetilde{t} \in [a_j + (z+\overline{k})T, b_j + (z+\overline{k})T)$. Because $b_j - T < t'$ and $\widetilde{t} \geq t'$, we infer that $z + \overline{k} \in \mathbf{N}$. It has resulted that $\widetilde{t} \in \bigcup_{k \in \mathbf{N}} ([a_1 + kT, b_1 + kT) \cup [a_2 + kT, b_2 + kT) \cup ... \cup [a_{k_1} + kT, b_{k_1} + kT))$. □

Example 56. *We show that in the previous Lemma, item b) we have in general $k_1 \neq p_1$ (unlike item a)) and we consider $x \in S^{(1)}$,*

$$x(t) = \chi_{[0,1)}(t) \oplus \chi_{[2,3)}(t) \oplus \chi_{[4,5)}(t) \oplus ...$$

In this case we have $\mu = 1, T = 2, \mathbf{T}^x_\mu = [0,1) \cup [2,3) \cup [4,5) \cup ...$ and

- for $t \in [-1,0] \cup [1,2] \cup [3,4] \cup ...$ we have $k_1 = 1$, for example $t = -0.5$ when $[a_1, b_1) = [0,1)$;

- for $t \in (0,1) \cup (2,3) \cup (4,5) \cup ...$ we have $k_1 = 2$, for example $t = 0.5$ when $[a_1, b_1) \cup [a_2, b_2) = [0.5, 1) \cup [2, 2.5)$.

Lemma 11. *For any $a, b \in \mathbf{R}$ with $a < b$ and any $\rho \in \Pi'_n$, the set $\{t | t \in [a,b), \rho(t) \neq (0,...,0)\}$ is finite.*

Proof. Let $\alpha \in \widehat{\Pi}'_n$ and $(t_k) \in Seq$ such that

$$\rho(t) = \alpha^0 \cdot \chi_{\{t_0\}}(t) \oplus \alpha^1 \cdot \chi_{\{t_1\}}(t) \oplus ... \oplus \alpha^k \cdot \chi_{\{t_k\}}(t) \oplus ...$$

We notice first that the set $\{t_k | k \in \mathbf{N}, t_k \in [a,b)\}$ is finite (we consider that the empty set is finite). Indeed, if we suppose against all reason that $\{t_k | k \in \mathbf{N}, t_k \in [a,b)\}$ is infinite, as (t_k) is strictly increasing, we infer that $\forall k \in \mathbf{N}, t_k < b$, contradiction with the fact that (t_k) is superiorly unbounded.

We can infer now, as far as the right hand set of the following inclusion

$$\{t | t \in [a,b), \rho(t) \neq (0,...,0)\} \subset \{t_k | k \in \mathbf{N}, t_k \in [a,b)\}$$

is finite, that the left hand set is finite too. □

References

[1] Dmitri V. Anosov, Vladimir I. Arnold (Eds.), "Dynamical systems I", Springer-Verlag, Berlin, (1988).

[2] David K. Arrowsmith, C. M. Place, "An introduction to dynamical systems", Cambridge University Press, New York, (1990).

[3] Michael Brin, Garrett Stuck, "Introduction to dynamical systems", Cambridge University Press, New York, (2002).

[4] Constanţa-Dana Constantinescu, "Haos, factali şi aplicaţii", editura Flower Power, Piteşti, (2003).

[5] Robert L. Devaney, "A first course in chaotic dynamical systems. Theory and experiment", Perseus Books Publishing, (1992).

[6] Robert W. Easton, "Geometric methods in discrete dynamical systems", Oxford University Press, (1998).

[7] Adelina Georgescu, Mihnea Moroianu, Iuliana Oprea, "Teoria Bifurcaţiei, Principii şi Aplicaţii", editura Universităţii din Piteşti, Piteşti, (1999).

[8] Boris Hasselblatt, Anatole Katok, "Handbook of dynamical systems", Volume 1, Elsevier, (2005).

[9] Richard A. Holmgren, "A first course in discrete dynamical systems", Springer-Verlag, New York, (1994).

[10] Jurgen Jost, "Dynamical systems. Examples of complex behaviour", Springer-Verlag, Berlin, Heidelberg, (2005).

[11] Rudolf E. Kalman, Peter L. Falb, Michael A. Arbib, "Teoria sistemelor dinamice", Editura tehnică, Bucureşti, (1975).

[12] Yuri A. Kuznetsov, "Elements of Applied Bifurcation Theory", Second Edition, Springer, New York, (1997).

[13] Mihaela Sterpu, "Dinamică şi bifurcaţie pentru două modele van der Pol generalizate", editura Universităţii din Piteşti, Piteşti, (2001).

[14] Mariana P. Trifan, "Dinamică şi bifurcaţie în studiul matematic al cancerului", editura Pământul, Piteşti, (2006).

[15] Ivan Tyukin, "Adaptation in dynamical systems", Cambridge University Press, Cambridge, New York, (2011).

[16] Serban E. Vlad, Boolean dynamical systems, *ROMAI Journal*, vol. 3, Nr. 2, 277-324, (2007).

[17] Serban E. Vlad, Universal Regular Autonomous Asynchronous Systems: Fixed Points, Equivalencies and Dynamical Bifurcations, *ROMAI Journal*, vol. 5, Nr. 1, 131-154, (2009).

[18] Serban E. Vlad, "Asynchronous systems theory", second edition, LAP LAMBERT Academic Publishing, Saarbrucken, (2012).

[19] Serban E. Vlad, Eventually periodic points of the binary signals: definition, accessibility and limit of periodicity, *ROMAI Journal*, vol. 10, Nr. 2, 213-224, (2014).

[20] Serban E. Vlad, Binary signals: the set of the periods of a periodic point, *Buletinul Institutului Politehnic din Iaşi*, Secţia Matematică. Mecanică teoretică. Fizică, Tomul LX (LXIV), Fasc. 4, 23-30, (2014).

[21] Serban E. Vlad, The double eventual periodicity of the asynchronous flows, Proceedings of the Third Conference of Mathematical Society of Moldova IMCS-50, August 19-23, Chişinău, Republic of Moldova, (2014).

[22] Serban E. Vlad, Binary signals: necessary and sufficient conditions of periodicity of a point, *Analele Universităţii din Oradea*, Fascicola Matematică, TOM XXII, Issue No. 1, 23-31, (2015).

[23] Stephen Wiggins, "Introduction to Applied Nonlinear Dynamical Systems and Chaos", Second Edition, Springer, New York, (2003).

About the Author

Dr. Serban E. Vlad
Department of Computers,
Oradea City Hall, Oradea, Romania

Serban E. Vlad was born in 1959 in Oradea, Romania and he works at present as analyst-programmer. He is a member of ROMAI and GAMM, the Romanian and the German societies of industrial and applied mathematics. He is the author of more than fifty papers and a chapter in a book. His research monograph Asynchronous Systems Theory was published in several editions. His main fields of interest are binary valued analysis and asynchronous systems theory. In binary valued analysis, he defined for the real argument, binary functions the notions of integral and derivative. He studied also the periodicity of the real argument and discrete argument binary functions.

In systems theory, Dr. Vlad was concerned with modeling the asynchronous circuits from electronics. He defined the concept of delay and he wrote differential equations that model the ideal latches and other circuits. He introduced the regular asynchronous systems that, by some analogy with the (real, usual) dynamical systems, are generated by Boolean functions which iterate their coordinates independently on each other.

Index

G

H

I

L

N

O

P